STO

EVOLUTION
OF THE ATMOSPHERE

EVOLUTION
OF THE ATMOSPHERE

James C. G. Walker

Macmillan Publishing Co., Inc.
NEW YORK
Collier Macmillan Publishers
LONDON

Macmillan Publishing Co., Inc.
866 Third Avenue, New York, N. Y. 10022

Collier Macmillan Canada, Ltd.

Library of Congress Catalog Card Number: 77-23796

Printed in the United States of America

printing number
1 2 3 4 5 6 7 8 9 10

Library of Congress Cataloging in Publication Data

Walker, James Callan Gray.
  Evolution of the atmosphere.

  Includes index.
  1. Atmosphere.   I.   Title.
QC861.2.W33   551.5   77-23796
ISBN 0-02-854390-4

To Ann, John, and Kate

Contents

Preface

The atmosphere interacts with organisms, rocks, oceans, and the interplanetary medium. An attempt to describe the evolution of the atmosphere therefore depends on evidence and understanding provided by a diversity of disciplines in addition to atmospheric science. Aspects of geology, planetology, paleobiology, microbiology, geochemistry, and oceanography are all involved in the subject matter of this book.

Recent advances in these disciplines have made it possible, perhaps for the first time, to present a coherent account of the evolution of the atmosphere. The subject is still so much in its infancy, however, that many areas of uncertainty remain. These should not discourage us. They represent opportunities for further research. I have tried to expose areas of uncertainty while not obscuring the plot with a tangle of alternatives. Where clear evidence is lacking, it has been necessary to guess what might have happened. Much of this account of atmospheric evolution is therefore speculative. Many of my guesses may be wrong, but if my version of atmospheric history stimulates the development of more accurate histories I shall be well pleased.

The book is devoted largely to the evolution of the atmosphere's chemical composition. Weather and climate are important properties of the atmosphere, but they depend on composition. An account of the evolution of composition must therefore precede an account of the evolution of climate. The atmospheres of the other planets receive some attention because they can tell us whether or not our speculative theories are leading us astray, but most of the book is devoted to earth.

Because of the diversity of disciplines that impinge on the study of atmospheric evolution, I have sought to write a book that can be understood by scientists with a diversity of backgrounds. I have therefore defined terms and explained concepts which seemed likely to be unfamiliar to at least part of the intended readership. I have used the convenient device of enclosing background information as well as detailed derivations in boxes, distinct from the text. These boxes can be studied or ignored in accordance with the needs and interests of the individual reader. I see my readers as having undergraduate educations in one of the natural sciences. The book could therefore be used as a text in a graduate or advanced undergraduate course. Because of the absence of review literature in the field of atmospheric evolution, however, the book is intended also to serve the research community as a monograph. Sources of data and ideas are therefore referenced in accordance with standard scientific practice.

The book is organized into three parts. The first chapter presents background information on the atmosphere, ocean, earth, and life as an introduction to concepts that are used repeatedly in the later chapters. The next three chapters describe the processes that control the composition of the present-day atmosphere. An understanding of these processes is essential if we are to deduce what the atmosphere was like in the past, or what it may be like in the future. Controlling processes fall naturally into three classes, to each of which I devote a chapter. In Chap. 2 I describe photochemical processes within the atmosphere and the way in which they govern the concentrations of minor atmospheric constituents. The discipline most heavily involved is chemistry. Chapter 3 deals with processes at the bottom of the atmosphere that control the abundances of the major atmospheric gases, oxygen, nitrogen, and carbon dioxide. Geochemical arguments are extensively used in this chapter. Processes occurring at the top of the atmosphere are taken up in Chap. 4, where we explore the loss of atmospheric gases to space. The subject matter of this chapter is mainly physics.

Our understanding of the contemporary atmosphere is, of course, more highly developed than our understanding of atmospheric evolution. Before undertaking to explore atmospheric history, moreover, we must seek to develop the fullest possible understanding of the processes that control atmospheric composition. The three chapters devoted to the contemporary atmosphere therefore place greater demands on the reader than does the rest of the book. For those readers who want just the essence of the argument, at least on a first reading, the major conclusions are summarized at the end of each chapter.

The material developed in Chap. 2, 3, and 4 provides the basis for discussion of the ancient atmosphere in the last three chapters of the book. In Chap. 5 I examine the origin of the atmosphere. In Chap. 6 I attempt to deduce the properties of the atmosphere prior to the origin of life. Finally, in Chap. 7, I advance a tentative survey of the history of the atmosphere both during the early stages when the atmosphere evolved in response to developments in microbial metabolism and during the later stages when the atmosphere evolved in response to changing geological conditions.

I am deeply indebted to colleagues from several disciplines who have read all or part of the manuscript and have attempted to correct my errors and clarify my exposition. They include E. A. Adelberg, S. J. Bauer, D. M. Hunten, G. E. Hutchinson, D. H. Stedman, and especially, L. Margulis. I have also enjoyed many fruitful discussions with A. L. McAlester and K. K. Turekian. The manuscript has been critically reviewed by students of geology and atmospheric science in a graduate course at Yale University. Most helpful were the suggestions of Steve Ashe, Bill Chameides, Gary Feulner, John Sans, and Karl Taylor.

I am grateful to Helen Harris, Gay Jarvinen, Linda Jensen, and Cindy Sidley for patient and accurate typing.

Much of this book was written while I was a member of the Department of Geology and Geophysics at Yale University and while I was on leave in South Africa, where I enjoyed the hospitality of the Department of Geology at the University of the Witwatersrand. Both of these departments provided excellent environments in which to pursue my research; what I learned while in them made it possible for me to write this book.

Part One

INTRODUCTION

Chapter One
Atmosphere, Ocean, Solid Earth
and Life

Much of this book is devoted to discussion of the processes determining the properties of earth's present atmosphere. The goal is to understand these processes well enough to be able to describe how they may have changed in the course of geological history and what changes in atmospheric properties may thereupon have resulted.

In this chapter we present elementary but essential background information drawn from the various disciplines that contribute to the understanding of atmospheric evolution. Patience is therefore requested of the many readers for whom much of this background information will be familiar. We start with a summary of the properties of the atmosphere that are most relevant to our subject. Then, after a brief description of the ocean, we review the composition and structure of the solid earth, as well as its history. We end the chapter with a review of basic biological and ecological concepts, an outline of the evolution of life, and a brief discussion of the influence of life on the evolution of the atmosphere.

ATMOSPHERE

Atmospheric Mass

The atmosphere is composed of a compressible fluid, air, bound to the earth by the force of gravity. At any point in the atmosphere the *pressure* is just equal to the downward gravitational force on all of the fluid in a

1

vertical column of unit cross-sectional area above the point in question. If the air pressure were not balanced in this way by the weight of the overlying gas, the air would simply expand or contract until balance was achieved. We can therefore derive the mass of the atmosphere from the value of the air pressure at the ground.

The average value of atmospheric pressure at sea level is 1.012×10^6 dyn cm^{-2} ($= 1012$ mb) (Verniani, 1966). In terms of the balance of forces we have just described, this is equal to the weight of all of the air in a vertical column with a cross section of 1 cm^2. The mass of this vertical column of air is obtained by dividing this atmospheric pressure by the acceleration due to gravity, which has an average value at sea level of 980 cm sec^{-2}. We thus find that the mass of the atmosphere in the column above each square centimeter of the earth's surface is 1.03×10^3 gm. Upon multiplying this sum by the total surface area of the earth, 5.10×10^{18} cm^2, the total mass of the atmosphere is found to be 5.27×10^{21} gm.

A more exact calculation of this quantity has been presented by Verniani (1966). The largest error in our estimate results from our use of sea-level pressure rather than the average surface pressure. Making allowance for the areas of land at different heights above the sea, Verniani finds that the average surface pressure is 984 mb, corresponding to an average height above sea level for the bottom of the atmosphere of 236 m. His value for the mass of the atmosphere is $(5.136 \pm 0.007) \times 10^{21}$ gm, where the uncertainty results largely from uncertainty in the height distribution of the land.

Barometric Law

Atmospheric pressure decreases with increasing altitude because the weight of the overlying atmosphere decreases. The variation of pressure with altitude may be calculated from the condition just described, of hydrostatic balance between pressure and the gravitational force on the overlying gas.

We may write the pressure at altitude z as

(1-1) $$p(z) = \int_z^\infty \rho(h) g(h) \, dh$$

where $\rho(h)$ is the density of the atmospheric gas at height h, and $g(h)$ is the acceleration due to gravity. The rate of change of pressure with altitude is obtained by differentiating this expression:

(1-2) $$dp/dz|_z = -\rho(z) g(z)$$

For atmospheric gases, pressure is related to density and temperature by the *ideal gas law*

(1-3) $$p = nkT$$

where n is the number of gas molecules per unit volume, equal to ρ/m where m is the average molecular mass of the gas, k is Boltzmann's

PRESSURE

Consider a vertical column of air of cross-sectional area A extending from the bottom of the atmosphere out into space. The force that this column of air exerts on the ground is just equal to its weight, given by the product of its mass and the acceleration due to gravity. Pressure is force per unit area, so the pressure at the ground is equal to the weight of all of the air in a vertical column of unit cross-sectional area. Further, in a fluid, such as air, pressure is isotropic, which means that it acts equally in all directions. Thus, the air at the bottom of our hypothetical column exerts the same pressure sideways on the surrounding air and upwards on the overlying air as it exerts on the ground.

In cgs units, which we use for most of the calculations in this book, pressure is expressed in dynes per square centimeter (dyn cm^{-2}), a unit which expresses directly the dimensions of pressure as force per unit area. Other common units are the *atmosphere* (not usually abbreviated), equal to 1.013×10^6 dyn cm^{-2}, and the *millibar* (abbreviated mb) equal to 10^3 dyn cm^{-2}.

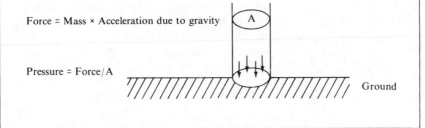

Force = Mass × Acceleration due to gravity

Pressure = Force/A

Ground

constant, equal to 1.38×10^{-16} erg deg^{-1}, and T is the absolute temperature. We use the ideal gas law to eliminate ρ in Eq. (1-2) and obtain

(1-4) $$dp/dz = -(mg/kT)p$$

We define a quantity with the dimensions of length called the *scale height:*

(1-5) $$H = kT/mg$$

Then

(1-6) $$dp/dz = -p/H$$

and

(1-7) $$p(z) = p(z_0) \exp[-\int_{z_0}^{z} (1/H)\, dh]$$

where $p(z_0)$ is the pressure at reference altitude z_0.

SCALE HEIGHT

The term scale height is used in various ways in atmospheric physics. Atmospheric scale height usually refers to the quantity computed according to Eq. (1-5) from known values of temperature, mean molecular mass, and gravitational acceleration. Sometimes, however, Eq. (1-6) is used to derive a scale height from a known variation of pressure with altitude:

$$H = -1/[(1/p)\,dp/dz]$$

Similarly, a *density scale height* can be defined as the negative reciprocal of the logarithmic derivative—i.e., the derivative of the logarithm to the base e $[d(\ln x)/dy \equiv (1/x)\,dx/dy]$—of density with respect to altitude:

$$H_n = -1/[(1/n)\,dn/dz]$$

A relationship between pressure scale height, H, and density scale height, H_n, can be derived from Eqs. (1-3) and (1-6):

$$H_n = 1/[1/H + (1/T)\,dT/dz]$$

An interesting and useful property of the scale height is that the product of density and scale height at any level is approximately equal to the integral of density over all heights above that level. This result can be derived by neglecting the variation of gravitational acceleration with altitude in Eq. (1-1). Then, using Eq. (1-3),

$$\int_z^\infty \rho(h)\,dh = p(z)/g = \rho(z)kT/mg = \rho(z)H$$

In words, if density did not vary with height above level z, the total thickness of the atmosphere above this level would be equal to the scale height. For this reason, the scale height is occasionally called the thickness of the equivalent homogeneous atmosphere.

For more information on the barometric law and its properties see, for example, Walker (1975a).

The gravitational acceleration, g, varies inversely as the square of the distance from the center of the earth. This variation is sufficiently small to be neglected in many atmospheric applications. Let us take g as constant and consider an isothermal atmosphere of uniform composition and therefore of constant m. For this atmosphere H is constant, and the barometric law, Eq. (1-7), simplifies to

(1-8) $$p(z) = p(z_0)\exp[-(z - z_0)/H]$$

Thus, atmospheric pressure in an isothermal atmosphere decreases exponentially with height, with each altitude increase of one scale height corresponding to a decrease in pressure by a factor of $1/e$.

In situations where H is a function of altitude, the integral in Eq. (1-7) may be evaluated either analytically or numerically. The variation of pressure with altitude will still be approximately exponential, so a plot of the logarithm of pressure against altitude will yield a fairly straight line.

Air near the ground consists principally of a mixture of molecular oxygen and nitrogen with a mean molecular mass of 29 amu (1 atomic mass unit $\equiv 1.66 \times 10^{-24}$ gm). The average temperature of surface air is 288°K. So, with $g = 980$ cm sec^{-2} and $k = 1.38 \times 10^{-16}$ erg deg^{-1}, we find $H = 8.4$ km as the value of the scale height. Thus, at a height of 8.4 km the atmospheric pressure is only $1/e = 37\%$ of its value at the ground. At a height of 84 km, pressure is reduced by a factor of $(1/e)^{10} = 1/22,000$ if the variation of temperature with altitude is neglected.

Thermal Structure

The barometric law described in the preceding section shows how temperature controls the rate of decrease of atmospheric pressure with altitude. Where temperature is high, the scale height is large and the exponential decrease of pressure is slow. Conversely, where temperature is low, the scale height is small and pressure decreases rapidly. The pressure can, in fact, be calculated at any height in the atmosphere by means of Eq. (1-7) if the temperature and mean molecular mass are known as functions of altitude. Atmospheric density at any altitude can be calculated in the same way because density is related to pressure and temperature by the ideal gas law, Eq. (1-3).

We shall return to the subject of the mean molecular mass, but for the present note that it is very nearly constant in the lowest 100 km of the atmosphere. This constancy means that a profile of temperature as a function of altitude contains most of the information we need to determine profiles of pressure and density. Figure 1-1 illustrates a typical temperature profile as well as pressure and density profiles. The figure shows that the rates of decrease of pressure and density are relatively small where the temperature is high and relatively large where the temperature is low, in accordance with the barometric law.

The maxima in the profile of temperature vs. altitude are reflections of the heating which results from the absorption of solar radiation of different wavelengths at different levels. Most of the energy incident on the earth from space is sunlight in the visible and near-infrared regions of the spectrum. Atmospheric gases are largely transparent to electromagnetic radiation at these wavelengths, so most of the sunlight is absorbed by the ground. The ground transfers heat to the lowest atmospheric layers in

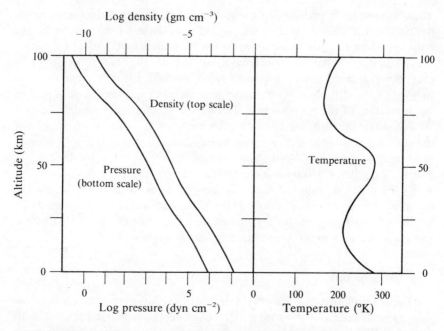

Fig. 1-1. The logarithm of pressure, the logarithm of density, and the temperature plotted against altitude in the earth's atmosphere. (From Goody and Walker, 1972. Copyright 1972 by Prentice-Hall, Inc., Englewood Cliffs, N.J. Used by permission of the publisher.)

the form of far-infrared radiation, latent heat of evaporation of water, and thermally conducted energy. This heat causes relatively high temperatures at the very bottom of the atmosphere, as Fig. 1-2 shows.

Solar radiation in the near-ultraviolet region of the spectrum, with wavelengths between 2000 A and 3000 A, does not penetrate through the atmosphere all the way to the ground. Instead, this radiation is absorbed by ozone, with the rate of absorption achieving a maximum at an altitude of about 50 km. The energy deposited in the atmosphere by solar near-ultraviolet radiation causes the temperature maximum at intermediate heights in Fig. 1-2.

Electromagnetic radiation in the extreme-ultraviolet region of the spectrum, with wavelengths shorter than 1000 A, is strongly absorbed by all atmospheric gases. Most of this component of the solar radiation is absorbed at heights above 100 km, causing high temperatures in the upper levels of the atmosphere, as shown in Fig. 1-2.

Atmospheric Nomenclature

The most commonly used names for the different levels of the atmosphere are based directly on the temperature profile (cf. Walker,

1967b). The bottom layer is called the *troposphere*. Temperature in the troposphere decreases steadily with increasing height up to the *tropopause*, where there is a marked discontinuity in the temperature gradient, the temperature becoming very nearly independent of height. The height of the tropopause varies from about 15 km in the tropics to about 10 km at high latitudes.

The atmospheric layer above the tropopause is called the *stratosphere*. Temperatures in the lower stratosphere are nearly constant; in the upper

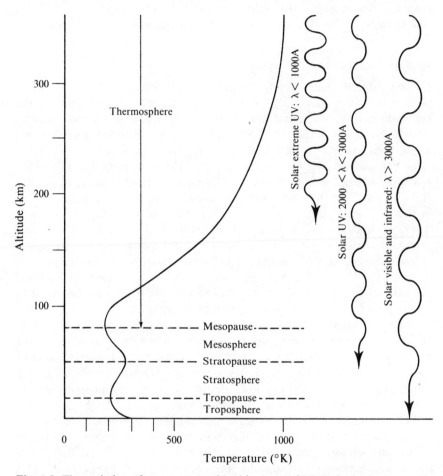

Fig. 1-2. The variation of temperature with altitude at middle latitudes in the earth's atmosphere. Layers of the atmosphere are called troposphere, stratosphere, mesosphere, and thermosphere; these are separated by features called tropopause, stratopause, and mesopause. The high-temperature regions of the atmosphere result from the absorption of solar radiation in certain wavelength ranges, as shown on the right. (From Goody and Walker, 1972. Copyright 1972 by Prentice-Hall, Inc., Englewood Cliffs, N.J. Used by permission of the publisher.)

stratosphere temperature increases to a maximum at a height of about 50 km. This temperature maximum defines the upper boundary of the stratosphere, the *stratopause*. Temperature decreases in the overlying layer, the *mesosphere*, until a temperature minimum, the *mesopause*, is reached at a height of about 85 km. Above the mesopause is the *thermosphere*, where temperature increases rapidly at first, and then more and more slowly until the atmosphere becomes nearly isothermal above 200 km. The altitude at which the temperature of the thermosphere ceases to rise is not clearly definable, but it is convenient to refer to the temperature of this isothermal region as the *thermopause temperature* (or the exospheric temperature).

Stability and Vertical Transport

The rate of change of temperature with altitude has an important influence on the ease with which air can move from one level of the atmosphere to another, and thus on the rapidity with which a constituent introduced at one level of the atmosphere is redistributed over adjacent levels. Imagine that a parcel of air is displaced to a higher altitude by some chance perturbation. If the air in the parcel is more dense than the surrounding atmosphere at the new level, the parcel tends to sink back down to the level from which it originated. If, on the other hand, the air in the displaced parcel is less dense than the surrounding atmosphere, buoyant forces tend to drive the parcel still further upwards. In the first case, the atmosphere is said to be *stable*, vertical motion is inhibited, and individual molecules of gas are carried relatively slowly from one altitude to another. In the second case, the atmosphere is *unstable*, vertical motions occur spontaneously—a phenomenon known as *free convection*—and mixing from one altitude to another is rapid.

As our imaginary parcel of air rises through the atmosphere it takes on, at each instant, the pressure of the surrounding gas. The temperature of the parcel may differ from that of the surroundings, however, because heat is transferred relatively slowly across the boundaries of the parcel. According to the ideal gas law, density is inversely proportional to temperature for gases at the same pressure, so whether the air in the parcel is more or less dense than the surrounding air depends on whether the temperature of the parcel is less than or greater than the temperature of the surrounding atmosphere. It is for this reason that the stability of the atmosphere depends on the rate of change of temperature with altitude.

As the parcel rises to levels of lower pressure it must expand. As it expands, the gas in the parcel performs mechanical work on the surrounding gas; this work draws on the internal, thermal energy of the parcel, causing the temperature to fall. Therefore, unless the temperature of the surrounding atmosphere also decreases with increasing altitude, the

BUOYANCY

Consider the vertical forces acting on a small parcel of air in which the density, ρ^*, differs from the density, ρ, of the surrounding atmosphere. For simplicity, assume that the parcel is a vertical cylinder with cross-sectional area A and height L. In addition to the gravitational force on the parcel, $-\rho^* ALg$ (the minus sign indicates a downward force), there is an upward force due to pressure of the ambient atmosphere on the bottom of the parcel and a downward force due to pressure on the top of the parcel. The difference of these two forces is $-A\,\Delta p = -AL\,dp/dz$, where Δp is the pressure difference between the top and bottom of the parcel and dp/dz is the pressure gradient in the ambient atmosphere. From the barometric law we have $dp/dz = -\rho g$. The sum of the pressure and gravitational forces on the parcel is therefore $(\rho - \rho^*)ALg$. We divide by the mass to obtain the acceleration, $(\rho/\rho^* - 1)g$. The parcel is accelerated upwards if $\rho^* < \rho$ and downwards if $\rho^* > \rho$.

parcel of displaced air will be cooler and more dense than the surrounding atmosphere; it will therefore tend to fall back to its original level.

In fact, for instability it is not enough for the temperature of the ambient atmosphere to decrease with altitude: only if it decreases more rapidly than the temperature in the displaced parcel of air, will free convection occur. The rate of decrease of temperature with altitude is called the *lapse rate*. In the displaced parcel of air, temperature decreases with height at approximately the *adiabatic lapse rate*. That is, as shown in Fig. 1-3, the atmosphere is stable if the ambient lapse rate is less than the adiabatic lapse rate; it is unstable if the ambient lapse rate exceeds the adiabatic lapse rate.

In the troposphere, the ambient lapse rate is generally close to the adiabatic lapse rate, and the air is relatively free to move vertically. This means that gaseous pollutants introduced near the ground are readily dispersed to higher levels of the troposphere. From time to time, however, an *inversion* occurs: The lapse rate is negative and temperature actually increases with increasing height. In this situation, vertical motion of the air is strongly inhibited by buoyant forces, and pollutants are trapped close to the ground. Severe air pollution episodes are always associated with temperature inversions.

In the stratosphere and thermosphere, temperature is either constant or increasing with altitude. The atmosphere in these regions is therefore stable, and the mixing of air from one level with air from another level is slow (hence the widespread concern over stratospheric pollution). In the

ADIABATIC LAPSE RATE

We wish to evaluate the rate at which temperature changes in a parcel of air moving adiabatically from one level of the atmosphere to another. An adiabatic change is one in which no energy is exchanged between the parcel and its surroundings. If the parcel moves from level z_1 in the figure, where its temperature is T_1, to level z_2, where its temperature is T_2, the lapse rate is

$$\Gamma = -(T_2 - T_1)/(z_2 - z_1)$$

The route followed by the parcel in getting from point 1 to point 2 does not affect the result, so we will imagine the parcel following the dashed line, a, b, c, in the figure rather than the straight line that connects the two points (see Goody and Walker, 1972).

On track a, the parcel is cooled to the absolute zero of temperature without any change in the pressure or altitude. The specific heat of a substance is the amount of heat that must be added to unit mass of the substance in order to raise its temperature by 1°. Since the specific heat of an ideal gas is independent of pressure and temperature, the amount of heat that must be extracted from the parcel in the cooling process is $c_p T_1 m$, where c_p is the specific heat of air at constant pressure, and m is the mass of the parcel. At absolute zero the parcel is very small and very dense, and it is effectively incompressible, like a solid or a liquid. The parcel therefore does not expand, performing work on its surroundings, when the pressure is reduced along track b. Work must, however, be performed on the parcel in order to raise it in the gravitational field of the earth. Along track b, therefore, the parcel gains an amount of gravitational potential energy equal to $mg(z_2 - z_1)$. Along track c, finally, heat is added to the parcel at constant pressure and altitude to bring its temperature to T_2. The amount of heat that is added is $c_p T_2 m$.

In an adiabatic change the parcel neither gains nor loses energy, so

$$-c_p T_1 m + mg(z_2 - z_1) + c_p T_2 m = 0$$

and

$$\Gamma = -(T_2 - T_1)/(z_2 - z_1) = g/c_p$$

mesosphere, the lapse rate is positive, and vertical mixing is relatively rapid.

Composition

The significance of vertical mixing to our investigation is that it influences the variation of composition with altitude. Mixing is a process

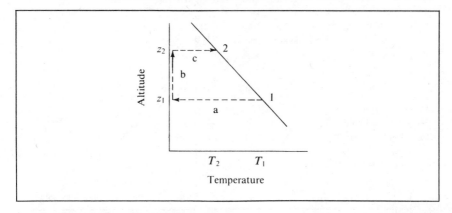

in which air flows from one altitude to another, carrying with it a complement of constituents characteristic of the level from which it originated. Imagine an atmosphere in which mixing is the only process affecting composition. After a period of time, the atmosphere will have been sufficiently stirred to yield a uniform composition, independent of altitude. Thus, a *well-mixed* atmospheric layer is one in which the relative proportions of the different constituents do not vary with altitude. Loosely defined, *mixing time* is the time required for initial inhomogeneities in composition to be largely eliminated by mixing. The mixing time is short in the troposphere and mesosphere and relatively long in the stratosphere and thermosphere.

Imagine, now, an atmosphere in which no mixing whatsoever occurs. In this situation, the action of gravity on atmospheric constituents of different molecular masses will lead to a marked variation of composition with altitude. We may regard Eq. (1-2)—

$$\mathrm{d}p/\mathrm{d}z = -\rho g$$

—as expressing a balance between the total-pressure gradient and the gravitational force on all of the atmospheric gases combined. When these two forces are balanced the atmosphere neither expands nor contracts. If gravitational attraction is the only process affecting the altitude distribution of the different gases, however, the forces on each individual gas must also be balanced:

(1-9) $$\mathrm{d}p_i/\mathrm{d}z = -n_i m_i g$$

where p_i is the partial pressure of constituent i, n_i is its number density, and m_i is its molecular mass. This balance of forces leads to the following expression for the partial pressure profiles of the different gases:

(1-10) $$p_i(z) = p_i(z_0) \exp[-\textstyle\int_{z_0}^{z} (1/H_i)\, \mathrm{d}h]$$

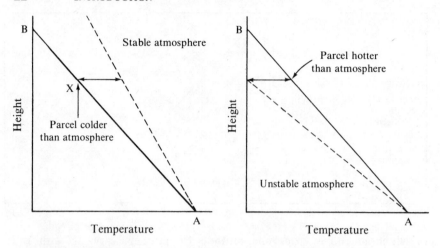

Fig. 1-3. The temperature profile in a stable atmosphere is shown by the dashed line in the left-hand diagram. The dashed line in the right-hand diagram shows the temperature profile in an unstable atmosphere. The solid line in each diagram has a negative slope equal to the adiabatic lapse rate; it is approximately the temperature of a parcel of air rising rapidly through the atmosphere. The parcel starts at the bottom point A in each case, where it has the same temperature as the surrounding atmosphere, and rises along the path AB. When the parcel on the left reaches point X, it is colder than the surrounding atmosphere and therefore denser. It tends to sink back to the level where it started. When the parcel on the right reaches point X, it is hotter and therefore less dense than the surrounding atmosphere. Buoyancy causes it to rise still further. (From Goody and Walker, 1972. Copyright 1972 by Prentice-Hall, Inc., Englewood Cliffs, N.J. Used by permission of the publisher.)

where

(1-11) $$H_i = kT/m_i g$$

is the scale height for constituent i. The scale height, H_i, is inversely proportional to molecular mass, so constituents with large molecular masses have small scale heights, and their partial pressures decrease relatively rapidly with height. Light constituents, on the other hand, have large scale heights and partial pressures that decrease slowly with increasing altitude. This effect is illustrated in Fig. 1-4, which shows representative density profiles in the terrestrial thermosphere.

In an unmixed atmosphere, we see that composition varies with altitude, with heavy gases becoming relatively less abundant and light gases becoming relatively more abundant as the altitude increases. In spite of this decrease of mean molecular mass with altitude, the total gas pressure still satisfies Eq. (1-7), as we may verify by summing Eq. (1-10) over all constituents i. If this condition were not satisfied, as we have already noted, the whole atmosphere would expand or contract.

Let us imagine an atmosphere in which mixing produces an initially homogeneous composition and then ceases altogether. In time, the action of gravity will redistribute the constituents in the manner we have described, but this redistribution will not occur instantaneously. The constituents have to diffuse through one another in order to achieve distributions that satisfy Eq. (1-10), the heavier constituents diffusing downwards while the lighter constituents diffuse upwards. The *diffusion time* at any altitude is the time required for this diffusive redistribution of constituents under the action of gravity to cause a substantial change in individual partial pressures at that altitude. After several diffusion times have elapsed, partial pressure profiles approach Eq. (1-10). When that equation is satisfied, the constituents are said to be in *diffusive equilibrium*.

The diffusion time is given, approximately, by $10^{-7}N$ sec, where N cm^{-3} is the total number density. At the ground N is 2.6×10^{19} cm^{-3}, so the diffusion time is about 3×10^{12} sec ($\sim 10^5$ yr) at the bottom of the atmosphere. This is very much longer than the mixing time in the troposphere (Sheppard, 1963), so we may conclude that mixing is much more important than gravitational separation in determining the variation of composition with altitude in the troposphere.

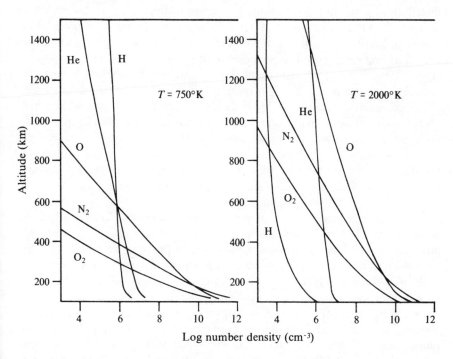

Fig. 1-4. Density profiles of the different constituents of the upper atmosphere for two values of the temperature. Densities decrease with altitude more rapidly in the cold atmosphere than in the hot one. (Data from Banks and Kockarts, 1973.)

Although the diffusion time is proportional to ambient atmospheric density, N, and therefore decreases exponentially with altitude, gravitational separation continues to be negligible compared with mixing throughout the stratosphere and mesosphere. In the thermosphere, however, the mixing time is long because temperature increases with altitude, and the diffusion time is relatively short because of the low ambient density. The two times are approximately equal at 101 km,, where N is 9×10^{12} cm^{-3}, and the diffusion time is 10^6 sec (10 d) (von Zahn, 1970). Above this transition level, called the *homopause*, diffusion is more important than mixing in determining the altitude profiles of the different atmospheric constituents; the lighter constituents become progressively more abundant relative to the heavier constituents. This behavior is illustrated in Figs. 1-4 and 1-5 and in Table 1-1.

We may conclude, provisionally, that the atmosphere is well-mixed and homogeneous in composition up to altitudes of about 100 km. There is, however, another factor that affects the variation of composition with altitude. Most atmospheric constituents are involved in chemical reactions that either produce or consume the constituent at rates that vary with altitude. Whether or not these chemical reactions influence the density distributions of the constituent in question depends on the relative magnitude of a third characteristic time, the *chemical lifetime*. The chemical lifetime of a given constituent is the time required for chemical reactions, operating in the absence of mixing or diffusion, to cause a significant change in the density of the constituent in question. Where the chemical lifetime is much longer than either the mixing or diffusion times, chemical reactions cause negligible departures from either the well-mixed or diffusive-equilibrium altitude profiles. For constituents with short chemical lifetimes, however, the departures may be substantial.

We are now in a position to understand qualitatively the profiles of the number densities of the different atmospheric constituents shown in Figs. 1-4 and 1-6. The major gases, oxygen and nitrogen, and the inert gases, helium, neon, and argon, are present in constant proportions at all levels up to the homopause. In this region of the atmosphere, which is sometimes called the *homosphere*, their distributions are dominated by mixing. Above the homopause, the homosphere gives way to the *heterosphere*, where diffusion is more rapid than mixing, and heavy gases such as molecular oxygen and nitrogen decrease in density more rapidly than lighter gases such as atomic oxygen, helium, and hydrogen.

With the exception of the inert gases, the trace constituents of the atmosphere are chemically active and throughout much of the homosphere their chemical lifetimes are shorter than the mixing time. The relative densities of these gases therefore vary with altitude in a manner that reflects local production and destruction in chemical reactions.

In the troposphere, on the other hand, most of the permanent

DIFFUSION TIME

We may estimate the order of magnitude of the diffusion time for a minor constituent, i, in the following way. Consider an isothermal atmosphere and suppose that the number density, n_i, of constituent i is initially independent of altitude. We assume that constituent i must diffuse through a stationary background gas of density N much greater than n_i. Each molecule of i collides with a molecule of the background gas on average every τ sec, where

$$\tau = 1/N\sigma v$$

In this expression, σ is the collision cross section and v is the mean thermal speed of the i molecules. We assume that the velocities with which i molecules emerge from collisions are completely random, so the average vertical velocity of i molecules after collisions is zero.

During the period between collisions, these molecules are accelerated downwards by the force of gravity; on average, they are displaced downwards a distance of $g\tau^2/2$ between each collision. Since the average time between collisions is τ, the average vertical drift velocity of the i molecules (defined to be positive upwards) is

$$U_i = -g\tau/2$$

and the flux is

$$\phi_i = n_i U_i = -n_i g\tau/2$$

This downward flux of i molecules causes the density to change at a rate given by minus the divergence of the flux:

$$\partial n_i/\partial t = -\partial \phi_i/\partial z = (n_i g/2)\, \partial \tau/\partial z = (n_i g/2\sigma v)\, \partial(1/N)/\partial z$$

Since the background gas is stationary, its density varies with altitude according to the barometric law, $N \propto e^{-z/H}$. Therefore

$$\partial(1/N)/\partial z = 1/NH$$

and

$$\partial n_i/\partial t = n_i g/2\sigma v H N$$

As a measure of the diffusion time we may take the reciprocal of $(1/n_i)\, \partial n_i/\partial t$, or

$$\tau_D = 2\sigma v H N/g$$

Collision cross sections for atmospheric gases are about 10^{-15} cm^2, and the mean thermal speed, $v = \sqrt{(3kT/m)}$, is about 5×10^4 cm sec^{-1}. Taking $H \sim 10$ km and $g = 980$ cm sec^{-2}, we find

$$\tau_D \sim 10^{-7} N \text{ sec}$$

TABLE 1-1. Representative model of the atmosphere

ALTITUDE (km)	TEMPERATURE (°K)	SCALE HEIGHT[a] (km)	$n(M)$[b] (cm^{-3})	$n(N_2)$ (cm^{-3})	$n(O_2)$ (cm^{-3})	$n(O)$ (cm^{-3})	$n(He)$ (cm^{-3})	$n(H)$ (cm^{-3})	MEAN MOLECULAR MASS (amu)
0	288.2	8.44	2.55(19)[c]	1.99(19)	5.34(18)				28.96
10	223.3	6.58	8.60(18)	6.71(18)	1.80(18)				28.96
20	218.9	6.45	1.70(18)	1.32(18)	3.56(17)				28.96
30	235.2	6.95	3.55(17)	2.77(17)	7.43(16)				28.96
40	268.2	7.95	8.11(16)	6.34(16)	1.70(16)				28.96
50	274.0	8.15	2.31(16)	1.80(16)	4.84(15)				28.96
60	252.8	7.54	7.17(15)	5.60(15)	1.50(15)				28.96
70	211.2	6.32	2.02(15)	1.58(15)	4.23(14)				28.96
80	177.2	5.32	4.30(14)	3.36(14)	9.00(13)				28.96
90	176.7	5.34	6.01(13)	4.67(13)	1.25(13)	3.00(11)			28.87
100	209.2	6.41	9.35(12)	7.05(12)	1.90(12)	3.20(11)			28.54
110	261.9	8.42	1.97(12)	1.44(12)	2.85(11)	2.30(11)			27.26
120	324.0	10.87	5.42(11)	4.00(11)	4.00(10)	1.00(11)	2.00(7)	2.41(6)	26.21
130	523.2	18.2	1.66(11)	1.16(11)	1.04(10)	4.01(10)	1.33(7)	1.01(6)	25.4
140	677.5	24.2	8.01(10)	5.28(10)	4.40(9)	2.29(10)	1.05(7)	5.90(5)	24.8
160	888.4	33.2	3.05(10)	1.81(10)	1.34(9)	1.11(10)	7.94(6)	2.94(5)	23.8

180	1015.9	39.6	1.54(10)	8.21 (9)	5.55 (8)	6.65 (9)	6.65(6)	1.85(5)	23.0
200	1095.5	44.5	8.90 (9)	4.23 (9)	2.63 (8)	4.41 (9)	5.83(6)	1.31(5)	22.2
220	1146.3	48.5	5.54 (9)	2.33 (9)	1.34 (8)	3.07 (9)	5.24(6)	1.01(5)	21.4
240	1179.4	51.9	3.62 (9)	1.33 (9)	7.11 (7)	2.21 (9)	4.78(6)	8.22(4)	20.7
260	1201.3	54.9	2.44 (9)	7.84 (8)	3.88 (7)	1.62 (9)	4.39(6)	6.99(4)	20.1
280	1216.1	57.6	1.69 (9)	4.69 (8)	2.16 (7)	1.20 (9)	4.05(6)	6.14(4)	19.5
300	1226.2	60.0	1.20 (9)	2.83 (8)	1.22 (7)	8.96 (8)	3.76(6)	5.54(4)	19.0
320	1233.1	62.3	8.58 (8)	1.73 (8)	6.93 (6)	6.74 (8)	3.49(6)	5.10(4)	18.5
340	1238.0	64.4	6.23 (8)	1.06 (8)	3.97 (6)	5.09 (8)	3.25(6)	4.77(4)	18.1
360	1241.5	66.3	4.58 (8)	6.57 (7)	2.29 (6)	3.87 (8)	3.03(6)	4.51(4)	17.7
380	1244.0	68.0	3.39 (8)	4.08 (7)	1.33 (6)	2.94 (8)	2.83(6)	4.30(4)	17.4
400	1245.9	69.6	2.53 (8)	2.54 (7)	7.76 (5)	2.24 (8)	2.64(6)	4.14(4)	17.1
450	1248.6	73.2	1.26 (8)	7.94 (6)	2.05 (5)	1.15 (8)	2.23(6)	3.82(4)	16.6
500	1249.9	76.4	6.44 (7)	2.53 (6)	5.54 (4)	5.99 (7)	1.90(6)	3.60(4)	16.1
550	1250.6	79.6	3.39 (7)	8.18 (5)	1.53 (4)	3.14 (7)	1.61(6)	3.42(4)	15.7
600	1250.9	83.2	1.83 (7)	2.70 (5)	4.29 (3)	1.67 (7)	1.38(6)	3.27(4)	15.3

After Banks and Kockarts (1973).

[a] Pressure scale height; corresponds to mean molecular mass.

[b] $n(M)$ is the total number density.

[c] $2.55(19) \equiv 2.55 \times 10^{19}$.

17

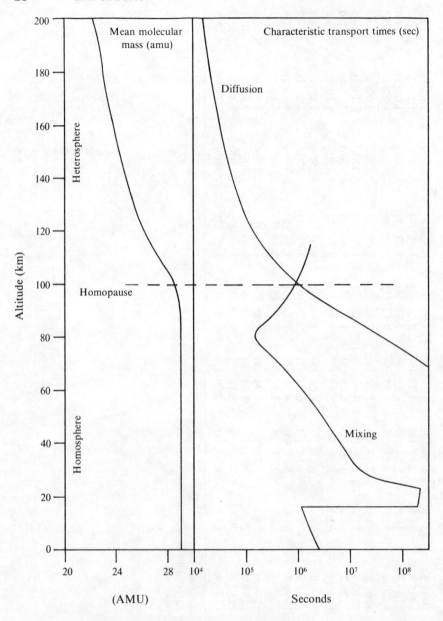

Fig. 1-5. Altitude profiles of the mean molecular mass and of the characteristic times for transport by molecular diffusion and by vertical eddy mixing. Below the homopause, at 101 km, mixing is more rapid than diffusion, and the mean molecular mass is constant. Above the homopause, diffusion is more rapid than mixing, and the mean molecular mass decreases with increasing altitude. (Data from Wofsy and McElroy, 1973; von Zahn, 1970; and Banks and Kockarts, 1973.)

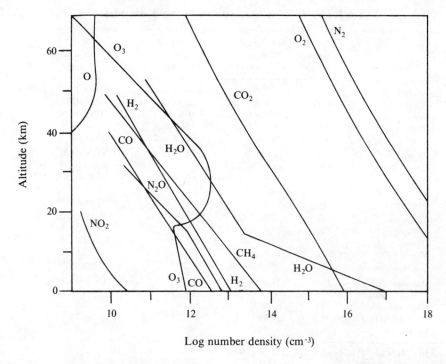

Fig. 1-6. Approximate altitude distributions of some minor atmospheric constituents.

atmospheric constituents have chemical lifetimes that are longer than the mixing time, so the composition of the troposphere is effectively homogeneous. The most important exception is water vapor, which shows a rapid decrease in relative density with increasing altitude. Water vapor is present in the troposphere at partial pressures that are typically 65% of the saturated vapor pressure. The concentration of water vapor therefore depends on temperature. As mixing carries saturated air aloft, the temperature falls, condensation occurs, and water is carried to lower levels in the form of precipitation.

Since the troposphere is relatively homogeneous and since it contains 80% of the mass of the atmosphere we may take the composition of ground-level air as a reasonably accurate reflection of the overall composition of the atmosphere. This composition is summarized in Table 1-2. Our goal, in this book, is to investigate the processes that may have affected the overall composition and mass of the atmosphere and to seek an understanding of the evolution of these properties in the course of geological history.

TABLE 1-2. Composition of the atmosphere

Constituent	Chemical Formula	Molecular Weight ($^{12}C = 12$)	Percent by Volume in Dry Air	Total Mass (gm)
Total atmosphere				$(5.136 \pm 0.007) \times 10^{21}$
Water vapor	H_2O	18.01534	variable	$(0.017 \pm 0.001) \times 10^{21}$
Dry air		28.9644	100.0	$(5.119 \pm 0.008) \times 10^{21}$
Nitrogen	N_2	28.0134	78.084 ± 0.004	$(3.866 \pm 0.006) \times 10^{21}$
Oxygen	O_2	31.9988	20.948 ± 0.002	$(1.185 \pm 0.002) \times 10^{21}$
Argon	Ar	39.948	0.934 ± 0.001	$(6.59 \pm 0.01) \times 10^{19}$
Carbon dioxide	CO_2	44.00995	0.0315 ± 0.0010	$(2.45 \pm 0.08) \times 10^{18}$
Neon	Ne	20.183	$(1.818 \pm 0.004) \times 10^{-3}$	$(6.48 \pm 0.02) \times 10^{16}$
Helium	He	4.0026	$(5.24 \pm 0.05) \times 10^{-4}$	$(3.71 \pm 0.04) \times 10^{15}$
Krypton	Kr	83.80	$(1.14 \pm 0.01) \times 10^{-4}$	$(1.69 \pm 0.02) \times 10^{16}$
Xenon	Xe	131.30	$(8.7 \pm 0.1) \times 10^{-6}$	$(2.02 \pm 0.02) \times 10^{15}$
Methane	CH_4	16.04303	$\sim 1.5 \times 10^{-4}$	$\sim 4.3 \times 10^{15}$
Hydrogen	H_2	2.01594	$\sim 5 \times 10^{-5}$	$\sim 1.8 \times 10^{14}$
Nitrous oxide	N_2O	44.0128	$\sim 3 \times 10^{-5}$	$\sim 2.3 \times 10^{15}$
Carbon monoxide	CO	28.0106	$\sim 1.2 \times 10^{-5}$	$\sim 5.9 \times 10^{14}$
Ammonia	NH_3	17.0306	$\sim 1 \times 10^{-6}$	$\sim 3 \times 10^{13}$
Nitrogen dioxide	NO_2	46.0055	$\sim 1 \times 10^{-7}$	$\sim 8.1 \times 10^{12}$
Sulfur dioxide	SO_2	64.063	$\sim 2 \times 10^{-8}$	$\sim 2.3 \times 10^{12}$
Hydrogen sulfide	H_2S	34.080	$\sim 2 \times 10^{-8}$	$\sim 1.2 \times 10^{12}$
Ozone	O_3	47.9982	variable	$\sim 3.3 \times 10^{15}$

After Verniani (1966) and Williamson (1973).

The Atmospheres of Venus and Mars

During our investigation of atmospheric evolution we shall have occasion to refer to the atmospheres of Venus and Mars. Of all the planets in the solar system, these two are closest to the earth in mass, bulk composition, and distance from the Sun. We can therefore expect some similarities in the histories of their atmospheres. If we are to understand atmospheric evolution, then, we must account for present-day differences in their atmospheres. Here we shall summarize, for use in later chapters, the properties of these two planets and their atmospheres, drawing on material presented by Goody and Walker (1972).

Relevant physical data concerning the planets are given below. We note that Venus is closer to the Sun than earth, and therefore presumably warmer, while Mars is further away. In fact, the surface temperature of Venus is about 750°K while the average surface

Physical data for the planets

PLANET[a]	MASS (10^{26} gm)	MEAN RADIUS (km)	MEAN DENSITY (gm cm^{-3})	GRAVITATIONAL ACCELERATION (cm sec^{-2})	AVERAGE DISTANCE FROM SUN (10^6 km)	LENGTH OF YEAR (Days)	INCLINATION OF EQUATOR TO ORBIT (Degrees)	ORBITAL ECCENTRICITY	PERIOD OF ROTATION (Days)
Mercury	3.35	2439	5.51	376	58	88	(0)[b]	0.206	58.7
Venus	48.7	6049	5.26	888	108	225	<3	0.007	−243[c]
Earth	59.8	6371	5.52	981	150	365	23.5	0.017	1.00
Mars	6.43	3390	3.94	373	228	687	25.2	0.093	1.03
Jupiter	19,100	69,500	1.35	2620	778	4330	3.1	0.048	0.41
Saturn	5690	58,100	0.69	1120	1430	10,800	26.8	0.056	0.43
Uranus	877	24,500	1.44	975	2870	30,700	98.0	0.047	−0.89[c]
Neptune	1030	24,600	1.65	1134	4500	60,200	28.8	0.009	0.53
Pluto	11	—	—	—	5900	90,700	—	0.247	(6.39)[b]

[a] The first four planets are similar in size, mass, density, and probably chemical composition. They are the *inner planets*. The remaining five are very different from earth, but apart from Pluto, are similar to one another. They are the *outer planets*.

[b] Data in parentheses are uncertain.

[c] Venus and Uranus rotate in the opposite sense to the other planets.

From Goody and Walker (1972); data from Kaula (1968) and Dole (1970). Copyright 1972 by Prentice-Hall, Inc., Englewood Cliffs, N.J. Used by permission of the publishers.

21

THE ATMOSPHERES OF VENUS AND MARS (continued)

temperature of earth is 288°K and of Mars, 210°K. We note, also, that the gravitational acceleration is much smaller on Mars than on Venus or earth.

The compositions of the atmospheres of Venus, earth, and Mars are compared below. Most striking is the fact that the atmospheres of Mars and Venus are composed principally of carbon dioxide, while the terrestrial atmosphere is mainly nitrogen. A complete theory of atmospheric evolution should explain this difference in atmospheric composition, as well as the differences in atmospheric masses and in surface temperatures.

Atmospheres of the inner planets

	VENUS	EARTH	MARS
Surface pressure (bar)	100	1	0.006
Atmospheric mass (gm)	5.3×10^{23}	5.3×10^{21}	2.4×10^{19}
Composition *(relative molecular abundance)*			
CO_2	>0.98	3×10^{-4}	0.96
N_2		0.78	0.025
^{40}Ar	<0.02	9.3×10^{-3}	0.015
^{36}Ar		3.1×10^{-5}	5.5×10^{-6}
O_2	$<10^{-6}$	0.21	2.5×10^{-3}
H_2O	10^{-6} [a]	up to 1	up to 10^{-3} [b]
HCl	6×10^{-7}		
HF	5×10^{-9}		
CO	5×10^{-5}	10^{-7}	10^{-3}

See Ingersoll and Leovy (1971), Hunten (1971), Walker (1975b), and Owen and Biemann (1976).
[a] Above the clouds. Below the clouds there may be a few tenths of a per cent of water vapor.
[b] On Mars, as on Earth, the water vapor abundance varies with season.

OCEAN

Introduction to the Ocean

The most important processes affecting the composition of the atmosphere involve the exchange of material between the atmosphere and the oceans or the solid earth; it is therefore appropriate to describe some of their properties. To see how important such exchange is likely to be, let us compare the mass of the atmosphere, 5.14×10^{21} gm, with the mass of the ocean, 1.39×10^{24} gm, and the mass of the solid earth, 5.98×10^{27} gm. We see that the atmosphere contains very much less matter than either the ocean or the solid earth. A small change in the overall composition of ocean or solid earth resulting from the exchange of material with the atmosphere can therefore cause a large change in the composition of the atmosphere. If, for example, surface temperature were to rise to the point where one third of the oceans evaporated, and if there were no accompanying change in the composition of the solid earth, the mass of the atmosphere would increase by a factor of 100 and the composition would become predominantly water vapor.

The oceans cover 71% of the surface of the earth to an average depth of 3729 m (cf. Turekian, 1968). The *salinity* of sea water is defined as the number of grams of dissolved salts in 1000 gm of sea water. In the open ocean salinity varies from 33 to 38 parts per thousand. Larger variations are encountered in waters close to shore. Salinity variations result from dilution by atmospheric precipitation, fresh stream water, and melting ice, as well as from concentration by evaporation and freezing (the ice that is formed when sea water freezes is nearly free of salt).

Although the salinity of sea water is variable, the relative proportions of the major elements dissolved in sea water are constant. These elements and their concentrations are listed in Table 1-3.

TABLE 1-3.   Concentrations   of   the major components of sea water with a salinity of 35‰

Component	Grams per Kilogram
Chloride	19.353
Sodium	10.76
Sulfate	2.712
Magnesium	1.294
Calcium	0.413
Potassium	0.387
Bicarbonate	0.142
Bromide	0.067
Strontium	0.008
Boron	0.004
Fluoride	0.001

From Turekian (1968). Copyright 1968 by Prentice-Hall, Inc., Englewood Cliffs, N.J. Used by permission of the publisher.

There are other constituents of sea water, however, with concentrations that vary markedly. The most important of these, for our purposes, are the elements involved in biological processes, called the *nutrient elements*. These are carbon, nitrogen, phosphorus, and silicon, and they are present in sea water principally as dissolved bicarbonate (HCO_3^-), nitrate (NO_3^-), phosphate ($PO_4^{\equiv}$), and silica (SiO_2). Organisms extract these elements from sea water and incorporate them into cell material. This activity is largely confined to the upper 100 m or so of the ocean, where there is enough light to permit photosynthesis to proceed. As a result, the nutrient elements are to varying degrees depleted in surface water. The depletion is nearly complete for phosphorus and nitrogen, as shown in Fig. 1-7. For the other nutrient elements it is less marked.

The nutrient elements are returned to solution in sea water when cell materials are degraded by respiration and decay. Most of the degradation occurs in the surface waters, but some of the solid organic matter settles to greater depths before it is destroyed. This downward transport of organic particles leads to an enhancement in the concentration of nutrient

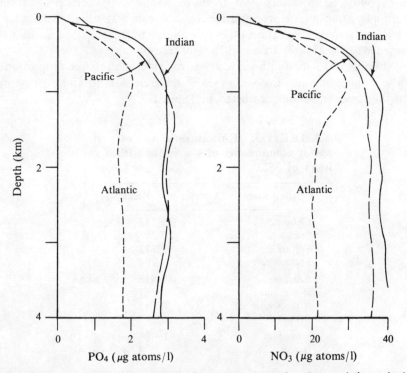

Fig. 1-7. Vertical distributions of the nutrient components, phosphate and nitrate, in the Atlantic, Pacific, and Indian Oceans. (From Turekian, 1968; data from Sverdrup et al., 1942. Copyright 1968 by Prentice-Hall, Inc., Englewood Cliffs, N.J. Used by permission of the publisher.)

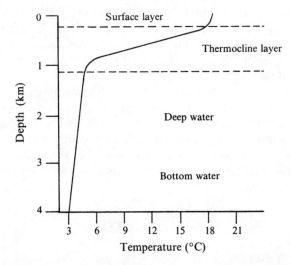

Fig. 1-8. A typical profile of temperature vs. depth in the ocean. (From Turekian, 1968. Copyright 1968 by Prentice-Hall, Inc., Englewood Cliffs, N.J. Used by permission of the publisher.)

elements in deep waters. Dissolved oxygen is consumed when the organic matter is degraded, so there is a corresponding depletion in the concentration of dissolved oxygen at deeper levels of the ocean.

Settling of organic particles results in a continual transport of nutrient elements from the surface waters to the deep waters. This downward transport is balanced, when averaged over the whole ocean, by upward transport of dissolved nutrients associated with the circulation of the ocean waters. Most important are the regions of upwelling, where the circulation brings deep water to the surface. The large supply of nutrients in these areas enables biological activity to flourish. Areas of upwelling are therefore biologically productive, and they support the world's major fishing industries.

Nutrient elements can also be carried to the surface by mixing processes similar to those we have discussed in the atmosphere. Vertical mixing is inhibited, however, by a strong increase in temperature between the deep water and the surface. In the ocean, as in the atmosphere, an upward increase in temperature corresponds to a stable situation in which buoyant forces oppose vertical motion.

A typical profile of temperature vs. depth in the ocean is shown in Fig. 1-8. There is a nearly isothermal, warm, surface layer, in which the temperature is determined largely by a balance between solar heating and loss of heat to the atmosphere by conduction, evaporation, and emission of radiation. Underlying this is the *main thermocline*, in which temperature decreases monotonically with increasing depth. This is the layer that is stable against vertical motions of the water. Underlying the main thermocline is a thick, cold layer of deep water and bottom water. This water is cold because it originates at high latitudes. Since it is cold, it has a relatively high density. It therefore settles to the bottom of the ocean basins and spreads around the world.

The general circulation of the ocean may be summarized schematically as follows. Cold water forms at high latitudes, sinks to the bottom of the ocean because of its high density, and spreads toward the equator. It is replaced by warm surface water that flows towards the poles. At high latitudes this surface water cools, sinks, and becomes deep water. The supply of surface water is maintained by upwelling of deep water accompanied by solar heating.

SOLID EARTH

Structure of the Earth

The material of which both the ocean and the atmosphere are composed was released originally from the solid earth. Before discussing the materials of which the earth is composed, let us briefly describe some of the physical properties of the solid earth.

Studies of the propagation of seismic waves have shown that there are three broad subdivisions of the solid earth with markedly different seismic velocities, densities, and presumably chemical compositions (cf. Clark, 1971). These are the *core*, which is presumed to consist mainly of iron, the overlying *mantle*, which contains 68% of the mass of the earth, and a thin layer on the surface called the *crust*. Some properties of these layers are listed in Table 1-4, along with equivalent data for the ocean and the atmosphere.

TABLE 1-4. The major subdivisions of the earth

	THICKNESS OR RADIUS (km)	MASS (gm)	MEAN DENSITY (gm/cm^3)
Oceanic crust	7	7.0×10^{24}	2.8
Continental crust	40	1.6×10^{25}	2.8
Mantle	2870	4.08×10^{27}	4.6
Core	3480	1.88×10^{27}	10.6
Oceans	4	1.39×10^{24}	1.0
Atmosphere	—	5.1×10^{21}	—
Earth	6371	5.98×10^{27}	5.5

From Clark (1971). Copyright 1971 by Prentice-Hall, Inc., Englewood Cliffs, N.J. Used by permission of the publisher.

The properties of the layers of the earth are not everywhere the same. The density of the mantle increases with increasing depth, and composition may change as well. Density also increases with depth in the core. The core, moreover, is divided into a liquid outer core and a solid inner core. The crust is particularly heterogeneous, as can be seen from the variety of rock types exposed at the surface.

The most important feature of crustal heterogeneity is not apparent to the casual observer, however. It is the distinction between oceanic crust and continental crust. The average thickness of the crust under the continents is about 40 km, while the thickness under the oceans is only

ISOSTASY

The thickness of the crust is not constant under the continents. In particular, it appears to increase under mountain ranges. The phenomenon was discovered in measurements of the earth's gravity field. The gravitational attraction exerted by the mountains is smaller than would be predicted on the basis of their bulk. According to the theory of isostasy, the excess mass of the mountain above the surface is balanced by a deficiency of mass below the surface. Mountains are underlain by low-density roots of crustal rock, and they effectively "float" in surrounding mantle rock of greater density, much as an iceberg floats in the ocean. Similarly, continents rise above the sea because continental crust is thicker than oceanic crust. In isostatic balance, the total mass per unit area above some level within the earth is a constant, regardless of the elevation of the surface.

An important consequence of isostasy is that buoyant forces on the roots of mountains and continents tend to push them upwards as mass is removed from the surface by erosion. The total thickness of material eroded can therefore be many times greater than the original height of the land.

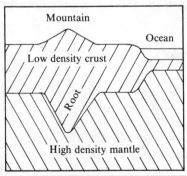

7 km. The average composition of continental and oceanic crusts is different also. Differences between continents and oceans have been detected beneath the crust, in the upper mantle, but horizontal in-homogeneity has not been detected at greater depths.

Temperatures increase with depth at a rate of about $20°K\,km^{-1}$ near the surface. The temperature gradient must decrease at greater depths, or the mantle would melt; but quantitative knowledge on temperatures in the deep interior is lacking. A probable value for the temperature at the center of the earth is 5000°K (Clark, 1971). Most of the heat that is conducted upwards to the surface of the earth is provided by the decay of radioactive elements incorporated into rock-forming minerals. There may also be a contribution from gravitational energy converted into heat when the earth was originally formed.

Minerals

We are now in a position to discuss the materials of which the earth is composed. Of these the most obvious are rocks, but rocks are made up of minerals. We shall therefore consider minerals first.

A mineral is a naturally occurring, crystalline solid with a definite chemical composition, or of a limited range of compositions. By crystal-line we mean that the atoms comprising the solid are chemically bonded in a systematic, three-dimensional array characteristic of a particular mineral. The orderly structure of a mineral is frequently reflected in the regular shapes of individual crystals of the mineral.

The most abundant minerals on earth are *silicates*, composed of silicon, oxygen, and other elements. The dominant structural component of a silicate mineral is a tetrahedron of four oxygen atoms with a silicon atom at the center. Different silicate minerals exhibit different spatial arrangements of these silicon–oxygen tetrahedra and different admixtures of other elements. The compositions and names of the more important silicate minerals are indicated in Table 1-5.

Nonsilicate minerals are diverse both in composition and in crystal structure. Table 1-6 gives the names and compositions of some common ones.

Rocks

This brief introduction to minerals enables us to turn to a discussion of rocks. Rocks are mixtures of minerals. In some rocks, the individual mineral grains are readily apparent to the eye. In other, fine-grained rocks, the individual mineral grains may be apparent only under the microscope. The relative proportions of the different minerals that make up a rock determine the overall chemical composition of the rock.

TABLE 1-5. Silicate minerals

NAME BASED ON STRUCTURE	RANGE OF COMPOSITION	NAME BASED ON COMPOSITION
Olivine	$Fe_2^{+2}SiO_4$ Mg_2SiO_4	fayalite forsterite
Pyroxene	$CaMg(SiO_3)_2$ $\updownarrow$ $CaFe^{+2}(SiO_3)_2$	diopside augite hedenbergite
Amphibole	Wide range of composition including Mg, Fe^{+2}, Ca, Na, Al, and OH, as well as Si and O.	
Micas and clay minerals	K, Al, OH, Si, O K, Mg, Fe^{+2}, Al, Si, O Al, OH, Si, O Al, OH, H_2O, Si, O Mg, Al, OH, Si, O	muscovite biotite kaolinite montmorillonite chlorite
Quartz	SiO_2	
Feldspar	$CaAl_2Si_2O_8$ $\updownarrow$ $NaAlSi_3O_8$ $\updownarrow$ $KAlSi_3O_8$	anorthite plagioclase albite alkali feldspar orthoclase

Based on Turekian (1972) and Ernst (1969).
$\updownarrow$ The vertical arrows indicate that composition varies between the limits indicated.
Fe^{+2} indicates that the iron is in the reduced, ferrous state rather than in the oxidized, ferric state.

The most basic classification of rocks depends not upon composition, however, but on mode of formation. *Igneous rocks* appear to have been formed by crystallization from a molten state. *Sedimentary rocks* consist of hardened deposits of material eroded from preexisting rocks. *Metamorphic rocks* are formed when igneous or sedimentary rocks are subjected to temperatures and pressures high enough to cause chemical and structural changes.

Igneous rocks. The molten material from which igneous rocks crystallize is called *magma*. Magma that flows out upon the surface of the earth is called *lava*; it solidifies to form an *extrusive* igneous rock. An *intrusive* igneous rock is formed when magma solidifies below the surface. Lavas tend to cool rapidly, allowing little time for the growth of large crystals of individual minerals. Extrusive igneous rocks therefore tend to be fine grained. Cooling at depth is typically slow, on the other hand, allowing large crystals to develop. Intrusive rocks therefore tend to be coarse grained. As shown in Fig. 1-9, igneous rocks with the same mineralogical makeup are given different names depending on whether

Oxidation and Reduction

In the course of our examination of the oxygen budget of the atmosphere, we will need to distinguish between oxidized minerals and reduced minerals. Oxidation and reduction are equivalent to a change in the valence state of the element concerned. In pyrite (FeS_2), for example, the iron is in its +2 valence state while the sulfur is in its −1 valence state. If pyrite is oxidized to hematite and sulfur dioxide—

$$4FeS_2 + 11O_2 \rightarrow 2Fe_2O_3 + 8SO_2$$

—oxygen is consumed while the valence of the iron changes to +3 and that of the sulfur changes to +4. Oxidation therefore corresponds to an increase in the positive charge on the element being oxidized. The ferric iron (Fe^{+3}) in hematite is more highly oxidized than the ferrous iron (Fe^{+2}) in pyrite. The iron in magnetite (Fe_3O_4) is intermediate, with an average valence of $+2\frac{2}{3}$. Magnetite can therefore be oxidized to hematite:

$$4Fe_3O_4 + O_2 \rightarrow 6Fe_2O_3$$

Oxygen is not necessarily involved in an oxidation reaction. What is needed is an element to accept the electrons being lost by the element undergoing oxidation (the *electron donor*). Oxygen serves as an *electron acceptor* in the reactions cited above. Its valence changes from 0 to −2. In the fermentation reaction which converts sugar to alcohol and carbon dioxide—

$$C_6H_{12}O_6 \rightarrow 2C_2H_5OH + 2CO_2$$

—the carbon serves as both electron donor and electron acceptor. The valence of the carbon changes from 0 in the sugar to −2 in the alcohol and +4 in the carbon dioxide. Some of the sugar is therefore oxidized to carbon dioxide while the rest of it is reduced to alcohol. Charge is conserved in the process.

they are intrusive or extrusive. The rocks richest in olivine, called peridotite and dunite, do not occur in extrusive forms.

Igneous rocks underlying the oceans and in midocean islands are almost all *basaltic* (also called *simatic* or *mafic*), that is, they are composed of olivine, pyroxene, and calcium-rich plagioclase. Basaltic rocks occur also on the continents in both extrusive and intrusive forms. Basaltic magmas originate at depths of about 50 km, in the layer of the earth called the upper mantle (cf. Turekian, 1972).

Granitic (or *sialic*) rocks, on the other hand, are almost entirely

TABLE 1-6. Common nonsilicate minerals

MINERAL NAME	CHEMICAL COMPOSITION	MINERAL NAME	CHEMICAL COMPOSITION
Graphite	C	Hematite	Fe_2O_3
Diamond	C	Chalcopyrite	$CuFeS_2$
Halite	NaCl	Sphalerite	ZnS
Calcite	$CaCO_3$	Gypsum	$CaSO_4 \cdot 2H_2O$
Aragonite	$CaCO_3$	Anhydrite	$CaSO_4$
Pyrite	FeS_2	Fluorite	CaF_2
Galena	PbS	Apatite	$Ca_5(PO_4)_3(OH, F, Cl)$
Magnetite	Fe_3O_4	Cassiterite	SnO_2

From Turekian (1972). Copyright 1972 by Holt, Rinehart and Winston, Inc., New York. Used by permission of the publisher.

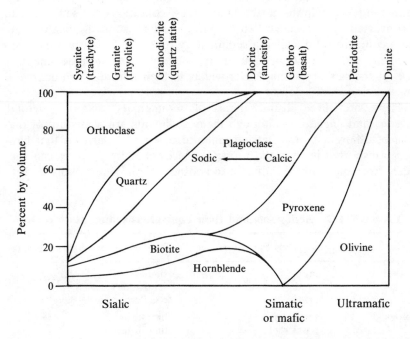

Fig. 1-9. The mineralogical makeup of common igneous rocks. The names given without parentheses are intrusive rocks. Names of equivalent extrusive rocks are given in parentheses. Groups of rocks with similar compositions are frequently described by the adjectives *sialic* (silica and aluminum), *simatic* (silica and magnesium), *mafic* (magnesium and iron), or *ultramafic*, as indicated below the figure. (From Turekian, 1972. Copyright 1972 by Holt, Rinehart and Winston, Inc., New York. Used by permission of the publisher.)

restricted to the continents. They are composed mainly of potassium feldspar (orthoclase for example), quartz, sodium-rich plagioclase, mica (biotite, for example), and amphibole (hornblende, for example). Granitic magmas are typically more viscous than basaltic magmas, and most granitic rocks are intrusive. Most granites have probably been formed by the partial melting of preexisting crustal rocks, both igneous and sedimentary. Granitic rocks thus differ markedly in origin from basaltic rocks, which come from the upper mantle.

Sedimentary rocks.   Rocks exposed at the surface of the earth are subject to degradation, both physical and chemical, by the weather and by the activities of organisms. The rocks are broken down by erosion and weathering to debris that can be transported considerable distances by running water, wind, or glaciers. Some of the rock material dissolves and is transported in solution. In due course the transported material is deposited somewhere to become *sediment.*

Burial of sediments under subsequently deposited sediments causes compaction and a moderate increase of temperature (temperature increases with depth in the earth). Under these conditions, the sediment can be converted by a process called *lithification,* from its unconsolidated state into a cohesive solid, called a sedimentary rock. Table 1-7 lists common sediments and the types of sedimentary rock that they form. Sedimentary rocks of the same type may differ widely in their hardness and degree of consolidation.

There are three distinct types of sedimentary material—*detrital, biogenic,* and *chemical*—although many sediments are a mixture of material of different types. Detrital material consists of particles that have been transported in the solid form. The sizes of these particles can vary widely from one detrital sediment to another. Their composition tends to

TABLE 1-7   Sediments and their equivalent sedimentary rocks

ORIGIN	TYPE OF SEDIMENT	TYPE OF SEDIMENTARY ROCK
Detrital	mud	shale
	sand	sandstone
	gravel	conglomerate
Biological	seashells	limestone
	coral reefs	limestone
	diatoms	chert (or flint)
	radiolaria	chert (or flint)
Chemical	salt	evaporite
	gypsum	evaporite

From Turekian (1972). Copyright 1972 by Holt, Rinehart and Winston, Inc., New York. Used by permission of the publisher.

depend on their size as well as on the rock from which they were eroded. Pebbles and larger particles are difficult to transport far from their place of origin. They suffer little chemical change during the processes of erosion and transportation. so they retain the composition of the parent rock. The sand-sized particles, on the other hand, consist largely of minerals with crystal structures that make them resistant to chemical degradation. Of these, quartz is the most common. Only the smallest detrital particles can be transported far enough from their sources to be present all over the ocean floor. Clay minerals (see Table 1-5) are abundant in the fine-grained material. They are formed by chemical changes in those silicate minerals that are not highly resistant to weathering. Different degrees of chemical change produce different clay minerals.

Most of the calcium carbonate deposited on the floor of the ocean consists of the skeletons of minute organisms, foraminifera, and coccolithophorides (Turekian, 1972). Sediments produced by the activities of organisms are called biogenic sediments. While calcium carbonate is the most common skeletal material, some organisms, mostly diatoms and radiolaria, make shells from the silica dissolved in sea water. These shells can accumulate to form siliceous sediments of biological origin, as distinct from siliceous sediments of detrital origin. (Detrital material consists of grains that have not dissolved.)

Most sea water is undersaturated with respect to the various salts in solution, and inorganic precipitation of these salts does not occur in the open ocean. When sea water is enclosed in a restricted basin and subjected to solar heating and evaporation, however, a brine of enhanced salinity is produced. Supersaturation with respect to certain salts may then occur, leading to inorganic precipitation and the formation of chemical sediments. Sedimentary rocks formed by chemical precipitation are called *evaporites.* Evaporite deposits consist mainly of the minerals anhydrite ($CaSO_4$), gypsum ($CaSO_4 \cdot 2H_2O$), and halite ($NaCl$).

Metamorphic rocks. When a rock is exposed to a new chemical environment or to conditions of temperature and pressure that differ from those under which the rock was formed, changes can occur in the mineral assemblage, the structure, and the texture of the rock (Ernst, 1969). An example of such change is the formation of sedimentary rocks at the surface by erosion, weathering, transportation, deposition, and lithification of material derived from preexisting rocks. At the low temperatures and pressures that exist at the surface, chemical reactions tend to be slow, and the mechanical stabilities of the rock-forming minerals have a large effect on the properties of the resulting rock. A contrasting example is the formation of magma by the partial or complete melting of preexisting rocks buried at depths within the earth where temperature and pressure are high. The igneous rocks that crystallize from the magma contain mineral assemblages governed by the principles of chemical equilibrium between crystals and molten material. The mechanical properties of the minerals are irrelevant.

The process of change is called metamorphism if it occurs under conditions intermediate between those characteristic of the surface (low temperature and pressure) and those so severe as to cause melting. Metamorphic rocks are produced by the metamorphism of preexisting igneous, sedimentary, or metamorphic rocks. Familiar examples are slate, formed by metamorphism of shale, and marble, formed by metamorphism of limestone. Metamorphism can involve the mechanical dislocation of individual crystals as a result of stress applied to the rock as well as chemical recrystallization. The minerals are in the solid state when metamorphic recrystallization occurs, although they may be in contact with a chemically active fluid.

Metamorphism need not cause any change in the overall chemical composition of the rock. However, certain constituents may be carried away in a fluid phase. Water may be lost from hydrous minerals, for example, or carbon may react with ferric oxide to produce carbon dioxide, which can escape. Other constituents may be introduced from fluids penetrating into the rock undergoing metamorphism. *Metasomatism* is the name given to metamorphism that leads to change in the bulk chemical composition of the rock.

The rock cycle. Crustal materials are continually cycled between the different types of rocks that we have discussed above, in what may be

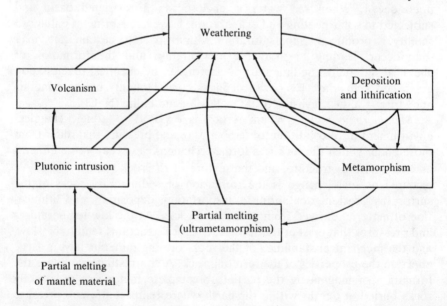

Fig. 1-10. The rock cycle. (From Ernst, 1969. Copyright 1969 by Prentice-Hall, Inc., Englewood Cliffs, N.J. Used by permission of the publisher.)

called the rock cycle (see Fig. 1-10). Igneous, metamorphic, and sedimentary rocks exposed at the surface of the earth are eroded and weathered to produce sedimentary rocks. In time, these sedimentary rocks may be subjected to temperatures and pressures that convert them to metamorphic rocks, or they may even be melted and converted to igneous rocks. Alternatively, they may be exposed at the surface and eroded to form new sedimentary rocks. Metamorphic rocks may be produced from igneous rocks or preexisting metamorphic rocks as well as from sedimentary rocks. They may be subjected to pressures and temperatures high enough to convert them to igneous rocks. Igneous rocks may be formed deep within the crust by the melting of sedimentary or metamorphic rocks or preexisting igneous rocks. Alternatively, they may be produced by partial melting of the upper mantle followed by upward migration of a basaltic magma.

The origin of basaltic magmas in the upper mantle indicates that material is transferred from the upper mantle to the crust. This transfer may mean that the crust is growing at the expense of the upper mantle. But there are processes that may transfer crustal material to the mantle. Such a process is discussed in the following section.

Plate Tectonics and Material Balance

We are not surprised by the fact that the flow of rivers into the sea does not cause the volume of the sea to increase. This flow is part of a cycle, the hydrologic cycle. Water evaporates from the oceans and returns to the land in the form of precipitation as fast, on the average, as it flows from land to sea. More perplexing is the fate of the detritus, derived from erosion of the land, that the rivers carry with them. Detritus is currently being added to the sea at a rate fast enough to fill the ocean basins with mud in a time of only 70 million years (Turekian, 1972). The present rate of erosion is fast enough to plane all of the continents down to sea level in

---

OCEAN RIDGES

A system of ridges extends through most of the oceans of the world. A typical ridge is about 2000 km wide (Clark, 1971) and rises to a height, at its crest, of about 2 km above the surrounding abyssal plains. At the crest of the ridge there is usually a steep-sided valley (called a *rift*) that resembles, in shape and dimensions, continental rift valleys such as those of East Africa. The topography is rugged near the crest, where sediments are absent. It becomes smoother on the flanks of the ridge where sediments fill the valleys and eventually also cover the hills.

a time as short as 10 to 20 million years. Since there is no evidence that erosion has not been occurring at approximately its present rate for hundreds or even thousands of millions of years, some mechanism must recycle the detritus, building the continents at the expense of sea floor sediments.

This conclusion is supported by a study of the sediments themselves. While we might expect the sea floor to be underlain by sediments dating back to the time when erosion began, an extensive program of exploration has failed to find any sediments older than 100 million years. What has been found is that sediments are thin and young near the crests of the oceanic ridges. As distance from the ridge crest increases, the sedimentary layer increases in thickness, and the sediments at the bottom of the layer increase in age (see Fig. 1-11). This pattern suggests that the sea floor is moving away from the crest of the ridge, carrying with it a blanket of sediments that becomes thicker as time passes and distance from the crest of the ridge increases. By sweeping sediments from the ridges to the edges of the oceans, this process of *sea-floor spreading* could accomplish the recycling of detrital material for which we have already noted a need.

Geophysical evidence has established the reality of sea-floor spreading beyond reasonable doubt, and has provided estimates of the velocities at which the sea floor moves. These velocities vary from about 1 cm yr^{-1} for the rate at which the floor of the Atlantic is moving away from the Mid-Atlantic Ridge to about 5 cm yr^{-1} for the motion of the Pacific floor away from the East Pacific Rise.

The motion of the sea floor away from the ridge leaves a gap at the crest which is filled by basaltic rocks derived from the upper mantle. Thus, sea-floor spreading requires that new oceanic crust be continually produced at the crests of the ridges. The area of new crust produced each year by all of the oceanic ridges is about 5 km^2. If the surface area of the

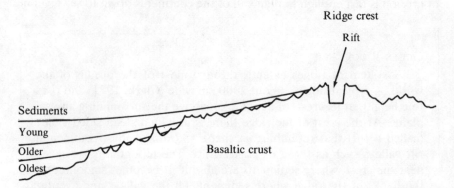

Fig. 1-11. Schematic representation of sea-floor sediments becoming thicker and older with increasing distance from the crest of a midocean ridge.

earth is to remain constant there must be places at which crust is consumed at an equal rate. These places have been identified as the *trenches*, where the oceans achieve their greatest depths. Trenches are topographical depressions caused by motion of the crust and the upper layer of the mantle downwards, at an angle of about 45° into the deeper mantle.

The phenomena of sea-floor spreading and creation and destruction of crust are related by the theory of *plate tectonics*. According to this theory, the surface of the earth consists of a number of rigid plates that are in motion with respect to one another. Plates move away from one another along ridges, towards one another along trenches, and past one another along transcurrent faults. Continents are carried along by the plates in which they are embedded. Thus, the theory of plate tectonics incorporates the older theory of continental drift.

The rigid plates together make up a layer of the earth called the *lithosphere*. It includes the crust and the top of the mantle. The thickness of the lithosphere is only imprecisely known. It may be from 50 to 100 km thick under the oceans, and perhaps 200 km thick under the continents (Clark, 1971). Under the lithosphere is a weak, lubricating layer, called the *asthenosphere*, over which the plates slide.

Of particular interest in connection with the balance of material entering and leaving the ocean are the trenches, in which one lithospheric plate turns down into the asthenosphere under another lithospheric plate (see Fig. 1-12). The descending plate warms up and gradually merges with the surrounding asthenosphere; but seismic evidence shows that the plate retains its strength to depths as great as 700 km. The regions of descending lithospheric plates are called *Benioff zones* or *subduction zones*.

Crustal rocks, including wet sea-floor sediments, are dragged down into the mantle in the subduction zones (Armstrong, 1971). The presence of water gives this material a low melting point, and much of it can return to the surface by way of volcanoes. Volcanoes and arcs of volcanic islands are therefore found over the descending lithospheric plate of a subduction zone.

Orogeny (the building of mountains) appears to be closely associated with subduction zones. Continental rocks and the sediments derived from them are too light to be dragged in large quantities into the mantle. When sea-floor spreading carries into a subduction zone the edge of a continent or the thick wedge of sediments (called a *geosyncline*) that accumulates off the shore of a continent, the lighter rocks remain at the surface where they are crumpled into mountain ranges. There was at one time a subduction zone between India and the rest of Asia, for example. The Himalayan Mountains formed when these two fragments of continent were brought together by the consumption of crust at this subduction zone.

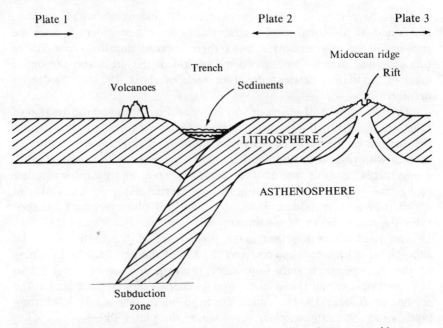

Fig. 1-12. Lithospheric plates moving together at a trench and apart at a ridge.

We would therefore expect orogenies to occur at the edges of continents, and this expectation is well supported by the geologic record. Except in cases where two continents have joined to form one, interiors of the continents tend to consist of *stable platforms*, or *cratons*, that have been protected from major deformation for a very long period of time. The platforms have not been far below the surface of the sea since they first formed. They have accumulated only a few thousand feet of sedimentary rocks, compared with the tens of thousands of feet that have accumulated in geosynclines. They have not been deeply eroded either, although numerous breaks in the sedimentary record indicate that they have frequently been above sea level. Evidently the platforms have remained at elevations close to sea level throughout much of geological history, while large vertical movements of the crust have occurred mainly at the continental margins.

For our attempts to understand the chemical interaction of the atmosphere, the ocean, and the crust, plate tectonics is important mainly because it provides a mechanism for the recycling of material. Oceanic sediments are swept into subduction zones from which they are returned to the continents by volcanism and orogeny. The oceans, therefore, do not fill with sediments and the continents are not eroded down to sea level. Cycling of material between crust and mantle is also possible, according to plate tectonics. Material that is lost to the mantle when new crust forms at the crests of the ocean ridges may be replaced by crustal material carried down into the mantle in subduction zones. We do not know whether the crust–mantle cycle is in balance, but the addition of basaltic rocks to the crust does not necessarily mean that the crust is growing at the expense of the mantle.

It is, indeed, possible that the entire system of atmosphere, ocean, crust, and mantle is in a state of *secular equilibrium.* By secular equilibrium we mean that there is, in the long term, no net transfer of material between these reservoirs. In other words, there is no systematic change with time in the overall composition and mass of any of the reservoirs, the atmosphere included. There may, however, be short-term fluctuations caused by temporary imbalances in the rates of transfer of material between reservoirs. In later chapters of this book we shall assume that such secular equilibrium presently exists. The assumption has the advantage of simplicity, and there is little evidence to contradict it.

It would be wrong, however, to assume that secular change has not occurred in the past. For one thing, the solar system, and with it the earth, had an origin, approximately 4.6 billion years ago. It is hardly likely that the earth formed with crust, oceans, and atmosphere already in place. It is more likely that there was a period, after the formation of the earth, when these features were accumulating and evolving, presumably as a result of gravitational differentiation of the material of the primitive earth. This phase of accumulation constitutes an important chapter in the history of the atmosphere, whether or not we assume that it has now ended.

Another source of secular change has been the origin and development of life. We shall show, in later chapters, that living creatures transfer large amounts of material into and out of the atmosphere. As organisms have changed during the course of time, we can expect the atmosphere to have changed as well.

LIFE

Biological Background

Here we shall be trying to extract just a few essential pieces of information from a large and complex subject. Much of our discussion will be grossly simplified. For an understanding of the complexities, the reader should consult a textbook on microbiology. Most of the factual information presented here, as distinct from the speculation, is from Stanier et al. (1970).

We are interested in organisms as agents of geochemical change. This means that we are mainly interested in *microorganisms* or *microbes.* Roughly speaking, these are organisms, such as bacteria and some algae, too small to be seen by the unaided human eye. Microorganisms are the important agents of geochemical change for several reasons. First, they are ubiquitous, being found in large numbers wherever life is possible on earth. Second, their small size gives them a large ratio of surface area to volume, permitting rapid exchange of material between the organism and its environment and, consequently, a high metabolic rate per gram of body weight. Third, they reproduce very rapidly, so that the population of a given microbe in a given environment can expand rapidly to take advantage of favorable conditions. Fourth, microbes, collectively, have a very much wider range of metabolic capabilities than larger organisms. They can flourish in hostile environments and derive energy from a remarkable variety of chemical compounds.

There are two functions of metabolism that need to be distinguished. One is *biosynthesis*, the synthesis of the organic molecules that make up the cell. Carbon is the key element in organic molecules, so we may think of biosynthesis as the process of building cell material from carbon compounds in the environment. Some organisms are able to synthesize cell material from inorganic carbon compounds such as carbon dioxide. These organisms are called *autotrophs*. All other organisms are dependent for growth on a source of organic carbon in their environment. They are called *heterotrophs*. In the modern world, all heterotrophs are sustained by autotrophs.

The second function of metabolism is to provide energy to the organism, to support its biological functions, including biosynthesis. Some organisms obtain energy by absorbing sunlight. These are *phototrophic* organisms. Since the absorbed energy is converted into chemical energy, generally by synthesis of an organic molecule, these organisms are also called *photosynthetic*. Provision of energy and biosynthesis are distinct functions, however, and they should not be confused. Most photosynthetic organisms can derive their carbon from inorganic compounds. They are therefore *photoautotrophs*.

All nonphotosynthetic organisms derive energy from chemical reactions of compounds that they extract from the environment. Most commonly they depend on organic compounds. These are *chemoheterotrophs*. There are some *chemoautotrophs*, however, that derive energy from the reaction of inorganic compounds, and use this energy to synthesize organic molecules, also from inorganic compounds.

The chemical reactions by which chemotrophic organisms derive energy invariably involve the *oxidation* of one element and the *reduction* of another. As described above these reactions can be considered in terms of the transfer of charge between an *electron donor*, which is oxidized, and an *electron acceptor*, which is reduced. Electron transfer can be involved in biosynthesis as well as in energy generation, depending on the oxidation state of the raw materials from which organic molecules are synthesized. For example, carbon is reduced when autotrophs synthesize carbohydrate, (CH_2O), from carbon dioxide; an electron donor is therefore required in order to conserve charge.

The oxidation and reduction reactions involved in metabolism are what make metabolic processes important for atmospheric evolution. Organisms can extract gases from the atmosphere, use these gases either in energy production or in biosynthesis, and return different gases to the atmosphere. *Methane bacteria*, for example, are chemoautotrophs that derive energy from the reaction

$$(1\text{-}12) \qquad\qquad CO_2 + 4H_2 \rightarrow CH_4 + 2H_2O$$

(Stanier et al., 1970, p. 672). They are *obligate anaerobes* (see box 1-10); today they are found in anaerobic sediments at the bottoms of lakes, where they make marsh gas, and in the rumens of herbivorous animals (cf. Deuser et al., 1973). In the early atmosphere, rich in hydrogen and devoid of oxygen, they may have been ubiquitous.

Biosynthesis is a little different from energy production in its effect upon the atmosphere, in that it converts atmospheric gases into nongaseous, organic compounds. In *green-plant photosynthesis*, for example, carbon dioxide and water are converted into carbohydrate and oxygen:

$$(1\text{-}13) \qquad\qquad CO_2 + H_2O \xrightarrow{\text{light}} (CH_2O) + O_2$$

Most of the carbohydrate is, in due course, converted back to carbon dioxide by the energy-yielding reaction of *respiration* [the reverse of reaction (1-13)]. If it were not so, the carbon dioxide of the atmosphere would soon be exhausted, and photosynthesis would come to a halt. So *mineralization*, which converts organic compounds into inorganic compounds, is as important a part of the biological cycle as biosynthesis.

Mineralization does not quite keep pace with biosynthesis, even with our modern, oxidizing atmosphere. Organic molecules are carried to the bottom of the sea adsorbed onto sediment grains or as organic particles.

AEROBES AND ANAEROBES

Some organisms cannot tolerate exposure to free oxygen. They are *obligate anaerobes*. They survive today in such anaerobic environments as deep-sea sediments, water-logged soils, and the intestines of animals. Other organisms are *obligate aerobes*. They cannot survive without free oxygen. All the organisms we see around us are in this class. *Microaerophils* can tolerate a little oxygen, but not too much. *Facultative* aerobic and anaerobic organisms can live either with or without free oxygen in the environment. Many of them change their metabolic processes in response to changes in the oxygen partial pressure to which they are exposed.

There they become incorporated into sediments and removed from the biological cycle, at least until the sediment is recycled by uplift and erosion. This drain of nutrient elements out of the biosphere may have been larger early in the history of life, when the atmosphere was weakly reducing and mineralization may have been less efficient.

An example of a nearly closed biological cycle that we have already mentioned is green-plant photosynthesis followed by aerobic respiration. It is worth presenting another example, which illustrates many of the features we have described in this section and which, in addition, does not require an oxidizing atmosphere. This closed biological system is called the *sulfuretum* (Postgate, 1968).

Any closed biological system must contain autotrophs, the *primary producers* of organic molecules. In the sulfuretum the primary producers are photosynthetic *purple* and *green sulfur bacteria*. These microbes use sunlight as a source of energy to synthesize organic molecules from carbon dioxide and water. An electron donor is required for the reduction of carbon dioxide to organic matter. The photosynthetic sulfur bacteria use hydrogen sulfide for this purpose, oxidizing it first to sulfur and then to the sulfate ion, $SO_4^=$. In the sulfuretum there are also heterotrophs that derive energy from the organic molecules synthesized by the photosynthetic sulfur bacteria. Most important are the *sulfate-reducing bacteria*. These microbes oxidize organic molecules in a process of *anaerobic respiration*. Respiration is anaerobic when the electron acceptor is not oxygen. The sulfate-reducing bacteria use sulfate as electron acceptor, converting the sulfate back to hydrogen sulfide. The real sulfuretum is more complicated than this, involving additional organisms and additional processes, but these are the essential elements of a closed biological cycle (see Fig. 1-13): autotrophic producers of organic molecules, heterotrophic consumers that convert organic molecules back to inorganic form, an element that can be oxidized by the producers and reduced by the consumers, and a source of energy, usually sunlight.

In Chap. 7 we shall argue that atmospheric evolution during much of geologic history was closely related to the evolution of microbial metabolism. We shall use these elements of a closed biological cycle as a framework for our speculation about the evolution of microbial metabolism. Here let us simply consider what caused microbial metabolism to evolve. According to Liebig's Law of the Minimum (cf. Hutchinson, 1973), microorganisms in a favorable environment will reproduce until some essential nutrient is exhausted. As life developed, therefore, there was always something that was in short supply, limiting further growth. When an organism developed a new metabolic capability that enabled it to substitute an abundant new nutrient for the one in short supply, it had an advantage over competing organisms and flourished. The way in which metabolic pathways involving many successive steps could have de-

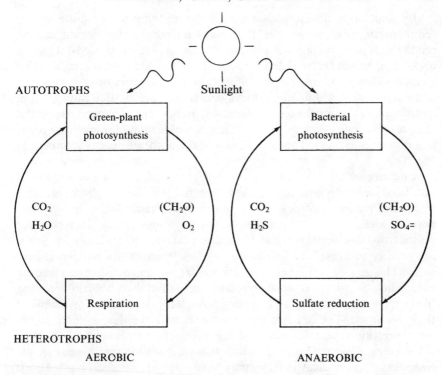

Fig. 1-13. Examples of closed biological cycles (with the drain of reduced organic matter to the sediments ignored).

veloped a step at a time, beginning with the later steps, is described by Miller and Orgel (1974).

Geologic Time and the History of Life

The major developments in the evolution of metabolism occurred more than a billion years ago. More recent biological evolution has involved increasing elaboration of multicellular organisms, rather than increasing metabolic capabilities of individual cells. This later biological evolution has provided the most useful means of distinguishing rocks of different ages. As organisms changed with time, so too did the fossils of these organisms, preserved in sedimentary rocks. Studies of the succession of fossils in layered sequences of rocks have enabled geologists to develop a relative time scale, in which the more recent history of the earth is divided into major eras and the eras are further divided into periods. The geologic time scale is illustrated in Fig. 1-14.

By and large, no method exists for determining the absolute age of sedimentary rocks, which are the ones that contain fossils and can be readily assigned to one or another of the geological periods. Igneous rocks can, however, be dated by studies of the decay of the radioactive elements they contain (Eicher, 1968). The relative ages of igneous rocks and surrounding sedimentary rocks can frequently be determined from stratigraphical considerations. Igneous rocks are younger than the sedimentary rocks they penetrate. They are older than sedimentary rocks laid down upon them. This is the basis of the absolute ages that appear in Fig. 1-14.

Life originated well before the beginning of the Cambrian, but it produced few obvious fossils. The absence of clearly distinguishable fossils in Precambrian rocks is the reason why Precambrian time has not been divided into eras or periods. At the beginning of the Cambrian, for reasons that are not understood, organisms started to make shells. Shells are readily preserved as fossils, so the fossil record has been rich ever since. This record, however, is restricted largely to shell-forming marine organisms. Soft-bodied creatures and those that live on land are less likely to produce fossils, and so they have left a less complete record of their evolution. The dates of important events in the development of life are therefore uncertain. Some of these events are shown in Fig. 1-15 (McAlester, 1968). The times when these organisms first appear in the fossil record are indicated. They may have originated earlier, but then left no fossils.

As far as the atmosphere is concerned, the most important lesson to be learned from the fossil record is that the evolution of the atmosphere was essentially complete by the end of the Precambrian. The multicelled organisms that are so abundantly represented in Phanerozoic rocks are

---

TAXONOMY

Organisms are classified into a hierarchy of groupings called *taxa.* The lowest-order taxon is the *species.* A species consists of a group of individual organisms that are potentially able to interbreed with one another. Closely related but distinct species are grouped into *genera.* Similar genera are combined in *families,* families in *orders,* orders in *classes,* and classes in *phyla.* Man, for example, is the species Sapiens of the genus Homo of the family Hominidae of the order Primates of the class Mammalia of the phylum Chordata. Most species survive on earth for a relatively short time, geologically speaking, but the higher-order taxa have long histories. There is no record of any phylum having become extinct since Cambrian times.

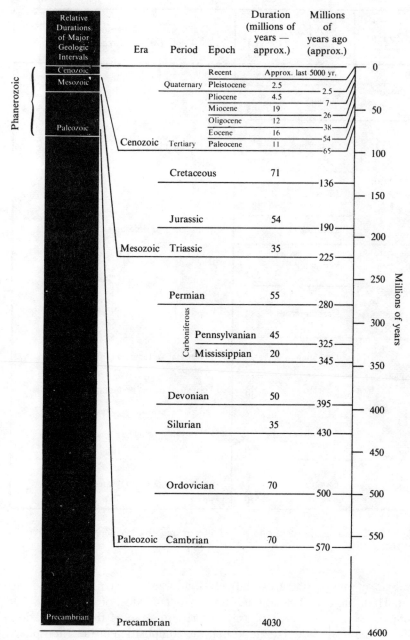

Fig. 1-14. Geologic time scale. (From Clark, 1971. Copyright 1971 by Prentice-Hall, Inc., Englewood Cliffs, N.J. Used by permission of the publisher.)

Era	Period	Age (10⁶ yr.)	First occurrence in the fossil record
Cenozoic	Quaternary	2.5	Man
	Tertiary	65	
Mesozoic	Cretaceous	136	Flowering plants
	Jurassic	190	Birds
	Triassic	225	Mammals and dinosaurs
Paleozoic	Permian	280	
	Carboniferous	345	Reptiles
	Devonian	395	Terrestrial vertebrates (amphibians)
	Silurian	430	Terrestrial plants and animals (arachnids)
	Ordovician	500	Vertebrates (fishes)
	Cambrian	570	Shell-forming invertebrates

Fig. 1-15. Events in the development of life during the Phanerozoic. (Data from McAlester, 1968.)

members of phyla that are relatively intolerant to changes in environment. Their metabolic capabilities are restricted. Most of the animals, for example, require a source of free oxygen to support their respiration. There is therefore little likelihood that the composition of the atmosphere or of the oceans could have changed markedly without these phyla becoming extinct. While atmospheric properties may well have fluctuated during the Phanerozoic, it is the Precambrian that must be examined for evidence concerning the growth and development of the atmosphere.

Unfortunately, the geological record of Precambrian time is far from complete. Occurrences of undisturbed Precambrian rocks are relatively rare. Most of them have either been eroded away, buried under younger rocks, or metamorphosed and deformed in later orogenetic events. The lack of a fossil record of evolving life in the Precambrian has prevented establishment of a global chronology comparable to that which exists for the Phanerozoic. Radioactive dating must be used to determine the relative ages of different occurrences of Precambrian rocks. Various names have been proposed for different times in the Precambrian. We shall use the system illustrated in Fig. 1-16. Note that no rocks at all have been found that date from the first billion years of earth's history.

Fossils are hard to find even in well-preserved Precambrian sediments because most Precambrian life consisted of single-celled, soft-bodied creatures (microbes). Only in late Proterozoic rocks have traces been found of multicelled animals (Metazoa). For reasons already described microbes have had much more effect on the development of the atmosphere than the larger creatures whose evolutionary history is preserved in the fossil record (Lovelock and Margulis, 1974; Margulis and Lovelock, 1974). Much of atmospheric evolution is therefore related to the evolution of microbial metabolic capabilities. Our understanding of microbial evolution, unfortunately, is largely speculative (Hall, 1971; Margulis, 1970, 1972).

Presumably, there was a period after the earth was formed before life

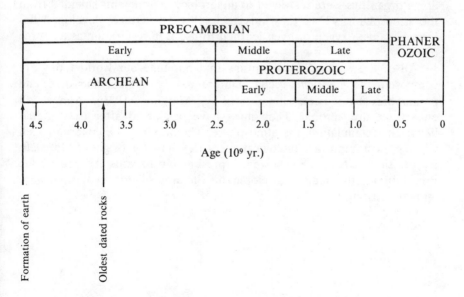

Fig. 1-16. Precambrian time scale. (After Schopf, 1972.)

had originated. During this time the atmosphere was accumulating and evolving as a result of the release of gases from the earth's interior and reaction between these gases and the rocks. Photochemical processes would have had an important effect on the composition of the atmosphere. Photochemical processes are also presumed to have led to the formation of organic compounds from which life developed.

The earliest organisms would have been single-celled creatures without intercellular organelles such as a cell nucleus. Such creatures survive today. They are called *prokaryotes*, and they include bacteria and blue–green algae. Their metabolic capabilities are extremely diverse. New capabilities presumably evolved one after another to take advantage of unexploited sources of energy in the environment. Atmospheric gases would have been consumed or produced by many of these energy-yielding metabolic reactions, so we expect that atmospheric composition would have changed in response to each new evolutionary development.

The most important development, as far as the atmosphere is concerned, was the origin of *green-plant photosynthesis*. Blue–green algae (Cyanophyta) were the first organisms with this capability. They derive energy from sunlight and release oxygen in the process. It appears that the atmosphere was largely devoid of oxygen before the onset of green-plant photosynthesis, but that the activities of the blue–green algae led to a gradual change from a reducing atmosphere to an oxidizing atmosphere. Since oxygen is toxic to many prokaryotes, the accumulation of oxygen in the atmosphere must have had a profound effect on the evolution of life. Some organisms were restricted to anaerobic environments shielded from the atmosphere, where they survive to this day. Others developed defenses against oxygen or even learned to use it in respiration as a source of energy.

All of the multicelled organisms that populate our world today are composed of *eukaryotic* cells. Differences between eukaryotes and prokaryotes have been summarized by Margulis (1970). Almost all eukaryotes are aerobic. They must have originated after oxygen had begun to accumulate in the atmosphere. The origin of the eukaryotic cell was the most important biological event to follow the origin of life itself. By the time it happened, however, presumably towards the end of the Precambrian, the major developments in atmospheric evolution were already drawing to a close.

Part Two

THE CONTEMPORARY ATMOSPHERE

Chapter Two

Atmospheric Chemistry and Its Effect on Minor Constituents

We begin our exploration of the processes that control atmospheric composition with those applying to the minor constituents, methane, hydrogen, carbon monoxide, and ozone. The atmospheric budgets of these constituents are fairly well understood and susceptible to quantitative treatment. The processes that control abundances of the major constituents, oxygen, nitrogen, and carbon dioxide, are less well understood; they are discussed in Chap. 3.

In this chapter we show how photochemical theory can be used to calculate the densities of many minor constituents of the atmosphere, provided the densities of the major constituents, the temperature of the atmosphere, and the rates at which some of the gases are produced by biological processes occurring at the surface are all known. In principle we could calculate the concentrations of the minor constituents at any stage of earth's history if these inputs to the calculation were known. At present, however, uncertainty regarding past values of these background parameters, particularly the biological production of minor gases, is a severe constraint. To illustrate the possibilities we present an application of photochemical theory to the recent geological past at the end of this chapter. Applications to the earliest history of the atmosphere appear in Chap. 6.

Atmospheric chemistry is a well-developed field with a complicated theory. The material presented in this chapter has been grossly simplified in order to expose the essential behavior of the atmospheric chemical system and to permit the derivation of approximate results without

49

recourse to complicated computer calculations. Experienced atmospheric chemists may feel that some of the simplifications are hard to justify. They may wish to go directly to the last section of the chapter, where the results are summarized. Newcomers to the field may find that even the simplified theory contains more than they want to know about atmospheric chemistry. They too are invited to skip the details and concentrate on the summaries that appear at the beginning and end of several of the following sections.

The chapter begins with background information on atmospheric chemistry for those newcomers to the field who want to understand the arguments that follow. There follow discussions of the penetration of sunlight into the atmosphere and the transport of minor constituents from one height to another—both topics important in photochemical theory. Next we describe the aspects of atmospheric chemistry that are reasonably well understood and not very controversial. This discussion is summarized in the opening paragraphs of the section entitled "Ozone in the Troposphere." Finally we take up aspects of atmospheric chemistry that are important for atmospheric evolution, but are not yet well understood.

We are concerned mainly with processes that affect the composition of the troposphere. Partly this is because most of the mass of the atmosphere is in the troposphere, so the composition of tropospheric air is close to the average composition of the whole atmosphere. But also we concentrate on the troposphere because this is where we live! Properties of the troposphere have determined the gaseous environment for the evolution of life.

For ozone, however, we cannot limit our attention to the troposphere. Ozone is the only one of the trace constituents that affects atmospheric properties other than composition. Ozone absorbs solar radiation in the near-ultraviolet region of the spectrum, from about 2000 A to 3000 A. The energy absorbed by ozone affects the energy budget of the middle atmosphere and causes the temperature maximum at the stratopause. There are no temperature maxima at intermediate altitudes on Mars and Venus, where ozone is much less abundant than on earth. Ozone is biologically important as well, because the near-ultraviolet radiation which it removes from sunlight is harmful to many organisms (Urbach, 1969). Most of the atmosphere's ozone is in the stratosphere; we therefore discuss processes that produce this constituent (*sources*) in the stratosphere together with processes that destroy it (*sinks*).

INTRODUCTION TO ATMOSPHERIC CHEMISTRY

We describe here some of the basic concepts of atmospheric chemistry, to which field so much of this chapter is devoted.

In order for two gas molecules to react they must collide. The rate at which molecules of one species collide with molecules of another species is proportional to the product of the densities of the two species. Molecules of a *species* are chemically identical to one another; N_2 and NH_3 are both species. We may therefore write

(2-1) $$R = K[A][B]$$

where R cm^{-3} sec^{-1} is the rate of collisions between species A and species B, $[A]$ cm^{-3} and $[B]$ cm^{-3} are the densities of the two species, and K cm^3 sec^{-1} is the proportionality constant, which we may call the *collision rate coefficient*. The collision rate coefficient is approximately equal to the product of the *cross section* for collisions—roughly speaking the average cross-sectional area of the colliding molecules—and the mean thermal speed of the molecules. Cross sections for collisions between atmospheric molecules are approximately 2×10^{-15} cm^2 and the mean thermal speed of the molecules in the troposphere is about 5×10^4 cm sec^{-1}, so the collision rate coefficient is about 10^{-10} cm^3 sec^{-1}.

As a rule only a fraction of all collisions between gas molecules result in their reaction. The reaction rate is therefore usually smaller than the collision rate, but proportional to it. Consider, for example, a reaction in which species A and B react to produce species C and D:

(2-2) $$A + B \rightarrow C + D$$

The rate of this reaction is given by Eq. (2-1), where R cm^{-3} sec^{-1} is now the rate at which molecules of A and B are destroyed by the reaction, equal to the rate at which molecules of C and D are produced, and K cm^3 sec^{-1} is now the *reaction rate coefficient*. Reaction rate coefficients are usually derived from laboratory experiments. Many of them depend on temperature. Few of them exceed the collision rate coefficient.

Some reactions can occur only when more than two molecules collide simultaneously. These are usually recombination reactions in which two molecules combine to form a single product. A reaction of the type

(2-3) $$A + B \rightarrow AB$$

is seldom possible in a gas because chemical energy is released when the combined molecule forms, and there is no way to dispose of this extra energy while conserving momentum.* If a third molecule takes part in the reactive collision, however, it can carry off the excess energy in the form of kinetic energy without doing violence to momentum conservation:

(2-4) $$A + B + M \rightarrow AB + M$$

---

* Collisions between gas molecules occur in isolation. The momentum and kinetic energy of a single reaction product are therefore determined by the requirement that momentum be conserved in the reaction. Chemical energy released by the reaction cannot therefore be converted into kinetic energy of a single product molecule.

Any species will function as the third molecule in such a *three-body reaction*, although not all molecules are equally effective in causing a reaction to occur. The symbol M is used to denote an atmospheric molecule in reactions where the identity of the molecule is of no concern.

The rate of a three-body reaction such as (2-4) is proportional to the product of the densities of the species that must collide simultaneously if a reaction is to occur. Thus for reaction (2-4) we have

(2-5) $$R = K[A][B][M] \, cm^{-3} \, sec^{-1}$$

where $K \, cm^6 \, sec^{-1}$ is now the three-body reaction rate coefficient.

Once we know the rates at which chemical reactions occur, we are able to evaluate their affects on the composition of the atmosphere. To calculate the density of a reactive gas we first identify the reactions that produce this gas. We sum the rates of these reactions to arrive at an expression for the chemical source of the gas. We next identify the reactions that remove the gas from the atmosphere and sum their rates to arrive at an expression for the chemical sink. By equating the expression for the source to the expression for the sink, we obtain an algebraic equation that can, in principle, be solved for the density of the gas under consideration.

Since the gas under consideration is a reactant in all of the reactions that destroy it, its density appears in the expressions for the rates of all of the destruction reactions. The total rate of destruction of the gas is therefore proportional to its density, and the density can be factored out of the expression for the chemical sink. Suppose, for example, that constituent X is destroyed in reactions with A, B, and C, with rate coefficients K_A, K_B, and K_C. The total chemical sink may be written

(2-6) $$S = [X]\{K_A[A] + K_B[B] + K_C[C]\}$$

or

(2-7) $$S = [X]L$$

or

(2-8) $$S = [X]/\tau$$

where $S \, cm^{-3} \, sec^{-1}$ is the rate of destruction of X by chemical reactions, $L \, sec^{-1}$ is an effective loss coefficient, and $\tau \, sec$ is the lifetime of X against destruction in chemical reactions. If $Q \, cm^{-3} \, sec^{-1}$ is the total chemical source of the gas, equating source to sink yields

(2-9) $$[X] = Q/L = Q\tau$$

We shall use this relationship between density, production rate, and lifetime frequently in our discussion of atmospheric chemistry.

As we have described it, Eq. (2-9) is valid only for gases that have very short chemical lifetimes. The chemical source of a gas at a given level of the atmosphere can be equated to the chemical sink at that level only if the gas reacts so rapidly that its density is not affected by transport of the gas from one level of the atmosphere to another. Gases for which this is true are said to be in *photochemical equilibrium*. There are gases like methane, hydrogen, and carbon monoxide, however, that play important roles in atmospheric chemistry, but which react so slowly that their chemical lifetimes are measured in years. Gases with lifetimes this long have local densities that are largely controlled by transport. Transport processes in the troposphere tend to produce a homogeneous chemical composition, so the constituents with long chemical lifetimes tend to have constant mixing ratios in the troposphere.

For these gases the role of chemistry is to control not the local density, but the total amount of the gas in the atmosphere. Equilibrium for these gases is achieved when the chemical production integrated throughout the atmosphere is equal to the chemical loss also integrated throughout the atmosphere. Equation (2-9) can still be used for long-lived gases, therefore, if we interpret $[X]$ to be the total amount of the constituent in the atmosphere and Q to be the total rate of production. In practice, it is frequently convenient to neglect horizontal variations in atmospheric properties, interpret $[X]$ cm^{-2} as the *column density* of the gas (the number of molecules in a vertical column of the atmosphere with a cross-sectional area of 1 cm^2), and take Q to be the height-integrated production rate. To convert column densities (in cm^{-2}) to local densities (in cm^{-3}) we need to know the altitude distribution of the gas. For a well-mixed gas with a constant mixing ratio, we may take the height distribution to be exponential with a scale height equal to that of the ambient atmosphere.

ABSORPTION OF SOLAR RADIATION IN THE ATMOSPHERE

Most atmospheric gases react with one another very slowly indeed at normally prevailing temperatures. Chains of rapid reactions are, however, initiated when atmospheric molecules are disrupted by solar radiation, producing chemically active *free radicals*. Solar photons, in fact, play such a key role in atmospheric chemistry that the subject is frequently called *photochemistry*. We must examine the penetration of solar photons into the atmosphere and derive the rates at which free radicals are produced by photodissociation (also called *photolysis*).

Atmospheric gases absorb radiation at some wavelengths and not at others. Radiation that is not absorbed by some gas or other has no effect on atmospheric chemistry. On the other hand, solar photons that are

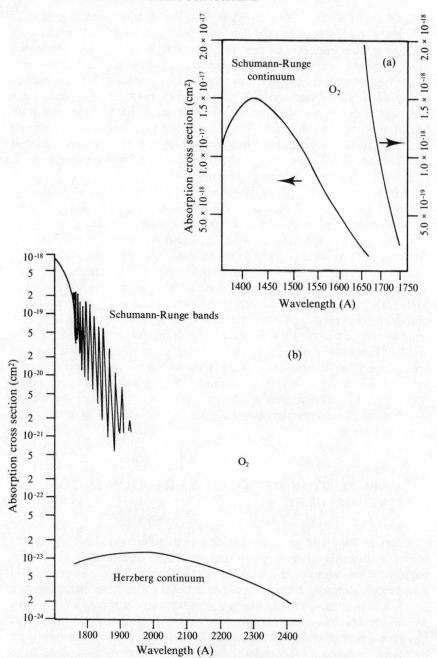

Fig. 2-1. Adsorption cross section of molecular oxygen. The names designate different wavelength regions of the absorption spectrum. (After Ackerman, 1971; Banks and Kockarts, 1973.)

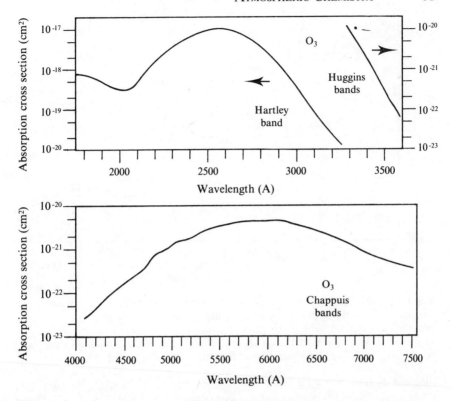

Fig. 2-2. Absorption cross section of ozone. (After Ackerman, 1971.)

absorbed at some level of the atmosphere are removed from the incoming beam of solar radiation and are not available for absorption at lower levels. We must therefore consider how the flux of solar photons passing downwards through the atmosphere varies with altitude and with wavelength.

The efficiency with which a given gas absorbs radiation is expressed by the *absorption cross section*, which is a function of wavelength. In the terrestrial atmosphere, the principal absorbers of radiation with wavelengths longer than 1400 A are molecular oxygen and ozone. Absorption cross sections of these gases are shown in Fig. 2-1 and 2-2. At wavelengths where the absorption cross sections are small, we can expect solar radiation to reach the ground with little attenuation. At wavelengths where the cross sections are large, on the other hand, we can expect absorption to occur fairly high in the atmosphere, with very little solar radiation penetrating to the ground or even into the troposphere. Roughly speaking, solar radiation of a given wavelength penetrates to the level of the atmosphere at which the overlying column density of absorbing gas is equal to the reciprocal of the absorption cross section. This

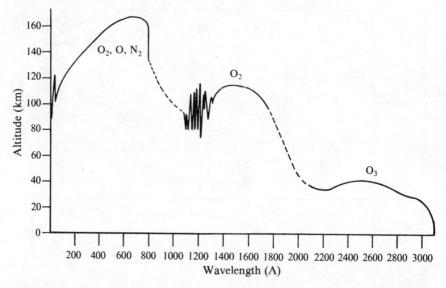

Fig. 2-3. Penetration of solar radiation into the atmosphere as a function of wavelength. The curve shows the altitude at which the rate of absorption of energy is a maximum when the sun is in the zenith. The principal absorbing constituents in each wavelength region are indicated. (From Friedman, 1960. Copyright 1960 by Academic Press Inc., New York, N.Y. Used by permission of the author and the publisher.)

altitude is sketched as a function of wavelength in Fig. 2-3. We see that solar radiation with wavelengths shorter than 3000 A is largely absorbed above the troposphere. Radiation at longer wavelengths reaches the ground with little attenuation.

For calculations of tropospheric photochemistry, however, we need quantitative information on the flux of solar radiation as a function of wavelength that penetrates into the troposphere. Figure 2-4 illustrates the incident solar spectrum as well as results of a particular calculation of the spectrum at the tropopause and at the ground.

With results such as these in hand, it is possible to calculate the rate of a specific photon absorption process. As an example, let us consider the rate of destruction of ozone molecules by photodissociation. The *dissociation cross section* $\sigma(\lambda)$ is defined by the statement that the rate of dissociation of ozone by radiation of wavelength λ is given by the product of the dissociation cross section, the ozone density, and the flux of photons at wavelength λ. (Because absorption processes other than dissociation are possible, the dissociation cross section may be smaller than the absorption cross section.) To obtain the total rate of ozone photolysis, we must integrate this product over all wavelengths for which dissociation occurs:

(2-10) $$Q = [O_3] \int_0^{\lambda'} \sigma(\lambda) F(\lambda)\, d\lambda$$

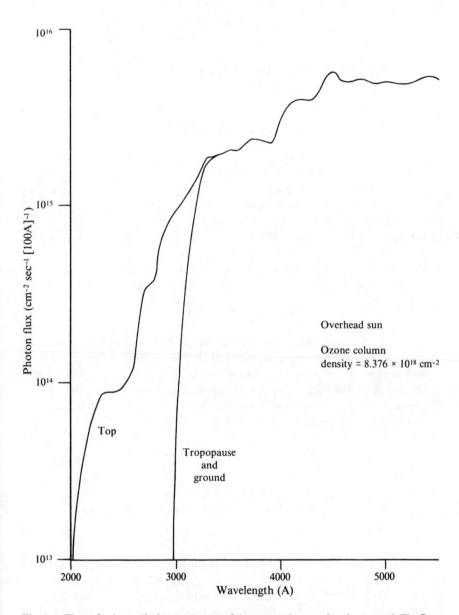

Fig. 2-4. Flux of solar radiation at the top of the atmosphere and at the ground. The flux at the tropopause cannot be distinguished from the flux at the ground on this scale. (From Chameides, 1974.)

HEIGHT DISTRIBUTION OF SOLAR PHOTON FLUX

Suppose that a beam of photons, all of one wavelength, suffers some attenuation in passing through an absorbing gas of density n cm^{-3}. Let F cm^{-2} sec^{-1} be the photon flux and dF be the number of photons absorbed when the beam travels a small distance, ds cm, through the absorbing gas. The absorption cross section, σ cm^2, is defined by the equation

$$dF = -F \, ds\sigma n$$

Absorption cross sections can be measured in the laboratory. Our problem is to derive the solar photon flux as a function of altitude and of wavelength from data on absorption cross sections and the altitude distributions of absorbing constituents.

First, we must allow for the possible existence of several absorbing gases with different densities, $n_i(z)$, and different absorption cross sections, $\sigma_i(\lambda)$. Since the absorption by one gas adds to the absorption by other gases, the equation for the reduction in the photon beam becomes

$$dF = -F \, ds \sum_i \sigma_i(\lambda) n_i(z)$$

Solar photons in the atmosphere usually travel downwards at an angle of θ with respect to the vertical (θ, the angle between the sun and the zenith, is called the *solar zenith angle*). We can therefore replace the path increment, ds, by a height increment, dz, using the relationship $ds = -dz/\cos \theta$. A simple differential equation for F as a function of z results. It can be integrated directly to give

$$F(\lambda, z, \theta) = F_\infty(\lambda) \exp[-\tau(\lambda, z)/\cos \theta]$$

(Chapman, 1926). In this expression, $F_\infty(\lambda)$ is the solar photon flux given as a function of wavelength incident on the top of the atmosphere, while $\tau(\lambda, z)$ is the zenith optical depth defined by

$$\tau(\lambda, z) = \sum_i \sigma_i(\lambda) \int_z^\infty n_i(h) \, dh$$

The integral on the right of this expression is the column density of constituent i above altitude z. It is equal to $n_i(z)H_i(z)$, where $H_i = kT/\bar{m}g$ if the atmosphere is well mixed, and $H_i = kT/m_i g$ if the atmosphere is in diffusive equilibrium (see Chap. 1 for discussion of the scale height H_i). We see that the optical depth is proportional to $n_i(z)$ and therefore decreases exponentially with increasing altitude. Since the optical depth appears in the exponent of the expression

HEIGHT DISTRIBUTION OF SOLAR PHOTON FLUX (continued)

for the photon flux, the flux is constant at high altitudes, where $\tau \ll 1$, and decreases rapidly at low altitudes, where $\tau > 1$. An indication of the altitude at which solar radiation is absorbed is provided by a plot against wavelength of the altitude where $\tau = 1$. Such a plot is presented in Fig. 2-3.

The expression for $F(\lambda, z, \theta)$ is not correct when θ is close to 90°, corresponding to the sun close to the horizon. Allowance must be made in this situation for the curvature of the earth's surface. An expression for the optical depth in a spherical atmosphere has been derived by Chapman (1931; cf. Smith and Smith, 1972). Since we are interested in radiation fluxes averaged over many days we need not concern ourselves with this refinement.

Absorption cross sections for the most important absorbing gases in the terrestrial atmosphere are shown in Fig. 2-1 and 2-2. The incident flux of solar radiation, $F_\infty(\lambda)$, is shown in Fig. 2-4. With these data and the altitude profiles of O_2 and O_3, the equations presented above can be used to calculate the flux of ultraviolet radiation, $F(\lambda, z, \theta)$, at any altitude and for any solar zenith angle. Some modifications to the calculation must be made to allow for absorption in the Schumann–Runge band system of molecular oxygen at wavelengths between 1750 A and 2030 A (Brinkmann, 1969; Kockarts, 1971). The absorption spectrum in this wavelength region consists of many narrow, closely spaced lines, and the structure is too fine to permit its resolution in any reasonable calculation of optical depth. An effective absorption cross section, averaged over many of these lines, must therefore be used. This effective cross section decreases as radiation penetrates deeper into the atmosphere because photons at the wavelengths of absorption maxima are removed from the incident beam at higher altitudes than photons at the wavelengths of absorption minima. Further corrections to the calculation must be made at low levels in the atmosphere to allow for Rayleigh scattering of the incident photons by atmospheric molecules as well as absorption and scattering by aerosol and dust particles (Leighton, 1961; Levy, 1972).

As an illustration we show in Fig. 2-4 the results of a calculation of the flux of photons as a function of wavelength at the ground. This spectrum was evaluated for an overhead sun and an ozone column density of $8.376 \times 10^{18} \text{ cm}^{-2}$. We see that the penetration of solar photons into the troposphere is negligible at wavelengths shorter than 3000 A. The short-wavelength photons are absorbed at higher levels in the atmosphere, as indicated in Fig. 2-3.

where λ_t is the longest wavelength at which dissociation is possible, and $F(\lambda)$ is the flux of solar radiation; Q, $[O_3]$, and $F(\lambda)$ are functions of altitude. It is convenient to define a *dissociation coefficient*,

(2-11) $$J = \int_0^{\lambda_t} \sigma(\lambda) F(\lambda)\, d\lambda$$

Then $Q = J[O_3]$, and the average lifetime of an ozone molecule before it is dissociated by sunlight is $1/J$.

The photodissociation coefficient is a function of solar zenith angle and therefore of latitude, season, and time of day. In the discussion that follows, we will not concern ourselves with diurnal variations of the photochemistry, but only with averages over a day of the densities of the chemically active trace constituents. We will therefore use photodissociation coefficients that have been averaged over 24 h with due allowance for the variation of solar zenith angle with time. Examples of average photodissociation coefficients appear in Table 2-1. The average lifetime

TABLE 2-1. Some photochemical reactions important in the troposphere

	REACTION	RATE COEFFICIENT[a]
R1	$O_3 + h\nu \rightarrow O_2 + O$	$J1 = 1.9 \times 10^{-4}$[b]
R2	$O + O_2 + M \rightarrow O_3 + M$	$K_2 = 1.1 \times 10^{-34} \exp(510/T)$
R3	$NO_2 + h\nu \rightarrow NO + O$	$J3 = 3.5 \times 10^{-3}$
R4	$NO + O_3 \rightarrow NO_2 + O_2$	$K4 = 9 \times 10^{-13} \exp(-1200/T)$
R5	$O_3 + h\nu_{UV} \rightarrow O_2 + O(^1D)$	$J5 = 9.2 \times 10^{-6}$
R6	$O(^1D) + M \rightarrow O(^3P) + M$	$K6 = 5 \times 10^{-11}$
R7	$O(^1D) + H_2O \rightarrow 2OH$	$K7 = 3.5 \times 10^{-10}$
R8	$OH + CO \rightarrow H + CO_2$	$K8 = 2.1 \times 10^{-13} \exp(-115/T)$
R9	$H + O_2 + M \rightarrow HO_2 + M$	$K9 = 3 \times 10^{-32}(273/T)^{1.3}$
R10	$HO_2 + NO \rightarrow OH + NO_2$	$K10 = 1.1 \times 10^{-11} \exp(-1000/T)$
R11	$OH + HO_2 \rightarrow H_2O + O_2$	$K11 = 2 \times 10^{-10}$
R12	$CH_4 + OH \rightarrow CH_3 + H_2O$	$K12 = 3.83 \times 10^{-12} \exp(-1830/T)$
R13	$H_2C = O + h\nu \rightarrow HCO + H$	$J13 = 1.5 \times 10^{-5}$
R14	$H_2C = O + h\nu \rightarrow H_2 + CO$	$J14 = 4.1 \times 10^{-5}$
R15	$H_2C = O + OH \rightarrow HCO + H_2O$	$K15 = 1.4 \times 10^{-11}$
R16	$H_2 + OH \rightarrow H_2O + H$	$K16 = 6.8 \times 10^{-12} \exp(-2020/T)$
R17	$HO_2 + O_3 \rightarrow OH + 2O_2$	$K17 = 1.0 \times 10^{-13} \exp(-1250/T)$
R18	$OH + NO_2 \overset{M}{\rightarrow} HNO_3$	$K18 = 8 \times 10^{-12}$
R19	$HNO_3 + h\nu \rightarrow OH + NO_2$	$J19 = 1.4 \times 10^{-6}$
R20	$HNO_3 + OH \rightarrow H_2O + NO_3$	$K20 = 6.0 \times 10^{-13} \exp(-400/T)$
R21	$NO_3 + NO \rightarrow 2NO_2$	$K21 = 8.7 \times 10^{-12}$

After Chameides, 1975.

[a] The units are $\sec^{-1}$ for photodissociation coefficients, $cm^3 \sec^{-1}$ for two-body reactions, and $cm^6 \sec^{-1}$ for three-body reactions.

[b] Photodissociation coefficients, designated J, were calculated by Chameides and Walker (1973) for ground-level air at 30°N latitude in July.

of a given constituent against photodissociation is the reciprocal of the average photodissociation coefficient. For ozone in the midlatitude troposphere, this lifetime is about 2 h. As we shall show, photodissociation of ozone in the troposphere is usually followed by reformation of ozone: so the photodissociation lifetime is not the same as the effective lifetime for permanent destruction of ozone.

CHARACTERISTIC TIMES FOR TRANSPORT

As described in Chap. 1, the relative importance of transport and photochemistry in determining the distribution of a reactive species may be assessed by comparing characteristic transport times with photochemical lifetimes such as the photodissociation times we have just derived. Transport times have been deduced from studies on a variety of atmospheric tracers (Junge, 1962; Jacobi and Andre, 1963). The subject has been reviewed by Sheppard (1963) and Reiter (1971) and results have been summarized by Pressman and Warneck (1970).

Consider first the rapidity of vertical transport in the troposphere. The vertical mixing time varies markedly with weather, season, and location, but an approximate average value is 25d. The important point is that constituents with photochemical lifetimes longer than about a month exhibit mixing ratios that do not vary with altitude in the troposphere. Water vapor is an exception because its mixing ratio in the troposphere is limited by condensation.

Within one hemisphere, horizontal mixing times in the troposphere are comparable to vertical mixing times. Horizontal distances are greater than vertical distances, but horizontal winds are greater than vertical winds. Mixing between northern and southern hemispheres is relatively slow, however, and takes about a year. Species with photochemical lifetimes between a month and a year are therefore likely to be homogeneously distributed in each hemisphere, but to have different concentrations in the two hemispheres. Species with photochemical lifetimes in excess of a year tend to have concentrations that are constant throughout the troposphere.

Because of its positive temperature gradient, the stratosphere is much more stable than the troposphere, and both horizontal and vertical mixing is slow. Studies of radioactive tracers show that horizontal mixing within one hemisphere takes from 1 to 2 y, and between hemispheres about 5 y (Pressman and Warneck, 1970). These studies also show that material remains in the lower stratosphere for 1 to 2 y before being transferred to the troposphere (cf. Gudiksen et al., 1968; Martell, 1970; Ehhalt and Heidt, 1973).

The average upward flux of tropospheric air into the stratosphere, which is the same as the average downward flux of stratospheric air into the troposphere, has been estimated by Newell (1970) from consideration of the energy budget of the atmosphere in the region of the tropical tropopause. His value is 10^{16} cm^{-2} sec^{-1}. By dividing this flux into the average column density of air in the stratosphere, 3.2×10^{24} cm^{-2} (SCEP, 1970), we obtain an estimate of the average length of time an air molecule spends in the stratosphere before being transported to the troposphere. This time is about 10 y. It corresponds to the residence time in the entire stratosphere, whereas the residence times derived from radioactive tracers refer to the lower stratosphere only. By dividing the flux into the column density of air in the troposphere, 1.8×10^{25} cm^{-2}, we derive a characteristic time of about 57 y for the transfer of air from the troposphere to the stratosphere. Photochemical lifetimes tend to be short in the stratosphere, but the tropospheric budgets of most reactive constituents are not affected by the stratosphere because of this long transport time.

FREE RADICAL DENSITIES IN THE TROPOSPHERE

The first few photochemical problems we consider are simple ones, susceptible to solution with pencil and paper. The reactions involved in these relatively simple photochemical problems are listed in Table 2-1, along with their rate coefficients. These reactions are numbered in the text as they are in the table. As we progress further into the subject, it will become increasingly necessary to appeal to the results of complex computer calculations of photochemical systems involving many more reactions than appear in Table 2-1. More complete tables of reactions important in the troposphere have been given by Chameides and Walker (1976) and Anderson (1976).

We begin with an illustrative calculation of the density of oxygen atoms produced by photolysis of ozone. Then we consider how this calculation should be modified to allow for the source of atoms provided by photolysis of nitrogen dioxide.

It turns out that oxygen atoms in their ground electronic state $O(^3P)$ are photochemically uninteresting. What we need to know is the density of electronically excited atoms, $O(^1D)$, produced by photolysis of ozone at shorter wavelengths. These atoms are in their first excited state, with an energy of electronic excitation equal to 1.96 eV. The state is *metastable*, which means that the atom hardly ever loses energy by emission of radiation. The excitation energy makes $O(^1D)$ atoms highly reactive. What gives them their importance in tropospheric chemistry is their reaction with water to produce hydroxyl radicals, OH. We shall therefore calculate the density of metastable oxygen atoms and then use this information to calculate the density of hydroxyl radicals. The OH density (see Fig. 2-7) is the main result of this section.

Photodissociation of ozone produces an oxygen atom,

(R1) $$O_3 + h\nu \rightarrow O_2 + O$$

which immediately combines with an oxygen molecule to reform ozone,

(R2) $$O + O_2 + M \rightarrow O_3 + M$$

where M is any atmospheric molecule. Taking 300°K as the average temperature at the ground at 30°N during July (U.S. Standard Atmosphere Supplement, 1966), the rate coefficient of this reaction (Table 2-1) is $K2 = 6.0 \times 10^{-34}\,cm^3\,sec^{-1}$. We can now estimate the lifetime of an oxygen atom at the ground in order to decide whether transport affects oxygen atom densities. The lifetime is

$$\{K_2[O_2][M]\}^{-1} = 1.4 \times 10^{-5}\,sec$$

Because it is very much shorter than the characteristic transport times discussed above, we conclude that the atomic oxygen density is not affected by transport. We can therefore estimate the local density by equating the local rate of production by reaction (R1) to the local rate of destruction by reaction (R2). This procedure is equivalent to multiplying the production rate by the photochemical lifetime.

For midlatitude summer conditions, the average ozone density at the ground is $7 \times 10^{11}\,cm^{-3}$ (Hering and Borden, 1964). We multiply this number by the photodissociation coefficient for reaction (R1), given in Table 2-1, to obtain $1.3 \times 10^8\,cm^{-3}\,sec^{-1}$ as the rate of production of oxygen atoms by photolysis of ozone. Multiplying the production rate by the photochemical lifetime derived above we find $1.8 \times 10^3\,cm^{-3}$ as the density at the ground of oxygen atoms produced by photodissociation of ozone.

We have presented this simple calculation in some detail in order to illustrate the methods that will be used throughout this chapter to arrive at densities of chemically active species. It is convenient to summarize the reactions involved in a calculation of this type in a diagram such as that of Fig. 2-5. In this diagram the arrows denote reactions that convert species in one circle into species in another. Species taking part in the reaction are shown alongside the arrow together with the number of the reaction. This diagram is very simple, but more complicated ones will follow. The diagram makes it clear that, in the reaction system we have examined, photolysis of ozone is always followed by reformation of ozone. The reactions are *cyclic*, providing no net production or destruction of ozone or atomic oxygen. We shall encounter many other reaction cycles.

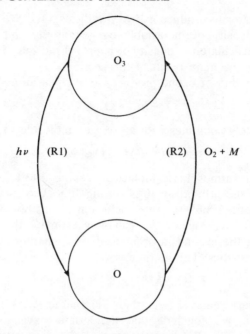

Fig. 2-5. The reaction cycle linking atomic oxygen and ozone.

Our calculation of the atomic oxygen density is incomplete because we have neglected a potentially important source of oxygen atoms provided by the photolysis of nitrogen dioxide:

(R3) $$NO_2 + h\nu \rightarrow NO + O$$

The nitric oxide molecule produced by this reaction is converted back into nitrogen dioxide by reaction with ozone:

(R4) $$NO + O_3 \rightarrow NO_2 + O_2$$

The ozone molecule destroyed in the process is replaced by a molecule formed by reaction (R2) from the oxygen atom produced by reaction (R3). The reactions are thus once again cyclic, with no net production or destruction of any of the participating species.

The combined reaction cycles are illustrated in Fig. 2-6. In photochemical equilibrium, considering just the reactions shown in this figure, the ratio of NO to NO_2 is given by

(2-12) $$[NO]/[NO_2] = J3/K4[O_3]$$

Although there are other reactions that produce and destroy NO and NO_2, reactions (R3) and (R4) are, in fact, by far the most important in the troposphere. Measurements by Stedman et al. (1975) have confirmed the accuracy of expression (2-12).

The oxygen atom density therefore depends on the NO_2 density as well as on the O_3 density. Measurements have shown that the concentration of NO_2 in tropospheric air varies with position and time by more than a factor of ten (Migeotte and Nevin, 1952; Junge, 1963; Lodge and Pate, 1966; Dalgarno, 1969; Robinson and Robbins, 1970a; Breeding et al., 1973; Lodge et al., 1974; Noxon, 1975). For purposes of illustration, we adopt a high value of the mixing ratio, 2.3×10^{-9}, corresponding to a density at the ground of $5.5 \times 10^{10} \, cm^{-3}$. The rate of production of O by photolysis of NO_2 is $J3[NO_2]$, equal to $1.9 \times 10^8 \, cm^{-3} \, sec^{-1}$. Adding this to the value of $1.3 \times 10^8 \, cm^{-3} \, sec^{-1}$ already obtained for production of O by photolysis of O_3 and multiplying by the lifetime previously calculated, we find $[O] = 4.6 \times 10^3 \, cm^{-3}$ at the ground. The calculation can be repeated at different altitudes leading to results like those shown in Fig. 2-7.

Atomic oxygen in its ground state is not sufficiently reactive to play an important role in tropospheric photochemistry. But radiation at wavelengths shorter than 3400 A dissociates ozone to produce electronically excited oxygen in the metastable ^1D level:

(R5) $$O_3 + h\nu(\lambda < 3400A) \rightarrow O_2 + O(^1D)$$

The photodissociation coefficient for the production of $O(^1D)$, $J5$, is given in Table 2-1. Photodissociation of NO_2 does not produce metastable

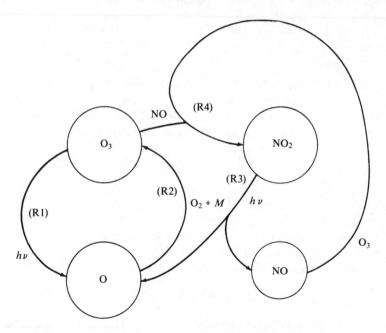

Fig. 2-6. Reactions controlling the O density and the NO-to-NO_2 ratio.

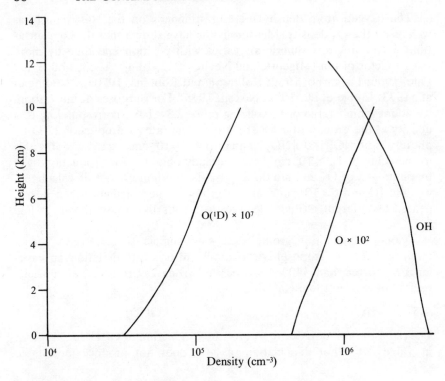

Fig. 2-7. Calculated daily average densities of OH, O, and O(^{1}D).

oxygen atoms. Taking 7×10^{11} cm^{-3} for the ozone density we calculate a rate of production of O(^{1}D) atoms of 6.4×10^6 cm^{-3} sec^{-1} at the ground.

The O(^{1}D) atoms produced by reaction (R5) are converted to ground-state O(^{3}P) atoms in collisions with nitrogen or oxygen molecules:

(R6) $$O(^1D) + M \rightarrow O(^3P) + M$$

The rate coefficient for this quenching reaction is 5×10^{-11} cm^3 sec^{-1} (Table 2-1). So the lifetime of O(^{1}D) at the ground is 8.3×10^{-10} sec. At the ground we find $[O(^1D)] = 5.4 \times 10^{-3}$ cm^{-3}. Representative densities as a function of altitude are shown in Fig. 2-7; the reaction cycle is illustrated in the upper portion of Fig. 2-8.

A small fraction of the O(^{1}D) atoms reacts with H$_2$O—

(R7) $$O(^1D) + H_2O \rightarrow 2OH$$

—to produce the very reactive hydroxyl radical. Using the rate coefficient in Table 2-1 we calculate that the production rate of OH at the ground is

$1.2 \times 10^6 \, \text{cm}^{-3} \, \text{sec}^{-1}$ for a water vapor mixing ratio of 0.027 and a total gas number density of $2.4 \times 10^{19} \, \text{cm}^{-3}$ (U.S. Standard Atmosphere Supplement, 1966).

A large number of reactions play a role in determining the density of OH in the troposphere. The photochemistry has been analyzed and discussed by Levy (1971, 1972, 1973a), McConnell et al. (1971), and Crutzen (1973, 1974a). We shall draw on Levy's results to present the

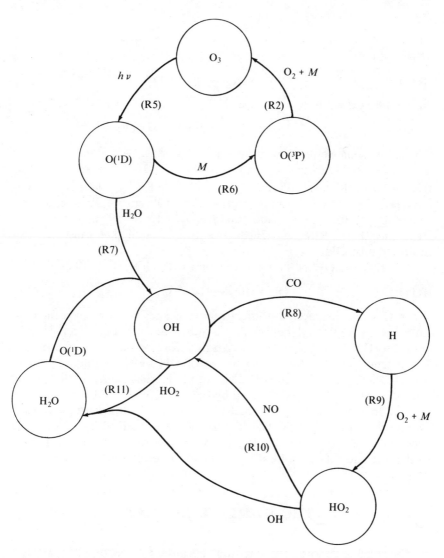

Fig. 2-8. Reactions controlling the densities of O(^{1}D), OH, and HO$_2$.

simplest possible model that leads to a reasonable estimate of the OH density. In doing this, we must neglect many important reactions. Those we consider are shown in Fig. 2-8.

The hydroxyl radical is converted to the hydroperoxyl radical by reaction with carbon monoxide,

(R8) $OH + CO \rightarrow H + CO_2$

followed by

(R9) $H + O_2 + M \rightarrow HO_2 + M$

The hydroperoxyl radical is converted back to hydroxyl by reaction with nitric oxide:

(R10) $HO_2 + NO \rightarrow OH + NO_2$

In equilibrium we have

(2-13) $[OH]/[HO_2] = K10[NO]/K8[CO]$

The rate coefficients are given in Table 2-1. At 300°K they are $K8 = 1.4 \times 10^{-13} \, cm^3 \, sec^{-1}$ and $K10 = 3.9 \times 10^{-13} \, cm^3 \, sec^{-1}$. With $[NO] = 1.7 \times 10^{10} \, cm^{-3}$—derived from Eq. (2-12) and $[NO_2] = 5.5 \times 10^{10} \, cm^{-3}$—and $[CO] = 2.88 \times 10^{12} \, cm^{-3}$ (Pressman and Warneck, 1970), we calculate $[OH]/[HO_2] = 1.6 \times 10^{-2}$ at the ground. Reactions (R8), (R9), and (R10) constitute a cycle that causes no net destruction of OH.

The sink for OH is the rapid reaction with HO_2:

(R11) $OH + HO_2 \rightarrow H_2O + O_2$

When this reaction occurs it effectively consumes two hydroxyl radicals, since the hydroperoxyl radical that is destroyed can no longer be converted to OH by reaction (R10). We therefore equate twice the rate of this reaction to the rate of production of OH by (R7), finding $[OH] = 6.9 \times 10^6 \, cm^{-3}$ at the ground. This value is reasonably close to the diurnally averaged values obtained by Levy (1973b) and McConnell et al. (1971). In what follows, we shall use Levy's (1973b) average altitude profile, shown in Fig. 2-7. The lifetime of OH, obtained by dividing the production rate into the density, is only a few seconds, so transport is negligible, and OH is in local photochemical equilibrium.

THE METHANE BUDGET

Hydroxyl is important to our study because it reacts rapidly with such permanent atmospheric constituents as methane, hydrogen, and carbon monoxide. Now that we have estimated the hydroxyl density we can examine the chemical reactions that these constituents undergo. We shall deal with methane first because, as we shall show, methane provides the source of a number of other trace gases in the troposphere.

In the calculations presented so far we have attempted to determine the concentrations of chemically active gases from the reactions that produce and destroy them. Our approach to methane and a number of the other gases will be different. The atmospheric concentrations of these gases have been measured, so we do not need to estimate their concentrations theoretically. Instead, we shall try to identify the processes that produce and destroy these gases (Junge, 1972) and to show that these processes are approximately in balance. These two approaches to a photochemical problem are equivalent. If we show that the measured concentration of methane, for example, leads to an atmospheric methane budget that is balanced, then we could use the methane budget to calculate a theoretical methane concentration that would agree with the measured concentration. We work from the known concentration to the budget because it is easier and because it reflects the actual course of research in this field.

A new kind of diagram is helpful for discussion of the budget of an atmospheric gas. In a *budgetary diagram* arrows denote the transfer of material between *reservoirs*, indicated by boxes. The rates of transfer are expressed in moles per year while the sizes of the reservoirs are expressed in *moles* (a mole is a mass in grams numerically equal to the molecular weight in atomic mass units). The *residence time* of material in a given reservoir is equal to the size of the reservoir divided by the rate of transfer of material into or out of the reservoir. A diagram of this type, illustrating the atmospheric methane budget, is presented in Fig. 2-9. The remainder of this section is devoted to a discussion of this budget.

Methane is destroyed by reaction with hydroxyl to form a methyl radical:

(R12) $$CH_4 + OH \rightarrow CH_3 + H_2O$$

With the hydroxyl density equal to $3.7 \times 10^6 \, cm^{-3}$ and $K12 = 8.6 \times 10^{-15} \, cm^3 \, sec^{-1}$ at the ground, the lifetime of a methane molecule is 3.1×10^7 sec, or about a year. Because of the temperature dependence of $K12$, the lifetime is longer at higher levels in the troposphere where temperatures are lower. Mixing times within the troposphere are shorter than this, so methane is homogeneously distributed.

We may roughly estimate the total rate of methane destruction by ignoring the increase of the methane lifetime with altitude. With a constant mixing ratio of 1.5×10^{-6} (Bainbridge and Heidt, 1966; Ehhalt, 1967) and an atmospheric scale height of 8.4 km, the column density of methane is $3 \times 10^{19} \, cm^{-2}$, and the height-integrated destruction rate is

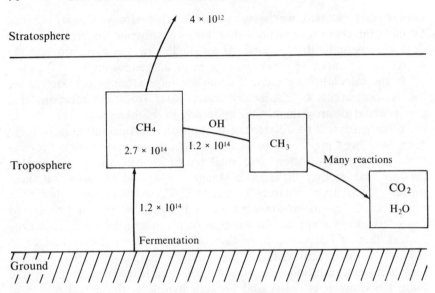

Fig. 2-9. Methane budget of the troposphere. The tropospheric reservoir is expressed in moles and the rates of transfer into and out of the reservoir are expressed in moles per year.

$9.8 \times 10^{11} \, \text{cm}^{-2} \, \text{sec}^{-1}$. A more accurate value for the height-integrated loss rate averaged over a day has been obtained by Levy (1973b) from a calculation of methane destruction as a function of altitude and time of day. His value is $4.5 \times 10^{11} \, \text{cm}^{-2} \, \text{sec}^{-1}$, close to the value obtained by McConnell et al. (1971) from a solution of the continuity and transport equations for methane. The average methane lifetime in the troposphere is then about 2.2 y. Our earlier discussion of transport from the troposphere to the stratosphere suggests that the flux of methane through the tropopause should be about $1.5 \times 10^{10} \, \text{cm}^{-2} \, \text{sec}^{-1}$, close to the values calculated by McConnell et al. (1971) and Wofsy and McElroy (1973). Loss to the stratosphere is therefore negligible compared with loss within the troposphere. This methane flux is, however, an important source of stratospheric hydrogen compounds. The flux of methane into the stratosphere is comparable to the flux of water vapor (Newell, 1970); and much of the methane is oxidized to water vapor in the stratosphere (McConnell et al., 1971; Wofsy et al., 1972; Ehhalt et al., 1972; Martell, 1973).

For the whole atmosphere a destruction rate of $4.5 \times 10^{11} \, \text{cm}^{-2} \, \text{sec}^{-1}$ equals destruction of $1.9 \times 10^{15} \, \text{gm yr}^{-1}$. There are no significant photochemical sources of methane. Surface sources are summarized in Table 2-2. The dominant sources are fermentation of organic matter in anaerobic soils in tropical forests and in swamps and paddy fields. Enteric fermentation in the intestines of domestic cattle and other large ungulates

TABLE 2-2.   Sources of atmospheric methane (gm yr^{-1})

Humid tropical areas	6.1×10^{14}	(Robinson and Robbins, 1968b)
Swamps	5.7×10^{14}	(Robinson and Robbins, 1968b)
Paddy fields	1.9×10^{14}	(Koyama, 1963)
Mines	0.8×10^{14}	(Koyama, 1963)
Ungulates	4.5×10^{13}	(Hutchinson, 1954)
Ocean	3.2×10^{12}	(Liss and Slater, 1974)
Total	14.9×10^{14}	

is also a source of methane (Hutchinson, 1954), but it appears to be minor. The release of methane from oil and gas wells and mines is less important than fermentation. Methane in surface sea water is almost in equilibrium with the atmosphere, so the open ocean is probably not a significant source (Lamontagne et al., 1973; Liss and Slater, 1974).

These estimates of surface sources are subject to considerable uncertainty. The agreement of the total with the rate of photochemical destruction of methane is close enough to suggest that the methane content of the atmosphere is determined by the balance between production at the surface and oxidation within the troposphere by hydroxyl radicals. The methane density is therefore proportional to the surface source divided by the OH density. The surface source has probably varied with time as a result of man's activities and as a result of changes in topography and climate which have changed the areas of anaerobic soils and swamps. So the methane content of the atmosphere has probably also varied with time. We shall consider this variability at the end of this chapter.

FORMALDEHYDE

The destruction of a methane molecule by reaction with hydroxyl initiates the chain of photochemical reactions illustrated in Fig. 2-10. The final products of this reaction chain are water and carbon dioxide, but intermediate products include formaldehyde, hydrogen, and carbon monoxide. The methane oxidation chain is, in fact, the major source of these constituents of the atmosphere. We shall therefore discuss it in this section.

The methyl radical produced by reaction (R12) combines immediately with oxygen to form a methylperoxyl radical:

$$(2\text{-}14) \qquad CH_3 + O_2 + M \rightarrow CH_3O_2 + M$$

The methylperoxyl radical may then react with nitric oxide to form a methoxy radical:

$$(2\text{-}15) \qquad CH_3O_2 + NO \rightarrow CH_3O + NO_2$$

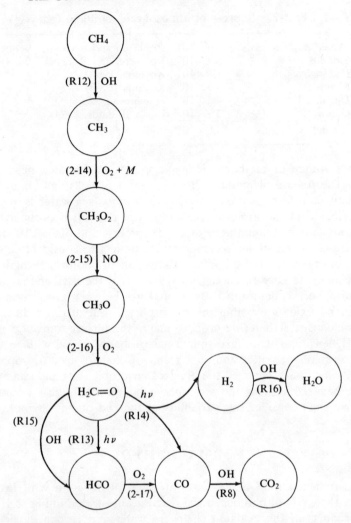

Fig. 2-10. Oxidation of methane.

Alternative reactions of the methylperoxyl radical have been discussed by Levy (1972, 1973a,b) and by Crutzen (1973). As in reaction (2-15), they result, ultimately, in the production of a methoxy radical.

The methoxy radical reacts with oxygen to form formaldehyde:

$$(2\text{-}16) \qquad CH_3O + O_2 \rightarrow H_2C{=}O + HO_2$$

The lifetimes of CH_3, CH_3O_2, and CH_3O are all so short that these species are in local photochemical equilibrium.

Reaction (2-16) is the principal source of formaldehyde in the atmosphere. Since the destruction of a methane molecule leads, by way of the reactions presented above, to the production of a formaldehyde molecule, the formaldehyde production rate is equal to the methane destruction rate. At the ground this rate is $1.2 \times 10^6 \, cm^{-3} \, sec^{-1}$; integrated through the troposphere it is $4.5 \times 10^{11} \, cm^{-2} \, sec^{-1}$.

Formaldehyde is removed by photolysis (Calvert et al., 1972):

(R13) $\qquad\qquad H_2C{=}O + h\nu \rightarrow HCO + H$

(R14) $\qquad\qquad H_2C{=}O + h\nu \rightarrow H_2 + CO$

and by reaction with hydroxyl:

(R15) $\qquad\qquad H_2C{=}O + OH \rightarrow HCO + H_2O$

With the photodissociation coefficients and rate coefficient in Table 2-1, and $[OH] = 3.7 \times 10^6 \, cm^{-3}$, we calculate a lifetime for formaldehyde at the ground of $9.3 \times 10^3 \, sec$, or 2.6 h. Transport does not affect the formaldehyde density; averaged over several days the density is in photochemical equilibrium, but the lifetime is long enough to cause departures from photochemical equilibrium during the course of a day. The average density we calculate is $1.1 \times 10^{10} \, cm^{-3}$ at the ground, close to the value deduced from measurements of formaldehyde in rain (Junge, 1963, p. 98).

The formyl radical, HCO, produced by reactions (R13) and (R15), is removed by reaction with oxygen:

(2-17) $\qquad\qquad HCO + O_2 \rightarrow HO_2 + CO$

Overall, one carbon monoxide molecule is produced for every methane molecule destroyed (McConnell et al., 1971). Molecular hydrogen is also produced (Levy, 1972) in the methane oxidation chain, but at a slower rate. Of the formaldehyde destruction reactions, only (R14) yields molecular hydrogen.

THE HYDROGEN BUDGET

With the information on the source of hydrogen provided by the methane oxidation chain we can now consider the budget of molecular hydrogen in the troposphere. This budget is summarized in Fig. 2-11. The principal source is photolysis of formaldehyde, while the principal sink is reaction with OH:

(R16) $\qquad\qquad H_2 + OH \rightarrow H_2O + H$

The photochemical lifetime at the ground is $3.3 \times 10^7 \, sec$, nearly equal to the methane lifetime. We therefore expect that hydrogen is well mixed in the troposphere. We may, as in the case of methane, obtain a rough estimate of the integrated rate of destruction by ignoring the variation of

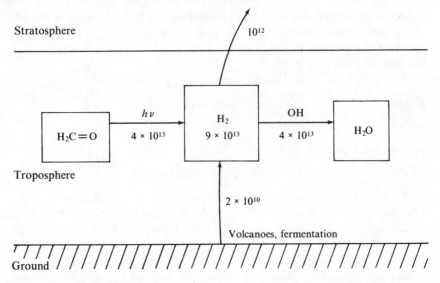

Fig. 2-11. Hydrogen budget of the troposphere. The reservoir is expressed in moles and rates of transfer in moles per year.

lifetime with altitude. The mixing ratio of H_2 in the troposphere is 5×10^{-7} (Scholz et al., 1970), so the integrated loss rate is 3.1×10^{11} cm^{-2} sec^{-1}. With allowance for the temperature dependence of the reaction rate coefficient the loss rate is reduced to 1.3×10^{11} cm^{-2} sec^{-1} (Levy, 1973b), corresponding to an average lifetime of 2.5 yr. This is equal to a loss rate for H_2 of 7.0×10^{13} gm yr^{-1} over the whole earth.

An estimate of the rate of release of hydrogen from volcanoes will be presented in Chap. 5. The value we obtain is 2×10^{10} gm yr^{-1}. Koyama's (1963) estimate of the rate of release of hydrogen as a result of fermentation in swamps and paddy fields is 1.8×10^{10} gm yr^{-1}. Although the uncertainties in these values are large, the estimates are so much smaller than the rate of photochemical loss that it is most unlikely that volcanoes and fermentation are significant sources of atmospheric hydrogen. A possible oceanic source (Williams and Bainbridge, 1973) is probably also small.

As we have already indicated, the principal source of hydrogen in the troposphere is photolysis of formaldehyde, reaction (R14). The hydrogen production rate is therefore $J14\,[H_2C{=}O]$. Using the photodissociation coefficient in Table 2-1 and $[H_2C{=}O] = 1.1 \times 10^{10}$ cm^{-3} we find a source of 4.5×10^5 cm^{-3} sec^{-1} at the ground. Because of the 2.5 yr lifetime of tropospheric hydrogen, production, like loss, must be integrated over altitude. Levy's (1973b) value for the total column production rate of hydrogen in the troposphere is 1.8×10^{11} cm^{-2} sec^{-1}. This value is close to the total loss rate, considering the uncertainties, so we may conclude that the hydrogen density in the troposphere is determined by balance between production in the photolysis of formaldehyde, reaction (R14), and destruction in the reaction with hydroxyl, reaction (R16).

As in the case of methane, the transport of hydrogen into the stratosphere has a negligible effect on the tropospheric budget. The hydrogen makes a minor but not negligible contribution to the budget of stratospheric hydrogen compounds (Martell, 1973).

THE CARBON MONOXIDE BUDGET

Like hydrogen, carbon monoxide is also produced as a byproduct of methane oxidation and destroyed by reaction with OH. The budget is illustrated in Fig. 2-12. There is some evidence for a source of carbon monoxide other than that provided by the methane oxidation chain. Let us examine this evidence.

Carbon monoxide is destroyed in the troposphere by reaction (R8) with OH. The photochemical lifetime at the ground is 1.9×10^6 sec. Because the rate coefficient is only weakly dependent on temperature (Table 2-1), the lifetime increases slowly with height, corresponding to the slow decrease in OH density (Fig. 2-7). Levy's (1973b) value for the average tropospheric lifetime is 0.1 yr, a value in harmony with results of radiocarbon measurements (Weinstock and Niki, 1972).

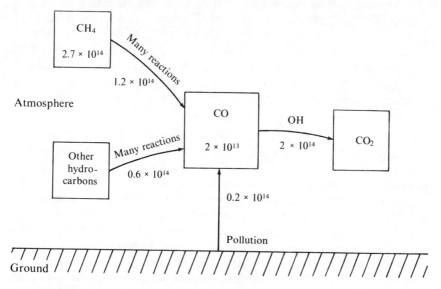

Fig. 2-12. The carbon monoxide budget. Reservoirs are expressed in moles and rates of transfer in moles per year.

SURFACE SOURCES OF CARBON MONOXIDE

Anthropogenic sources of carbon monoxide have been summarized by Jaffe (1973). His results are shown in the table. Automobiles are the largest source by far, but the total of all manmade sources is only 3.6×10^{14} gm yr^{-1}. Since the photochemical sink is 5.9×10^{15} gm yr^{-1}, pollution is not a major source of carbon monoxide on a global basis, although it may be dominant in restricted areas.

The oceans provide a natural source of carbon monoxide (Linnenbom et al., 1973). Several marine organisms secrete the gas (Wittenberg, 1960; Loewus and Delwicke, 1963; Chapman and Tocher, 1966; Pickwell et al., 1964), and it appears to be produced also by the photochemical oxidation of dissolved organic matter in sea water (Wilson et al., 1970). As a result, surface sea water can be supersaturated with respect to the partial pressure of carbon monoxide in the overlying atmosphere by a factor as large as 40 (Lamontagne et al., 1971). The average global flux of carbon monoxide from the ocean to the atmosphere has been estimated by Linnenbom et al. (1973) as 2.2×10^{14} gm yr^{-1}, and by Liss and Slater (1974) as 4.3×10^{13} gm yr^{-1}. This flux may be almost as large as the anthropogenic source, but it is substantially less than the photochemical sink.

The estimate of the volcanic source of carbon monoxide that we obtain in Chap. 5 is a negligible 2×10^{12} gm yr^{-1}. A number of other surface sources, mostly biological, have been mentioned by Jaffe (1973). They appear to be negligible, as does the production of carbon monoxide by lightning and other phenomena involving charged particles (Green et al., 1973). The total surface source is therefore probably less than 6×10^{14} gm yr^{-1}, equal to a flux of 8×10^{10} cm^{-2} sec^{-1}.

Comparing the lifetime with transport times presented earlier, we conclude that carbon monoxide densities can vary from one part of the troposphere to another, depending on the distribution of the source (Junge, 1974). Measurements of carbon monoxide have been reviewed by Jaffe (1973). Strong variations are indeed detected, with high concentrations in polluted, urban air and low concentrations over the southern oceans. As a typical mixing ratio in air remote from sources of pollution we shall use 1.2×10^{-7} (Pressman and Warneck, 1970; Jaffe, 1973; Linnenbom et al., 1973), although substantial variations apparently occur even in remote locations. With this value of the mixing ratio, the height-integrated loss rate is 7.9×10^{11} cm^{-2} sec^{-1} (Levy, 1973b), corres-

Estimated global anthropogenic CO sources for 1970

Sources	World Fuel Consumption $(10^{12}$ gm yr$^{-1})$	World CO Emission $(10^{12}$ gm yr$^{-1})$
Mobile		
Motor vehicles	439	199
Gasoline		197
Diesel		2
Aircraft (aviation gasoline, jet fuel)	84	5
Watercraft		18
Railroads		2
Other (nonhighway) motor vehicles (construction equipment, farm tractors, utility engines, etc.)		26
Stationary		
Coal and lignite	2983	4
Residual fuel oil	682	<1
Kerosene	69	<1
Distillate fuel oil	411	<1
Liquefied petroleum gas	34	<1
Industrial processes (petroleum refineries, steel mills, etc.)		41
Solid waste disposal (urban and industrial)	1130	23
Miscellaneous (agricultural burning, coal bank refuse, structural fires)		41
Total anthropogenic CO		359

From Jaffe (1973), *J. Geophys. Res.*, **78**, 5293–5305, copyrighted by American Geophysical Union.

ponding to a destruction rate for CO of 5.9×10^{15} gm yr^{-1} in the whole atmosphere.

Carbon monoxide is produced in the course of methane oxidation by reactions (R14) and (2-17). Since, with the reaction chain we have outlined, methane destruction is always followed by one or other of these reactions, we may equate the carbon monoxide production rate to the methane destruction rate. The production rate is therefore

$4.5 \times 10^{11} \, \text{cm}^{-2} \, \text{sec}^{-1}$, much larger than the total surface source of $8 \times 10^{10} \, \text{cm}^{-2} \, \text{sec}^{-1}$, but somewhat smaller than the photochemical sink.

Thus, although the methane oxidation chain is the largest source of tropospheric carbon monoxide, it appears that it is not the only important one. The additional source has not been unequivocally identified, but it is probably provided by the oxidation of gaseous hydrocarbons other than methane by reactions similar to those that oxidize methane. Kummler and Baurer (1973) have shown that this mechanism may reasonably provide the additional source required to bring the carbon monoxide budget into balance.

OZONE IN THE TROPOSPHERE

The material that has been presented so far has been relatively straightforward and not controversial. We are about to take up a number of subjects in which the uncertainties are greater and there is, as yet, little consensus among researchers in the field of atmospheric chemistry. Before doing this, let us review the discussion up to this point.

Tropospheric photochemistry is initiated by the photolysis of ozone molecules to produce metastable ^1D oxygen atoms. Metastable oxygen atoms react with water vapor to produce hydroxyl radicals. Hydroxyl radicals oxidize methane, hydrogen, and carbon monoxide. Methane is produced by biological processes in anaerobic environments at the surface. The methane content of the atmosphere is determined by a balance between the surface source and destruction in the reaction with hydroxyl. Destruction of methane initiates a sequence of reactions called the methane oxidation chain. Reactions in the methane oxidation chain provide the largest sources of formaldehyde, hydrogen, and carbon monoxide in the atmosphere. The concentrations of these gases are therefore related to the concentration of methane; they depend on the biological source of methane.

The situation of ozone in the troposphere is more complicated than that of these other gases. Recent research indicates that tropospheric ozone densities are determined by a combination of local photochemical production and destruction, transport downwards from the stratosphere, and destruction in reactions with reduced material at the ground. We shall first review the photochemical reactions that produce and destroy ozone in the troposphere and then present an approximate budget.

Ozone is actually formed by the three-body recombination of molecular and atomic oxygen, reaction (R2). But as we have already noted, almost every oxygen atom produced in the troposphere enters into this reaction. Therefore, the rate of production of ozone is equal to the rate of production of atomic oxygen. However, production of oxygen by photo-dissociation of ozone, reaction (R1), obviously does not count as a net source of ozone. Neither does production of oxygen by photolysis of NO_2, reaction (R3), if the NO_2 has been formed in the oxidation of NO by ozone (see Fig. 2-6). We must look for sources of atomic oxygen that do not involve the simultaneous destruction of ozone. This amounts, in practice, to sources of NO_2 other than reaction (R4).

As an example of such a source consider the oxidation of carbon monoxide by hydroxyl, reaction (R8). The reaction produces a hydrogen atom which combines to form HO_2 by reaction (R9). The HO_2 can oxidize nitric oxide to nitrogen dioxide, reaction (R10), restoring the original OH to the atmosphere. The nitrogen dioxide is then photodis-sociated, reaction (R3), restoring the original nitric oxide and producing an oxygen atom which combines to form ozone. The overall effect of this sequence of reactions, illustrated in Fig. 2-13, is

$$(2\text{-}18) \qquad CO + 2O_2 \xrightarrow{\text{light}} CO_2 + O_3$$

One ozone molecule can therefore be produced for every carbon monox-ide molecule oxidized.

There are several other reactions in the methane oxidation chain (Fig. 2-10) that can lead to the production of NO_2 and thus of O_3. In principle, the yield could be as high as four ozone molecules for each methane molecule destroyed, but detailed calculations by Chameides and Walker (1976) and Chameides and Stedman (1977) indicate that the yield is generally much lower.

The reactions that destroy ozone in the troposphere have been re-viewed by Chameides and Walker (1976). The most important are

$$(R17) \qquad HO_2 + O_3 \rightarrow OH + 2O_2$$

and reaction (R5),

$$O_3 + h\nu_{UV} \rightarrow O(^1D) + O_2$$

followed by reaction (R7),

$$O(^1D) + H_2O \rightarrow 2OH$$

If metastable $O(^1D)$ is converted to ground state O by reaction (R6), of course, O_3 is restored to the atmosphere by reaction (R2). Reactions involving NO and NO_2 can also destroy ozone at a significant rate when the densities of these gases are high.

The contributions of photochemistry to the tropospheric ozone budget are illustrated in Fig. 2-14, where the contributions of transport from the stratosphere and reactions at the ground are also shown (Chameides and

Stedman, 1977). The importance of photochemistry to the budget depends strongly on the concentrations of NO and NO_2. The values appearing in Fig. 2-14 correspond to a mixing ratio of NO and NO_2 combined of 0.4 ppb (an intermediate value).

Calculations of the ozone budget as a function of altitude have shown that ozone densities in the upper troposphere are determined largely by a balance between transport and photochemical destruction, while photochemical production contributes mainly in the lower troposphere. The lifetime of ozone against photochemical destruction varies from 10 d at low altitudes to 100 d at a height of 5 km, while the lifetime obtained by dividing the ozone density by the photochemical production rate varies from 20 d at low altitudes to 1 y at 5 km. The difference between the production and destruction lifetimes reflects the effect of vertical transport.

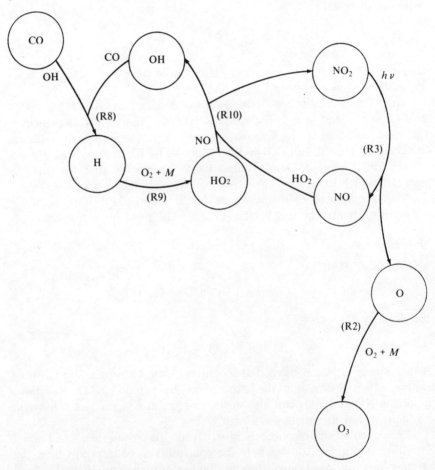

Fig. 2-13. Ozone production resulting from oxidation of carbon monoxide.

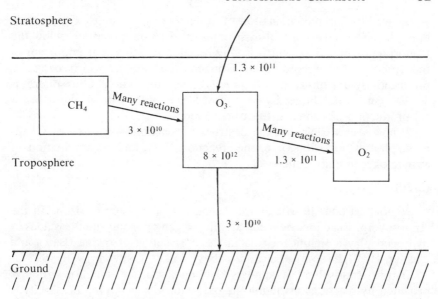

Fig. 2-14. Ozone budget of the troposphere. The reservoir is expressed in moles and rates of transfer in moles per day.

Since we are interested mainly in ozone densities near the ground, we shall use a purely photochemical model of tropospheric ozone at the end of this chapter, where we discuss possible variations in tropospheric composition during the course of geological time. Chameides and Walker (1973, 1976) have shown that the neglect of transport does not produce marked disagreement between theoretical and measured ozone densities at low altitudes.

OZONE IN THE STRATOSPHERE

As Fig. 1-6 shows, ozone is more abundant in the stratosphere than in the troposphere. Stratospheric ozone affects the temperature structure of the atmosphere and it also shields the troposphere from biologically harmful solar ultraviolet radiation. We shall therefore discuss the budget of stratospheric ozone in this section. The reactions responsible for the abundance of ozone in the stratosphere were first described by Chapman (1930). Our present understanding of the subject has been summarized by Crutzen (1973) and by McElroy et al. (1974).

Ozone is produced in the stratosphere and above by the photodissociation of molecular ozygen—

$$(2\text{-}19) \qquad \qquad O_2 + h\nu \rightarrow O + O$$

—followed by the recombination of atomic and molecular oxygen, reaction (R2). The radiation that dissociates O_2 does not penetrate below the stratosphere (Fig. 2-3), which is why reaction (2-19) is not important in the troposphere. According to Crutzen, the rate at which ozone is produced by reactions (2-19) and (R2) in the whole atmosphere is 3×10^{13} cm^{-2} sec^{-1}, larger by a factor of 500 than the rate of photochemical production of ozone in the troposphere.

It now seems clear that the destruction reactions suggested by Chapman, photodissociation of ozone, reaction (R1), and recombination of atomic oxygen and ozone—

(2-20) $$O_3 + O \rightarrow 2O_2$$

—are not fast enough to balance the rate of production of ozone in the stratosphere. Reactions between atomic oxygen or ozone and the oxides of hydrogen play a significant role in the destruction of ozone (Bates and Nicolet, 1950), but only at altitudes above 45 km (Crutzen, 1970, 1971).

In the stratosphere it appears that an important sink for ozone is provided by reactions involving the oxides of nitrogen. One possibility is the reaction between NO and O_3, reaction (R4), followed by

(2-21) $$NO_2 + O \rightarrow NO + O_2$$

Crutzen (1973) has described other reaction chains involving higher oxides of nitrogen. It is possible that reactions involving chlorine and its oxides are also important (Molina and Rowland, 1974; Cicerone, Stolarski and Walters, 1974; Stolarski and Cicerone, 1974; Wofsy and McElroy, 1974; Crutzen, 1974b).

The details of the destruction of ozone in the stratosphere are uncertain, partly because of insufficient knowledge of the photochemistry of the nitrogen and chlorine oxides, and partly because their mixing ratios in the stratosphere are not known. The subject has generated considerable interest because of the possibility that pollution of the stratosphere could affect the abundance of ozone and therefore the penetration of solar ultraviolet radiation to the ground (Hammond and Maugh, 1974).

OXIDES OF NITROGEN

The oxides of nitrogen, NO and NO_2, play important roles in the photochemistry of both the troposphere and stratosphere. They are probably involved in the destruction of ozone in both of these regions of the atmosphere, and in the troposphere they are active also in the production of ozone through their involvement in the methane oxidation

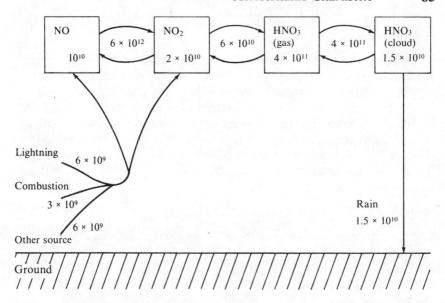

Fig. 2-15. The budget of odd-nitrogen oxides in the troposphere. Reservoirs are in moles and rates of transfer in moles per day.

chain and in the maintenance of the balance between OH and HO_2. In this section, we shall therefore examine the processes that control the densities of the nitrogen oxides in the atmosphere.

Our understanding of the atmospheric budget of the nitrogen oxides is far from complete. A tentative budget is given in Fig. 2-15. It appears that the source of these gases is provided by lightning and combustion, with a possible additional contribution of uncertain origin. Once nitrogen oxides have been released to the atmosphere, they take part in a number of rapid photochemical cycles that maintain a balance between NO, NO_2, NO_3, HNO_2, N_2O_5, and HNO_3. [One example of such a cycle is the reaction that converts NO to NO_2 followed by the reaction that converts NO_2 back to NO (Fig. 2-6).] These six species are referred to collectively as the *odd-nitrogen oxides*. The ones with the longest photochemical lifetimes and therefore the greatest densities are HNO_3, NO_2, and NO. Odd-nitrogen oxides are removed from the atmosphere when nitric acid dissolves in raindrops to provide the nitrate ions of rainwater. Before discussing this rainout process, let us provide a few examples of the reactions that convert one odd-nitrogen oxide into another. Nitrogen dioxide is converted to nitric acid by a number of reactions (Crutzen, 1973; Levy, 1973a; McConnell and McElroy, 1973), of which the most important may be

(2-22) $$OH + NO_2 \rightarrow HNO_3{}^*$$

ODD-NITROGEN

Many of the more chemically active species of the atmosphere contain an odd number of nitrogen, oxygen, or hydrogen atoms. These species are produced by the photolytic, chemical, or biological decomposition of the more stable diatomic molecules, N_2, O_2, and H_2. Due to their chemical activity, these molecules are of great interest to atmospheric chemists, and they are frequently referred to as "odd-molecules" (Bates and Nicolet, 1950). Here, for example, we use the term "odd-nitrogen" to refer to NO, NO_2, NO_3, $N_2O_5(NO_2$—O—$NO_2)$, HNO_2, and HNO_3.

---

followed immediately by quenching of the excited molecule,

(2-23) $$HNO_3^* + M \rightarrow HNO_3 + M$$

At tropospheric pressures, reaction (2-22) is the rate-limiting step, and the two reactions behave as a single reaction,

(R18) $$OH + NO_2 \xrightarrow{M} HNO_3$$

with an effective two-body rate coefficient. The lifetime of NO_2 before it is converted to HNO_3 is about half a day. Nitric acid is converted back to nitrogen dioxide by reactions such as

(R19) $$HNO_3 + h\nu \rightarrow OH + NO_2$$

and

(R20) $$HNO_3 + OH \rightarrow H_2O + NO_3$$

followed by

(R21) $$NO_3 + NO \rightarrow 2NO_2$$

These reactions lead to a photochemical lifetime for HNO_3 of about 6 d (Levy, 1973a).

This is the longest photochemical lifetime of any of the odd-nitrogen oxides by at least an order of magnitude, which implies that nitric acid is at least ten times as abundant as any of the other odd-nitrogen oxides. The lifetimes are so short that the odd-nitrogen oxides should be in photochemical equilibrium, with concentrations that depend on the concentration of nitric acid. The total abundance of nitric acid in the troposphere is determined by a balance between the sources of odd-nitrogen oxides and the sink provided by rainout of nitrate ions. The nitric acid abundance has not been measured in the troposphere, but it can be estimated using an argument presented by Chameides (1975).

Rainwater is undersaturated with respect to atmospheric nitric acid by many orders of magnitude, which suggests that the solution of nitric acid in cloud drops is limited by the rate at which nitric acid molecules can diffuse to the surface of a droplet. The growth of the droplets themselves is limited by the rate at which water molecules can diffuse to their surfaces. Since the diffusion processes should be similar for water molecules and nitric acid molecules, we expect that water and nitric acid are added to growing cloud drops at relative rates that are proportional to their relative concentrations in the ambient air. The ratio of nitric acid to water vapor in the lower troposphere is therefore approximately equal to the nitrate concentration in rainwater. From measurements of this concentration in unpolluted areas by Eriksson (1952) and Pearson and Fisher (1971), Chameides (1975) has deduced that the nitric acid concentration is about 4×10^{-7} times the water vapor concentration. Near the ground a typical value for the water vapor mixing ratio is 3×10^{-2} (U.S. Standard Atmosphere Supplement, 1966), leading to a nitric acid concentration of $3 \times 10^{11} \, cm^{-3}$.

Chameides (1975) has pointed out that the interaction of nitric acid and water should cause the nitric acid mixing ratio to decrease with increasing altitude in the troposphere much as does the water vapor mixing ratio. The water vapor mixing ratio decreases because decreasing temperatures cause water to condense into clouds. As the cloud drops gather water vapor at high altitudes, they gather nitric acid also. When they fall to lower levels, where temperatures are higher, they usually evaporate, resulting in a downward transport of both water and nitric acid. On the average, a nitric acid molecule probably survives for about a day in the gaseous form before dissolving in a cloud drop (Chameides, 1975). Most cloud drops do not turn into rain drops; instead they evaporate, restoring the nitric acid molecule to the atmosphere. The average lifetime of a cloud is about an hour (Mason, 1975).

A small fraction of the cloud drops do turn into raindrops, however, and fall to the surface carrying nitric acid out of the atmosphere. This appears to be the only significant sink for odd-nitrogen oxides in the troposphere. Its magnitude may be estimated from measurements of the nitrate concentration in rainwater. A rainout rate for nitrogen of $75 \times 10^{12} \, gm \, yr^{-1}$ has been derived by Robinson and Robbins (1970a) from data presented by Eriksson (1952). This value corresponds to an average rate of loss of odd-nitrogen oxides of $2 \times 10^{10} \, cm^{-2} \, sec^{-1}$. According to the model of Chameides (1975), the total abundance of odd-nitrogen oxides in the troposphere is $4.4 \times 10^{16} \, molecules \, cm^{-2}$, so the residence time is $2.2 \times 10^6 \, sec$ or about 20 d.

We must now seek a source of odd-nitrogen oxides in the troposphere of comparable magnitude to this sink. Nitric oxide and nitrogen dioxide are produced whenever air is exposed to sufficiently high temperatures, as

TABLE 2-3.   Estimated global anthropogenic sources of nitrogen compounds

COMPOUND	SOURCE	EMISSION $(10^{12} \text{ gm yr}^{-1})$
NO$_2$	Coal combustion	24.4
	Petroleum refining	0.6
	Gasoline combustion	6.8
	Other oil combustion	12.8
	Natural gas combustion	1.9
	Other combustion	1.4
NO$_2$	Total	48.0
NH$_3$	Combustion	3.8

After Robinson and Robbins (1970a).

in furnaces and engines. This pollution source has been estimated by Robinson and Robbins (1970a). Their results are presented in Table 2-3. The total production of odd-nitrogen oxides is $1.45 \times 10^{13} \text{ gm yr}^{-1}$ of nitrogen, only one-fifth of the rainout rate. Evidently pollution is a minor but not a negligible source of odd-nitrogen oxides.

Lightning is a natural source of high-temperature air. The rate of production of odd-nitrogen oxides by lightning is most uncertain, but it may be twice as large as the rate of production by pollution (Chameides et al., 1976).

A possible photochemical source has been suggested involving the oxidation of ammonia (Georgii, 1963; Crutzen, 1974a; McConnell and McElroy, 1973). Another possible source is the release of NO and NO$_2$ from soils (Junge, 1963; Bartholomew and Clark, 1965). Data are not available to permit useful direct estimates of the magnitudes of these sources. The best we can do is guess that the total production by lightning, combustion, soils, and NH$_3$ oxidation is equal to the total destruction by rainout.

NITROUS OXIDE AND ODD-NITROGEN IN THE STRATOSPHERE

In the preceding section we discussed the sources and sinks of odd-nitrogen in the troposphere. As already noted, however, NO and NO$_2$ play an important role in catalysing the recombination of ozone in the stratosphere as well as in the troposphere. Since most atmospheric ozone is in the stratosphere, it is stratospheric photochemistry that determines the total abundance of ozone. We shall therefore examine the stratospheric budget of odd-nitrogen oxides in this section. Although this subject has received much attention in connection with possible pollution of the stratosphere by high-flying aircraft and nuclear weapons, it is not well understood and is still controversial.

It appears that odd-nitrogen is produced in the stratosphere mainly by reaction between nitrous oxide and metastable oxygen atoms:

(2-24) $$N_2O + O(^1D) \rightarrow 2NO$$

(McElroy and McConnell, 1971b), and is removed by downward transport into the troposphere, where the total odd-nitrogen mixing ratio is less (McConnell and McElroy, 1973; McElroy et al., 1974). The budget is illustrated in Fig. 2-16. If this description is correct, the abundance of odd-nitrogen in the stratosphere depends on the abundance of nitrous oxide.

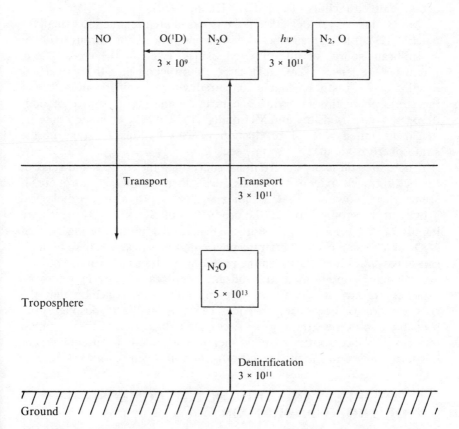

Fig. 2-16. Tentative budgets of nitrous oxide and stratospheric odd-nitrogen. The reservoir is expressed in moles and the rates of transfer in moles per year (of N_2O).

Atmospheric nitrous oxide is a product of microbial activity in anaerobic soils and ocean sediments (Arnold, 1954; Wijler and Delwiche, 1954; Alexander, 1961; Bates and Hays, 1967), as well as in oxygen-deficient waters of the open ocean (Hahn, 1974). The process, in which microorganisms reduce nitrate ions to nitrogen gas or nitrous oxide, is known as *denitrification* because it removes nitrogen from the soil. Nitrous oxide is relatively inert and does not, as far as we know, take part in chemical reactions in the troposphere. It is, however, subject to photodissociation by solar ultraviolet radiation in the upper stratosphere. The products are usually nitrogen molecules and metastable oxygen atoms. If we assume that photodissociation is the only process that destroys nitrous oxide and that the rate of production by denitrification is equal to the rate of destruction by solar ultraviolet radiation, we obtain the budget shown in Fig. 2-16. This budget is very tentative, however. There may well be sources and sinks of nitrous oxide that have not yet been identified (Schütz et al., 1970; Hahn, 1974).

Metastable oxygen densities in the troposphere are much too small for reaction (2-24) to be a significant destruction process for nitrous oxide or a significant source of odd-nitrogen oxides. Even in the stratosphere, where $O(^1D)$ densities are much larger, destruction of nitrous oxide by reaction (2-24) is still negligible compared with photodissociation. But in the stratosphere this reaction appears to be the major source of odd-nitrogen oxides. McElroy and McConnell (1971b) have estimated that the height-integrated rate of production of NO by this reaction in the stratosphere is about $2 \times 10^7 \, \text{cm}^{-2} \, \text{sec}^{-1}$.

There are still many uncertainties concerning the sources and sinks of the odd-nitrogen oxides in both the troposphere and the stratosphere. It seems clear, however, that a major source is provided by biological activity at the ground, both in the production of NO and NO_2, which are important for tropospheric photochemistry, and in the production of N_2O, which is important for stratospheric photochemistry. These biological sources must have varied in the past as a result of the evolution of life and changing conditions in the soil and the oceans. They may also be expected to vary in the future. We may therefore conclude that the concentration of odd-nitrogen oxides can vary in the troposphere and probably also in the stratosphere. A complete discussion of the evolution of atmospheric composition should therefore include the consequences of possible changes in the mixing ratios of odd-nitrogen oxides.

POSSIBLE VARIATION IN THE COMPOSITION OF TROPOSPHERIC AIR

In this chapter we have examined the processes that control the concentrations of photochemically active trace gases in the troposphere.

Our goal has been to understand these processes well enough to be able to predict how the concentrations may have varied during the course of geologic history. This is the subject we shall now take up. There has, as yet, been very little research in this area. Indeed, until very recently, we knew too little about tropospheric photochemistry to make such research profitable. This situation has now changed. We understand the photochemistry well enough to calculate, with some confidence, how the densities of methane, hydrogen, carbon monoxide, and ozone might respond to changes in external factors. What is difficult is to determine how the external factors have varied with time.

The factors that need to be considered may be deduced from the material of this chapter. First, there is atmospheric temperature, since many of the reaction rate coefficients depend strongly on temperature. Temperature also controls the water vapor content of the atmosphere, and water vapor is the source of the OH radicals that are so active photochemically. Past variations in mean global climate may have caused variations in trace gas concentrations. Next, there is the concentration of odd-nitrogen oxides, which depends, according to our present understanding, on lightning and rainfall and possibly also on the rate of release of NO and NO_2 from the ground. The ozone content of the stratosphere is important also, because it determines the penetration of solar ultraviolet radiation into the troposphere, and ultraviolet radiation initiates photochemical activity by producing metastable oxygen atoms. It appears that the stratospheric ozone content depends on the concentration of odd-nitrogen in the stratosphere, and this may depend on the rate of release of N_2O from the soil. A final external factor that must be considered is the flux of methane from the ground to the atmosphere. Hydrogen, carbon monoxide, and ozone are all products of the methane oxidation chain.

Chameides and Walker (1975) have used an elaborate photochemical model to examine the sensitivity of tropospheric photochemistry to variations in some of these factors. Figure 2-17 illustrates their results for the densities of methane, hydrogen, carbon monoxide, and ozone in the lower troposphere as functions of the methane flux from the ground. This quantity may have varied markedly with time, because methane is produced in swamps, and the extent of swamps should have varied in response to changes in climate, topography, and the level of the sea relative to the continents.

The variation in the density of ozone is particularly interesting because, of all the gases in the troposphere, it is closest to being present in toxic amounts. It is at least possible that past increases in the methane flux have led to ozone densities high enough to injure susceptible organisms. What is needed is an estimate of how the methane flux has varied with time. Such an estimate would make it possible to identify periods of

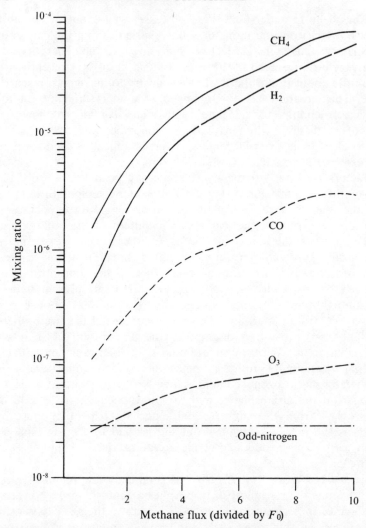

Fig. 2-17. Mixing ratios of various gases in the troposphere as a function of the methane flux from the ground. The present-day flux is approximately $\overline{F_0} = 5.3 \times 10^{11}$ cm^{-2} sec^{-1}. The odd-nitrogen mixing ratio is assumed to be constant. (From Chameides and Walker, 1975. Copyright 1975 by *American Journal of Science*, New Haven, Conn. Used by permission of the publisher.)

high ozone density and perhaps to correlate these periods with times when the fossil record shows evidence of high biological stress.

In the absence of information on the history of the methane flux, however, Chameides and Walker (1975b) chose to reverse the argument and use paleontological evidence to set limits on the methane flux. *Pinus Ponderosa* is a conifer that is particularly susceptible to ozone damage.

An increase in the average ozone density by a factor of about three would probably cause this species to become extinct. Such an increase in ozone density could result from an increase in the methane flux by a factor of six to eight, depending on assumptions concerning changes in temperature and odd-nitrogen densities. Fossil evidence indicates that *Pinus Ponderosa* has existed since at least the middle Oligocene, or for about 30 million years. Appropriate limits can therefore be set on the ozone density and the methane flux as well as on the densities of hydrogen, carbon monoxide, and methane during the last 30 million years.

Much more research should be possible on the interaction between minor atmospheric constituents and organic evolution. There is no reason to suppose that minor-constituent densities have not varied in the past, and it is quite possible that their variations have affected the succession of living species. The possible variations in minor-constituent densities depend largely on organisms, however, because biological activity is the source of the methane and some of the oxides of nitrogen. Examples of the interaction of life and the atmosphere have been suggested by Lovelock and Margulis (1974) and Margulis and Lovelock (1974).

Chapter Three

Major Constituents and Their Interaction with Solid Earth, Ocean, and Life

In Chap. 2 we showed how the densities of many minor constituents are controlled by photochemical reactions within the atmosphere. For the major constituents, O_2, N_2, and CO_2, photochemistry plays no role; their abundances are determined by interactions at the surface with solid earth, ocean, and life. These interactions are by no means as susceptible to detailed quantitative analysis as the chemical processes described in the previous chapter. This chapter therefore contains fewer equations and more appeals to reason—or faith?—than Chap. 2. It is also much more speculative. Most of the arguments are geochemical in nature, and some are quite involved.

OXYGEN

We shall begin our discussion of atmospheric oxygen by presenting its budget, describing the processes that contribute to this budget, and estimating the magnitudes of the various components of the budget. We shall then argue that the cyclical processes of photosynthesis followed by respiration and decay do not control the amount of oxygen in the atmosphere. Control is exercised, instead, by the much slower cycle of burial of reduced organic matter in sediments and the erosion and weathering of reduced matter in sedimentary rocks. We shall then argue that the weathering part of this cycle, which consumes oxygen, does not depend strongly on the amount of oxygen in the atmosphere and therefore does not determine this amount. The production of oxygen, which corresponds to the burial of organic matter in sediments, does depend on the oxygen content of the atmosphere, however, and we shall argue that this is the process that determines the content. Finally, we shall present a simple model of the controlling process that shows how the oxygen content of the atmosphere may depend on the phosphorus content of the sea. The discussion follows closely a paper by Walker (1974).

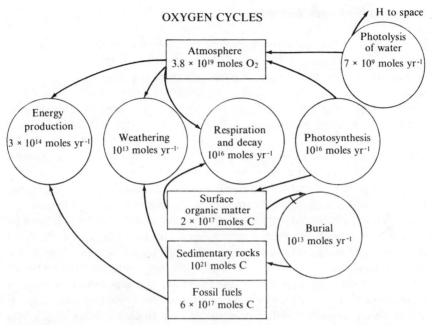

OXYGEN CYCLES

Fig. 3-1. Processes (circles) and reservoirs (boxes) affecting the oxygen content of the atmosphere. (From Walker, 1974. Copyright 1974 by the *American Journal of Science,* New Haven, Conn. Used by permission of the publisher.)

Since the earth is reducing in overall composition, the presence of free oxygen in the atmosphere must be considered in terms of processes that separate oxygen from its compounds and sequester the reducing material that is produced (Redfield, 1958; Sillén, 1966). Thus, the budget in Fig. 3-1 includes not only processes that directly affect the atmospheric reservoir of oxygen, but also processes that affect the reservoirs of reduced matter with which oxygen might combine. Since the arguments that follow are stoichiometric in nature, it is convenient to express both reservoirs and rates in terms of moles of oxygen and moles of reduced material. Finally, it should be noted that many of the values presented in Fig. 3-1 are subject to large uncertainty, as the discussion will show.

In an effort to keep the model as simple and comprehensible as possible, in Fig. 3-1 and in the discussion that follows we have ignored many processes which affect atmospheric oxygen but which do not appear to play a dominant role in determining the oxygen abundance. Thus, for example, we have lumped all of the reduced species in sedimentary rocks together and called them *organic carbon,* rather than treating explicitly the oxidation of sulfide and ferrous iron and the reduction of sulfate and ferric iron. These processes are important, but not essential. For an oxygen budget in which they are evaluated individually see Schidlowski et al. (1975).

Escape of Hydrogen to Space

The dissociation of water vapor by solar ultraviolet radiation initiates a chain of photochemical reactions leading either to the reformation of water vapor or to the escape of hydrogen from the upper atmosphere to space. Escaping hydrogen atoms leave behind the oxygen with which they were at one time associated, constituting a net source of atmospheric oxygen (cf. Kuiper, 1949). We shall discuss this process in later chapters and evaluate its rate. Here we take the present rate to be 7×10^9 moles yr^{-1}. It is completely negligible compared with the biological sources of oxygen to be discussed below.

Most of the processes we shall discuss in this chapter are cyclic, however, with oxygen being released to the atmosphere at one point of the cycle and consumed at another. Photolysis of water vapor followed by escape of hydrogen to space is noncyclic; it causes a net gain in the level of oxidation of the atmosphere, crust, and upper mantle. There are several noncyclic sinks of oxygen also, among them the oxidation of primitive igneous rocks and volcanic gases and the accretion of hydrogen from space (Van Valen, 1971). We shall examine these sinks in Chap. 5. While the evidence is not conclusive, it is probable that they are less considerable than the source provided by the escape of hydrogen. The surface layers of the earth are therefore probably becoming gradually more oxidized. The noncyclic processes are too slow, however, to affect the stability of atmospheric oxygen over time scales of a few hundred million years or less. We shall therefore follow Holland (1973a) in concentrating on the more rapid cyclic processes.

Photosynthesis

For present purposes we may think of photosynthesis as a process in which green plants add reduced organic material and oxygen at equal molar rates to the surface organic reservoir and the atmospheric reservoir (cf. Hutchinson, 1954). We may keep track of the organic material with adequate accuracy simply by referring to the reduced organic carbon. Approximately 1 mole of organic carbon is produced for each mole of oxygen added to the atmosphere (cf. Riley and Chester, 1971, p. 230).

Because of substantial differences in the *productivities* of different plant communities it is difficult to estimate with precision the total rate at which photosynthesis produces organic carbon. The figure we use in Fig. 3-1 is that of Bowen (1966, p. 50; cf. Gilbert, 1972). Bowen's estimate of production by oceanic plants differs little from previous values and agrees with the more recent result, based on a large body of radiocarbon data, of Koblentz-Mishke et al. (1970; cf. Ryther, 1969). Bowen, basing his estimate on a review by Westlake (1963), concludes that land plants contribute approximately four times as much to total production as do oceanic plants, a value that is higher than previous estimates, partly because of earlier neglect of plant roots in estimating the total masses of plant communities and partly because of the absence, until recently, of data on tropical forests. We shall argue below that production by land plants has little effect on the oxygen balance of the atmosphere, so the uncertainty in this figure is of little concern.

Respiration and Decay

We may think of respiration and decay as the reverse of photosynthesis, consuming surface organic carbon and atmospheric oxygen at equal rates. From the values presented in Fig. 3-1 we can calculate that the average time spent by organic carbon in what we have called the surface organic reservoir is only 20 yr. Our value for the size of this reservoir is taken from Bowen (1966, p. 52). The reservoir includes living and dead organic matter on the land, in the soil, and in the ocean. About one quarter of the mass consists of living organisms, mostly plants. About 80% of the remainder is in solution in sea water. Some portion of the organic carbon contained in the surface layers of ocean sediments should probably be included in this reservoir since this carbon is subject to oxidation on a relatively short time scale. The inclusion of this component would increase the size of the reservoir by approximately a factor of two. The arguments that follow, however, depend only on the fact that the surface organic reservoir of carbon is very much smaller than the atmospheric reservoir of oxygen, so we will ignore this correction.

From the short residence time of organic carbon in the surface organic reservoir and the rapid rate of photosynthesis compared with all of the other processes in Fig. 3-1, we may conclude that respiration and decay must almost exactly balance photosynthesis (Van Valen, 1971; Holland, 1973a) over times longer than about 100 yr (burial of organic carbon is a much slower process, which we shall consider below). Thus we arrive at the rate of respiration and decay shown in Fig. 3-1.

From the relatively small size of the surface organic reservoir some useful conclusions can be drawn. Suppose, for example, that the rate of photosynthesis were to be increased slightly by some change in world climate or by efficient agricultural techniques. The surface organic reservoir would increase at a rate proportionately very much greater than the rate of increase of atmospheric oxygen. We may assume that the rate of respiration and decay would increase as the surface organic mass increased, so we may predict that a new equilibrium would be achieved, in due course, with an increase in the amount of organic carbon in the surface reservoir, but with no significant change in atmospheric oxygen (Van Valen, 1971).

We can see, therefore, that the cycle of photosynthesis followed by respiration and decay, which links a large reservoir of atmospheric oxygen to a small reservoir of surface organic carbon, serves to control not the size of the oxygen reservoir, but the size of the carbon reservoir (Junge, 1972). This conclusion indicates that the oxygen content of the atmosphere is not directly affected by changes in the rate of photosynthesis such as might be caused, for example, by changes in the partial pressure of carbon dioxide or, through the Warburg effect (cf. Turner and Brittain, 1962; Gibbs, 1970), in the partial pressure of oxygen (Broecker, 1970b; Holland, 1973a). The relatively small size of the surface organic reservoir implies, also, that changes in the rate of photosynthesis and thus in the amount of carbon at the surface are possible, if they occur slowly, without causing any significant depletion in the supply of the raw materials of photosynthesis, carbon dioxide (3×10^{18} moles in the oceans) and water (10^{23} moles).

The time scale of these hypothetical changes is important as far as the supply of carbon dioxide is concerned. Photosynthesis removes carbon dioxide from the atmosphere and the surface layer of the ocean. The atmosphere contains 5.6×10^{16} moles of carbon dioxide and the surface ocean contains about 5×10^{16} moles, mostly in the form of bicarbonate ions (Broecker et al., 1971); together they could be exhausted in 10 yr if photosynthesis were not balanced by respiration and decay. A much larger amount of carbon dioxide, 3.2×10^{18} moles, resides in the deep ocean, and the rate of transfer of carbon from the deep ocean to the surface is sufficiently rapid to replenish the surface–atmosphere reservoir in about 50 yr (Broecker et al., 1971). Thus, substantial growth in the size of the organic reservoir over times of a few decades or less will be limited by the supply of carbon dioxide. Changes over times longer than a few centuries can draw on the relatively large reserve of carbon dioxide in the deep sea.

Burial of Organic Carbon

Even at equilibrium there is a small imbalance between the rates of photosynthesis and of respiration and decay which arises because some of the organic carbon is preserved by burial in sediments. Corresponding to this *"fossilization"* of organic carbon is a small net source of atmospheric oxygen (Van Valen, 1971; Holland, 1973a; Schidlowski et al., 1975).

The rate of fossilization of organic carbon can, in principle, be estimated from data on the carbon content and the rate of accumulation of recent sediments. Since both of these quantities exhibit substantial variation with position, a reliable estimate is not possible without a very much more comprehensive compilation of data than is presently available. We shall assume, rather arbitrarily, that the net source of oxygen due to burial of organic carbon is equal to the rate of consumption of oxygen by weathering, to be discussed below. Very rough estimates of the rate of burial of carbon (Walker, 1974) show that this assumption is at least reasonable. Most of the burial occurs as hemipelagic "blue mud" on the continental slopes (Garrels and Mackenzie, 1971, p. 219), where carbon contents and sediment accumulation rates are both high.

The rate of burial of organic carbon in lake sediments is negligibly small compared with oceanic sediments and the rate of burial in subaerial sediments is probably also negligible (Redfield, 1958). We may not, however, conclude without further examination that land plants do not contribute to the net source of atmospheric oxygen that corresponds to the carbon buried. It is possible that organic material particularly resistant to decay is synthesized on land and washed into the sea, making a significant contribution to the carbon in terrigenous sediments. There is a difference between the isotopic composition ($^{13}C/^{12}C$) of organic carbon produced by land plants and by *phytoplankton* (photosynthetic microbes living in near-surface waters) that makes it possible to evaluate this hypothesis (Sackett, 1964). The evidence, as reviewed by Degens (1969) and Williams (1971), is that while the presence of land-derived organic detritus is clearly discernible in river estuaries, it becomes undetectable in carbon isotope data outside the immediate environs of the river mouth. This evidence is not universally accepted; nevertheless, we shall assume that the net source of atmospheric oxygen corresponding to burial of organic carbon is almost entirely derived from photosynthesis by phytoplankton rather than by land plants.

From this it follows that the cycle of photosynthesis followed by respiration and decay on land is to all intents and purposes closed. Although the total rate of photosynthesis on land exceeds that in the ocean, photosynthesis on land makes no significant contribution to the oxygen budget of the atmosphere. Oxygen is released to the atmosphere at a rate corresponding to the rate of burial of organic carbon in sediments, and most of the organic material that is buried is produced by oceanic plants, not by land plants. If this was true also in the past, then the development of land plants in the late Silurian was not a significant stage in the evolution of atmospheric oxygen.

Fossilized Organic Carbon

A substantial amount of reduced organic carbon has accumulated in sedimentary rocks as a result of burial over the ages. Estimation of the total amount is subject to considerable uncertainty in the volumes of various types of rocks. The value shown in Fig. 3-1 for the size of the sedimentary carbon reservoir is that of Ronov and Yaroshevskiy (1967; cf. Garrels and Perry, 1974); a smaller estimate was given by Rubey (1951). But the arguments that follow depend only on the fact that the amount of fossilized carbon is considerably larger than the amount of atmospheric oxygen, and this is true of Rubey's estimate as well as of the one we use. In any event, these estimates refer only to material in the crust of the earth. It is, however, possible that the crust is not isolated from the mantle and that significant quantities of material have been exchanged between the two throughout geological history (Armstrong, 1968, 1971; Armstrong and Hein, 1973). The true size of the fossil carbon reservoir may therefore be much larger than the value given here, and our estimate of the residence time of material in this reservoir may be too small.

Most of the fossil carbon occurs in highly dispersed form, with an average concentration in sedimentary rocks of only 0.5% (Ronov and Yaroshevskiy, 1967, 1969). The proportion of the total that occurs in sufficient concentrations to be considered a recoverable fossil fuel is very small indeed (Hubbert, 1969).

By dividing the rate of burial of organic carbon into the fossil carbon reservoir we may calculate that carbon spends on the average 10^8 yr in sedimentary rocks.* Sediments do not accumulate indefinitely on the floor of the ocean. In due course, processes of sea-floor spreading and tectonism recycle the sediments, either by way of volcanism, or by raising them above the surface of the sea where they undergo erosion and weathering. When this happens, the reduced material that the sediments contain is subject to oxidation, constituting a sink for atmospheric oxygen.

Consumption of Oxygen in Weathering

We follow Holland (1973a) in estimating this quantity by calculating the total rate at which rivers are carrying material from the continents into the sea today and the average composition of the eroded material. The calculation, presented in the box, leads to an oxygen consumption rate of 10^{13} moles yr^{-1}. There is disagreement over some of the numbers that have been used in this calculation, but the result is close to estimates obtained in similar calculations by Holland (1973a) and Garrels and Perry (1974).

The weathering of igneous rocks plays a very minor role compared to weathering of sedimentary rocks, so we may think of weathering as the process that closes a cycle which commenced with the burial of reduced organic material produced by photosynthesis. Carbon is the most important reducing constituent of the sediments, and it is on carbon that we

---

* This estimate may be too small (Garrels and Mackenzie, 1972).

CALCULATION OF WEATHERING RATES

Turekian (1971) has discussed the amount of material carried into the sea by rivers. With an average suspended load of $330 \, mg \, l^{-1}$, an average bed load approximately 10% as large, and dissolved solids contributing $100 \, mg \, l^{-1}$—correction for bicarbonate ion contributed by atmospheric carbon dioxide is not necessary at this level of approximation—he concludes that $500 \, mg \, l^{-1}$ is the average amount of terrigenous material in river water. With an average runoff of $3.6 \times 10^{16} \, l \, yr^{-1}$, the total flux into the sea is $1.8 \times 10^{16} \, gm \, yr^{-1}$ (cf. Holeman, 1968; Gregor, 1970). Man's influence on this quantity is probably no larger than a factor of two or three (Judson, 1968), and it may be much smaller. If windblown material is neglected (Garrels and Mackenzie, 1971, p. 112), the river-borne flux must, in the long run, be equal to the rate of erosion; so we adopt this figure for the erosion rate.

We consider the principal reducing species in the eroded rocks to be organic carbon, sulfide, and ferrous iron. And we assume that the carbon is oxidized, upon erosion, to carbon dioxide, the sulfide to sulfate, and the ferrous iron to ferric (Holland, 1973a). Later in the chapter we shall examine the possibility that not all of the reduced material is oxidized during its exposure to the air before it is once again buried in sediments at the bottom of the sea. For the average compositions of continental rocks we use the compilation of Ronov and Yaroshevskiy (1967, 1969). As shown in the table (next page), sedimentary rocks consume 1.86 gm of oxygen per 100 gm of rock weathered while granitic and basaltic rocks consume an average of 0.87 gm.

From considerations of the areas of igneous and sedimentary rocks exposed at the surface, Gilluly et al. (1970) conclude that sedimentary rocks contribute 85% of eroded material in the United States, while igneous rocks contribute 15%. We shall assume that these proportions hold throughout the world. Then, for 100 gm of rock weathered, sedimentary rocks consume 1.52 gm of oxygen and igneous rocks consume 0.13 gm of oxygen, for a total of 1.65 gm of oxygen per 100 gm of eroded material. With the figure already presented for the total rate of erosion, we calculate that weathering consumes atmospheric oxygen at a rate of $3 \times 10^{14} \, gm \, yr^{-1}$ or $10^{13} \, moles \, yr^{-1}$, as shown in Fig. 3-1.

Oxygen consumed in the weathering of different rocks

	SEDIMENTARY ROCKS		GRANITIC ROCKS		BASALTIC ROCKS	
	Content (% by weight)	Oxygen Consumption[a] (gm)	Content (% by weight)	Oxygen Consumption[a] (gm)	Content (% by weight)	Oxygen Consumption[a] (gm)
Reduced carbon	0.47	1.25	0.17	0.45	0.11	0.29
Sulfide	0.15	0.30	0.04	0.08	0.03	0.06
FeO	2.82	0.31	2.86	0.32	4.78	0.53
		1.86		0.85		0.89

[a] Per 100 gm of rock weathered.

focus our attention. The other reducing constituents are largely produced by the oxidation of buried organic carbon by chemical reactions occurring within the sediments (cf. Berner, 1971a), so most of the reduced sulfur and iron in sediments has taken the place of a stoichiometrically equal amount of reduced carbon that was in the sediments when they were first deposited. It is for this reason that it is not necessary to consider the oxidation and reduction of sulfur explicitly in a simple model of the oxygen budget. Although there is isotopic evidence of substantial changes in the relative sizes of the reservoirs of reduced and oxidized sulfur over geological time (Holser and Kaplan, 1966; Holland, 1973c), these changes have probably been accompanied by equivalent changes in the reduced carbon reservoir, not in the abundance of atmospheric oxygen (cf. Garrels and Perry, 1974).

Photosynthesis followed by burial of organic carbon adds oxygen to the atmosphere and carbon to the sedimentary reservoir, while weathering closes the cycle by consuming atmospheric oxygen and sedimentary carbon. In Fig. 3-1 we have shown this cycle in balance, but the data are by no means good enough to determine whether this is true at the present time.

Volcanism

The rate of consumption of atmospheric oxygen by reaction with reduced gases brought to the surface by volcanoes and magmas may be estimated from arguments to be presented in Chap. 5. Hydrogen is the major reducing constituent of volcanic gases, and we shall estimate that the production rate is 2×10^{11} gm yr^{-1}. The corresponding oxygen consumption rate is 5×10^{10} moles yr^{-1}. This is negligible compared with the rate of consumption of oxygen in the weathering of sedimentary rocks.

Energy Production by Combustion of Fossil Fuels

In terms of the cyclical processes we have been discussing, we may think of the combustion of fossil fuels as an accelerated form of weathering. The rate at which man is currently using up atmospheric oxygen in this process (Hubbert, 1969) is substantially larger than the rate of oxygen consumption by natural weathering, as Fig. 3-1 shows. The fossil fuel reservoir, however, is small. From the values in Fig. 3-1 we can see that the present rate of fossil fuel consumption can be maintained for only 2000 yr. Combustion of all of the fossil fuel reserves would reduce the oxygen content of the atmosphere by less than 2%. While energy production by combustion of fossil fuels may cause many environmental problems, the disappearance of atmospheric oxygen is not one of them (Machta and Hughes, 1970; Broecker, 1970b).

The Role of Photosynthesis in Maintaining the Oxygen Content of the Atmosphere

Some insights into the workings of the oxygen cycles shown in Fig. 3-1 are obtained by imagining what would happen if photosynthesis were to cease altogether, possibly as a result of man's activities. If we were to kill all the green plants, photosynthesis would no longer add oxygen to the atmosphere and carbon to the reservoir of surface organics. Respiration and decay would continue to consume oxygen and carbon, however, so the atmospheric oxygen content and the surface organic carbon content would decline. The decline would continue until the surface carbon reservoir was completely exhausted, after a time of the order of 20 yr. At this time, the amount of oxygen in the atmosphere would have decreased by less than 1% (Broecker, 1970b; Van Valen, 1971). Atmospheric carbon dioxide would increase initially, but most of the excess would be taken up by the deep sea in 100 yr or so.

With no carbon left in the surface organic reservoir, the burial of organic carbon in sediments would now cease, but weathering would continue. The oxygen content of the atmosphere would therefore continue to decline, but much more slowly than before. It would take approximately 4 million years for weathering to consume all of the oxygen in the atmosphere. Consumption of all of the atmospheric oxygen would reduce the reservoir of sedimentary organic carbon by only 4%. We may conclude that the oxygen crisis that would result from the cessation of photosynthesis would, from man's point of view, occur in the very remote future.

Control of the Oxygen Content of the Atmosphere

With the sources and sinks of atmospheric oxygen enumerated and evaluated, we are now in a position to consider what processes determine the amount of oxygen there is. It is clear, from the "thought experiments" discussed above, that control is not exercised directly by the rapid cycle of photosynthesis followed by respiration and decay, which links the atmospheric oxygen reservoir to the surface organic carbon reservoir. Because the surface carbon reservoir is so much smaller than the atmospheric reservoir, this cycle controls the mass of surface carbon, not the mass of atmospheric oxygen. Although the atmospheric reservoir of carbon dioxide is smaller than the surface carbon reservoir, it is probable that the abundance of carbon dioxide averaged over times of the order of 10^6 yr is controlled by processes of weathering and precipitation of carbonate minerals (Broecker, 1971), and not by the cycle of photosynthesis followed by respiration and decay.

Therefore, since other processes we have examined are quantitatively insignificant, control of atmospheric oxygen must be exercised by the cycle of burial of organic carbon followed by weathering, which links the reservoir of atmospheric oxygen to a much larger reservoir of sedimentary organic carbon (Holland, 1973a). The rates are such that imbalances in this cycle, if they were to occur, could cause substantial fluctuations in the oxygen content of the atmosphere in times of the order of 4 million years.

As Van Valen (1971) and Holland (1973a) have pointed out, 4 million years is a short time compared with the history of terrestrial life. Mammals have been successful inhabitants of the earth for approximately 65 million years, since the beginning of the Cenozoic, and mammals are sensitive to the oxygen content of the atmosphere. Multicelled animals with oxidative metabolisms have been abundant for a period ten times as long, since before the beginning of the Paleozoic. It is hard to draw quantitative conclusions from these observations, but we can probably say quite definitely that oxygen has not been absent from the atmosphere in the last 600 million years at least; and we could possibly argue further, that atmospheric oxygen has not fluctuated by factors larger than two or three since the advent of mammals.

For oxygen to have survived in the atmosphere for so long a period in the presence of processes that replace it every 4 million years, there must be a negative feedback mechanism that provides a measure of stability (Holland, 1973a). Where, in the geochemical cycle we have described, does this feedback mechanism operate?

Variation of the Weathering Sink with Oxygen Partial Pressure

An obvious possibility is that the rate at which atmospheric oxygen is consumed by weathering increases as the oxygen content of the atmosphere increases (Holland, 1973a). We shall now show that this does not appear to be the case under present conditions.

We note, first, that there is an upper limit to the rate at which oxygen can be consumed by weathering, which is governed by the rate at which reduced material is exposed to the atmosphere by erosion. Sedimentary organic carbon cannot be oxidized faster than it is uncovered. With this in mind, let us think of the weathering sink as the product of three factors. The first factor is the erosion rate, which is the rate at which material is carried from the continents into the sea. Among the circumstances that might influence the erosion rate are topography, climate, and biological activity; but over a long period of time the erosion rate must be equal to the rate at which tectonic activity raises up new material to replace that which has worn away. The erosion rate, therefore, does not depend directly on the oxygen content of the atmosphere.

The second factor entering into the expression for the weathering sink is the content of reducing material in the eroded rocks. As shown above, this is principally sedimentary organic carbon. For present purposes, we need simply note that this factor does not depend on the oxygen content of the atmosphere.

The third factor is the fraction of the organic carbon (and iron and sulfur) that suffers oxidation during the period between erosion and reburial when the reducing material is exposed to the atmosphere. This factor does depend on the oxygen content of the atmosphere, at least at low oxygen levels (Holland, 1973a). If there were no oxygen in the atmosphere, essentially all of the organic carbon would survive its journey from the mountains to the sea, to be incorporated, once again, in new sediments. The fraction oxidized and the weathering sink would both be close to zero.

Increasing atmospheric oxygen would lead to the oxidation of an increasingly large fraction of the exposed material until, at sufficiently high oxygen levels, oxygen would cease to be limiting, essentially all of the exposed carbon would be oxidized before new sediments were laid down, and the weathering sink would become independent of the oxygen content of the atmosphere (Van Valen, 1971) while remaining dependent on the erosion rate. This behavior is sketched in Fig. 3-2 (cf. Holland, 1973a). The question we must now consider is whether we are at the low oxygen end of this figure, where the oxygen sink depends on the oxygen partial pressure, as Holland (1973a) has suggested, or at the high oxygen end where the sink is constant.

The isotopic composition of organic carbon in sediments on the sea floor differs significantly from that in ancient sedimentary rocks (Degens, 1969). The implication is that these new sediments do not contain a large proportion of fossilized carbon that has been recycled. Indeed, as we have already noted, the isotopic composition suggests that most of the organic matter has been freshly synthesized by phytoplankton. Since the average carbon content of new sediments is approximately equal to the average carbon content of the ancient sediments undergoing erosion (Holland, 1973a), this observation suggests, in turn, that most of the fossilized carbon is oxidized upon erosion.

We can argue, moreover, that each time a given volume of sediment passes through the cycle of erosion and deposition it receives a new charge of freshly synthesized organic matter. If this new material is not all oxidized during the subsequent cycle of erosion and deposition, the carbon content of the sediment will increase with the passage of time. Broecker (1970a), however, has presented an argument, based on the invariance with time of the isotopic composition of carbonate carbon, which shows that the sedimentary reduced carbon reservoir accumulated before the opening of the Paleozoic and has not been increasing at a significant rate since then.

The argument is based on the observation that reduced organic carbon

Sink = (Erosion rate) × (Carbon content) ×
(Fraction oxidized)

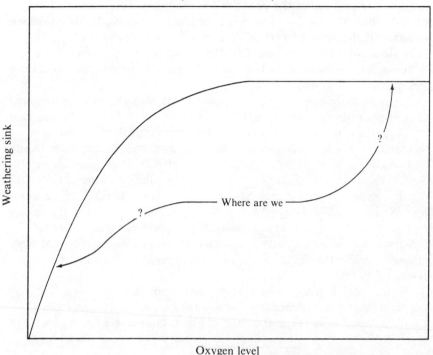

Fig. 3-2. Schematic representation of the rate of consumption of oxygen in weathering as a function of the oxygen partial pressure. (From Walker, 1974. Copyright 1974 by the *American Journal of Science,* New Haven, Conn. Used by permission of the publisher.)

is isotopically lighter than the oxidized carbon in the sea and the atmosphere. The difference arises because plants assimilate the lighter isotope, ^{12}C, more rapidly than the heavier isotope, ^{13}C. Imagine what happens as the reservoir of reduced organic carbon grows at the expense of the reservoir of isotopically heavier oxidized carbon, assuming that the total amount of carbon in the surface layers of the earth is fixed. As ^{12}C is removed from the oxidized reservoir more rapidly than ^{13}C, the oxidized carbon that remains must grow isotopically heavier. Since the difference between the isotopic compositions of reduced and oxidized carbon is approximately constant, the organic carbon must grow isotopically heavier also. This behavior is shown in Fig. 3-3. The average isotopic composition of all of the carbon in the crust remains constant, but the compositions of both reduced and oxidized reservoirs become heavier as the light, reduced reservoir grows at the expense of the heavy, oxidized reservoir.

Figure 3-4 shows the isotopic composition of sedimentary carbonate rocks formed at different times in the past. These rocks preserve the isotopic composition of the ocean at the time they were formed. There is no indication of a secular increase in the $^{13}C/^{12}C$ ratio such as would have occurred if the reduced carbon reservoir had accumulated at a constant rate since the beginning of the Paleozoic. From this fact Broecker deduces that the organic carbon reservoir grew to its present size before the opening of the Paleozoic.

Organic carbon in Phanerozoic sediments also shows no systematic change with time in isotopic composition (Degens, 1969)—a finding that supports the carbonate evidence for little or no growth in the reservoir of sedimentary reduced carbon. In fact, data from the Precambrian (Keith and Weber, 1964; J. W. Schopf et al., 1971; Oehler et al., 1972; Schidlowski et al., 1975) suggest that the accumulation of the sedimentary organic carbon reservoir may have occurred more than 3.3 billion years ago. Some additional support for the conclusion that the sedimentary carbon reservoir is not now growing is provided by data, shown in Fig. 7-5c, on the average carbon content of sediments of different ages (Ronov, 1958); these data show no clear evidence of a secular increase (Gregor, 1971).

If the sedimentary carbon reservoir is not now growing, we may conclude that any sedimentary carbon that is recycled is very old (Precambrian) and inert (Sackett et al., 1974); it has survived many cycles of

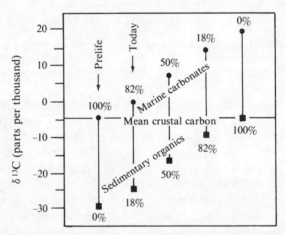

Fig. 3-3. Change in the $^{13}C/^{12}C$ ratios in the carbonate portion and in the organic portion of carbon stored in sediments as a function of the ratio of these two species (assuming that the organic carbon always has on the average 25 parts per thousand less ^{13}C relative to ^{12}C than the carbonate carbon and that mean crustal carbon has a $^{13}C/^{12}C$ ratio 4.5 parts per thousand less than that in the standard. (From Broecker, W.S., A boundary condition on the evolution of atmospheric oxygen. *J. Geophys. Res.* 75, 3553-3557, 1970. Copyrighted by American Geophysical Union.)

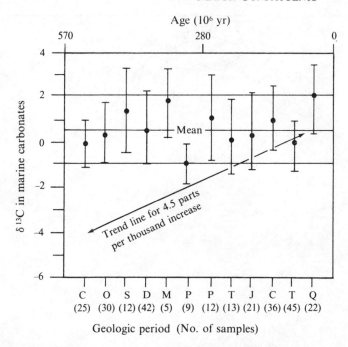

Geologic period (No. of samples)

Fig. 3-4. Variation in $^{13}C/^{12}C$ ratios in marine carbonates as a function of geologic period. The dots are averages computed by Keith and Weber (1964) for well-preserved marine limestones. The number of samples for each period is listed under its letter designation. The error bars represent the standard deviation of the individual results from the mean. The dashed line represents the trend in $^{13}C/^{12}C$ results expected if the organic reservoir had accumulated at a uniform rate since Cambrian time. (From Broecker, W.S., A boundary condition on the evolution of atmospheric oxygen. *J. Geophys. Res.* 75, 3553-3557, 1970. Copyrighted by American Geophysical Union.)

erosion and deposition without suffering oxidation. All of the reduced matter that is susceptible to oxidation is, in fact, oxidized upon erosion. Therefore, we are on the flat portion of the curve in Fig. 3-2, and the weathering sink for present-day oxygen is independent of the oxygen content of the atmosphere (cf. Van Valen, 1971). This being the case, we must look elsewhere for the feedback mechanism that stabilizes the oxygen content of the atmosphere. Since it is not associated with the sink of oxygen, it must be associated with the source.

Burial of Organic Carbon as a Source of Atmospheric Oxygen

We now wish to examine more closely the processes that lead to burial of organic carbon in sediments, thereby providing the net oxygen source that, in the long run, balances the oxygen sink caused by weathering. Since the sedimentary rocks that consume most of the oxygen upon

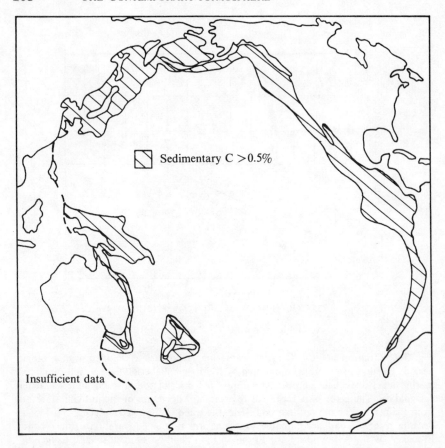

Fig. 3-5. Areas in the Pacific Ocean where recent sediments contain more than 0.5% organic carbon. (From Walker, 1974; data from Romankevich, 1968. Copyright 1974 by *American Journal of Science,* New Haven, Conn. Used by permission of the publisher.)

weathering contain, on the average, about 0.5% carbon, and since the mass of the sediments is largely conserved during erosion, weathering, and transport to the sea (Garrels and Mackenzie, 1971), it is evident that the deposition of new sediments containing less than 0.5% carbon results, overall, in a net loss of atmospheric oxygen. The important oxygen source regions are therefore those where the new sediments contain more than 0.5% carbon. Figure 3-5 shows for the Pacific that these regions are very restricted indeed (Romankevich, 1968), occurring only in narrow belts along the ocean margins. Ancient sediments also show a concentration of carbon in near-shore marine deposits (Ronov, 1958). In spite of the great disparity in the areas of the carbon-rich and carbon-poor sediments it is possible for balance to be maintained in the oxygen and carbon budgets because the sediment accumulation rate is approximately ten times as large in the near-shore areas as in the deep sea.

There are two factors that appear to contribute to the relatively high carbon content of the near-shore sediments (Romankevich, 1968). The first of these is the rapid accumulation of sediments. Organic detritus lies on the sea floor or in the surface layers of the sediments for a time that is inversely proportional to the sediment accumulation rate. During this time it is exposed to oxygenated water and is subject to attack by *benthic* (bottom-dwelling) organisms and aerobic microorganisms. The more rapidly the detritus is buried, the more likely it is to escape oxidation.

The second factor is the supply of organic matter to the sediments, which is much greater along the coasts than in the open ocean because the primary organic productivity is higher (Koblentz-Mishke et al., 1970; Riley and Chester, 1971). A large supply of carbon to the sediments can obviously lead directly to a high carbon content of the sediments, but there is an indirect effect of large carbon supply that is probably more important. A large supply of organic carbon promotes the establishment of anaerobic conditions.

As a simple model of what happens to the carbon in newly deposited sediments, consider a sample of material that has been completely isolated by subsequent deposition. If this sample contains more moles of oxygen than of organic carbon dissolved in the water in its pores, most of the carbon will in due course be oxidized, the sediment will remain aerobic, and little or no carbon will be fossilized. If, on the other hand, because of a greater initial supply of carbon, the sample contains more carbon than oxygen, the oxygen will be depleted in due course and the sediment will become anaerobic. The remaining organic material will still be subject to oxidation by microorganisms utilizing dissolved nitrate or sulfate ions as their sources of oxygen (fermentation does not change the amount of reduced material unless methane escapes from the sediment). But these organisms work much more slowly than aerobic organisms (Thimann, 1963) and are less able to attack resistant organic substrates (Postgate, 1968). Sulfate reduction, moreover, does not as a rule lead to a decrease in the total content of reduced material in the sediments. Carbon is oxidized, but the sulfate is reduced to hydrogen sulfide, which reacts with iron (Berner, 1970, 1972). Burial of sedimentary iron sulfide is equivalent to burial of organic carbon as far as the net production of atmospheric oxygen is concerned (Redfield, 1958). Only if the sediments are deficient in reactive iron and the hydrogen sulfide escapes into the sea does oxidation of organic carbon by sulfate reduction lead to a decrease in the oxygen source.

Our description of the processes affecting the decay of organic carbon in isolated sediment samples is, of course, greatly simplified. A more realistic model would allow for diffusion of dissolved constituents and for the activities of burrowing organisms (Berner, 1971a; Rhoads, 1973). Our conclusion, nevertheless, is valid. A large supply of organic carbon to the sediments promotes the establishment of anaerobic conditions and these, in turn, enhance the preservation of the carbon. In the world ocean today it is in the near-shore sediments that anaerobic conditions are found.

Given the highly restricted areas in which significant carbon burial occurs and the complex and nonlinear interactions that determine the carbon content of sediments it would appear that the total rate of carbon burial and thus the net source of atmospheric oxygen would vary quite markedly with time. Changes in the distribution of the continents, in world climate, and in the circulation of the oceans will lead to new areas of high primary productivity, a high sediment accumulation rate, and a high rate of burial of organic carbon; alternatively, areas of significant carbon burial may disappear. Thus it seems most probable that the oxygen content of the atmosphere can vary (Holland, 1973a), possibly by a substantial amount. Nevertheless, as we have already argued, there must be a feedback mechanism that prevents oxygen from disappearing altogether.

The Feedback Mechanism

Control appears to be exercised by the relative rates of supply of organic matter and of oxygen to the deeper levels of the ocean (Redfield, 1958; Broecker, 1971; Holland, 1973a). If the oxygen content of the atmosphere were to decrease, there would be larger areas of the ocean in which the supply of organic material would exceed the supply of oxygen. Anaerobic conditions would become more widespread, increasing the rate of burial of organic carbon and thus the net production of atmospheric oxygen; this mechanism would counteract the initial decrease.

What factors determine the rates of supply of dissolved oxygen and reduced carbon to the deeper levels of the ocean? In order to answer this question in the simplest way possible, let us imagine a horizontally stratified, one-dimensional ocean. In other words, let us ignore or average over the marked horizontal variations in oceanic properties that result from the varied shapes of the ocean basins, circulation, and latitudinal gradients.

In this hypothetical ocean, imagine that an isolated sample of water is brought from the deep sea to the surface. At the surface photosynthesis proceeds until the limiting nutrients are exhausted. The amount of organic material synthesized is proportional to the initial nutrient concentration in the sample. At the same time, the oxygen content of the sample comes to equilibrium with the atmosphere. Thus, the concentration of dissolved oxygen is proportional to the partial pressure of oxygen in the atmosphere.

Now imagine that this sample of sea water with its burden of organic matter and oxygen is carried back down to the depths, where photosynthesis is impossible because of the absence of light. Respiration and decay consumes the organic matter, restoring nutrient elements to the sea water. Oxygen is consumed in the process. That this model of the biogeochemical circulation of the ocean contains some elements of the truth is shown by the representative profiles of nutrient concentration and dissolved oxygen as functions of depth shown in Figs. 1-7 and 3-6. Over most of the oceans the limiting nutrient elements are phosphorus and nitrogen. Figure 1-7 shows that they are both exhausted at the surface. The oxygen concentration, on the other hand, is high at the surface, where equilibrium is maintained with the atmosphere, and is strongly depleted at the depths where respiration and decay release phosphorus and nitrogen from organic matter.

The question we want to consider for our hypothetical average ocean is whether respiration and decay exhausts the supply of organic matter or of oxygen in the deep water. If oxygen is present in excess, almost all of the organic matter is consumed, little carbon survives to be buried in sediments, and the net oxygen source is small. If organic carbon is in excess, however, the deep waters become anaerobic, and the rate of

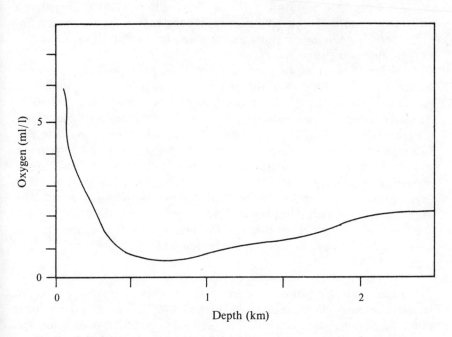

Fig. 3-6. Representative profile of dissolved oxygen as a function of depth in the Pacific Ocean. (After Riley and Chester, 1971.)

fossilization of organic carbon is large. The answer to the question depends on the nutrient content of average deep-sea water and the carbon-to-nutrient ratio in average organic matter as well as on the partial pressure of oxygen in the atmosphere and the average solubility of oxygen in surface sea water. The calculation that follows was given by Redfield (1958). We shall use his values for the necessary quantities averaged over the ocean.

Consider, first, the phosphorus-to-nitrogen ratio in average sea water and in average *plankton* (microbes in near-surface waters make up most of the mass of living creatures in the sea). According to Redfield (1958), average sea water contains 15 nitrogen atoms for every phosphorus atom, while average plankton contain 16 nitrogen atoms per phosphorus atom. Hence, given the range of uncertainties within which we are operating, we may take the phosphorus-to-nitrogen ratio as being the same in water and plankton. Thus, as the plankton consume nutrients at the surface, concentrations of both phosphorus and nitrogen in the water decline in proportion to one another, and both nutrient elements are exhausted simultaneously, as shown in Fig. 1-7. We shall use phosphorus as the reference nutrient, remembering that nitrogen is proportional to phosphorus.

Average sea water contains 1000 carbon atoms, mostly as bicarbonate ions, for every phosphorus atom, while average plankton have a carbon-to-phosphorus ratio of only 106. Thus, the growth of plankton in surface sea water exhausts the phosphorus supply long before the carbon supply is significantly depleted. In other words, phosphorus and nitrogen are the limiting nutrients rather than carbon.

In order to maintain its average composition, the average planktonic organism must consume 16 nitrogen atoms and 106 carbon atoms for every phosphorus atom it consumes. The nitrogen atoms are present in sea water largely as nitrate ions, (NO_3^-), and in plankton largely as amino radicals, (NH_2). The hydrogen in the amino radicals comes from water, so 64 oxygen atoms are released when 16 nitrogen atoms are incorporated into plankton. The incorporation of carbon into the plankton leads to the release of 212 oxygen atoms for every phosphorus atom consumed. Adding these two contributions we find that photosynthesis releases 276 oxygen atoms for each phosphorus atom consumed. When respiration and decay convert the plankton back to dissolved inorganic constituents, the same number of oxygen atoms are consumed for each phosphorus atom restored to the water.

Since average deep-sea water contains 70×10^{-6} gm l^{-1} of phosphorus, we find that 10^{-2} g l^{-1} of oxygen are released when photosynthesis at the surface consumes all of the dissolved phosphorus. When the water returns to the depths, the same amount of oxygen is consumed as respiration and decay converts the organic matter back to inorganic forms.

We have found that the decay of organic matter in deep-sea water creates a demand for dissolved oxygen of 10^{-2} gm l^{-1}. We must now evaluate the supply of dissolved oxygen in order to determine whether average deep-sea water is anaerobic or aerobic. The oxygen supply is governed by equilibrium between surface sea water and the atmosphere. According to Redfield (1958) average surface water saturated with oxygen contains 10^{-2} gm l^{-1}. The oxygen supply is therefore just large enough to balance the supply of organic matter, leaving neither oxygen nor organic matter in excess in the deep sea. If there were less oxygen in the atmosphere, however, the supply of oxygen to the deep sea would be inadequate, the water would become anaerobic, and organic matter would survive to be buried.

This is a greatly oversimplified description of a complex set of interactions. For one thing, most sea water, when it leaves the surface, does not have all of its limiting nutrients incorporated in organic material. Deep water is produced at high latitudes, where primary productivity may be limited by illumination rather than by nutrient supply. Thus, water descending to the depths may have its full load of oxygen, but may not carry as much organic material as it would have if photosynthesis had exhausted the nutrients. For this reason, the deep waters of the world ocean are not entirely depleted in oxygen, although the depletion is very substantial for a large volume of the Pacific, as shown by Duedall and Coote (1972).

The model we have described also neglects the transport of nutrients and oxygen by eddy diffusion and the transport of organic matter in particulate form by settling. In addition, there are very marked variations in the nutrient content of deep water in different parts of the ocean as a result of the interaction of biological processes and ocean circulation (Redfield et al., 1963). Therefore, the coincidence we have described between the nutrient content of sea water and the oxygen content of the atmosphere exists only in the sense of an ill-defined average.

The average is nonetheless significant. As we have just described, there is a suggestive coincidence between the oxygen content of the atmosphere, which governs the oxygen content of saturated sea water, and the amount of oxygen that is released when photosynthesis exhausts the nutrients in sea water (equal to the amount consumed when decay restores these nutrients to inorganic form). Consequently, if the oxygen content of the atmosphere were decreased or the nutrient content of sea water increased, anaerobic conditions would be created at depth, not everywhere, but in situations where circulation and productivity were favorable. The formation of these anaerobic areas would cause an increase in the rate of burial of organic carbon and thus in the net source of oxygen. Since the rate of consumption of oxygen in weathering would not be changed by small perturbations of the kind we are considering, the abundance of atmospheric oxygen would increase. Similarly, if the atmosphere contained too much oxygen relative to the nutrient content of average sea water, anaerobic environments would be rare; there would be little burial of organic carbon, and so little net production of oxygen. Oxygen consumption, however, would not be changed, so the level of atmospheric oxygen would decline.

This general mechanism for stabilizing the oxygen content of the atmosphere by means of a net oxygen source that varies inversely with oxygen partial pressure has been described by Broecker (1971) and Holland (1973a). The present model draws on the work of Redfield (1958) to make this general idea more quantitative. According to the model, the equilibrium level of oxygen depends in a calculable way on the nutrient content of sea water, the composition of plankton, and the solubility of oxygen.

Possible Variations in Atmospheric Oxygen

As Holland (1973a) has pointed out, the existence of a negative feedback mechanism does not imply that the oxygen content of the atmosphere has not varied with time. The equilibrium level can be expected to have varied as a result of changes in surface temperature and therefore in the solubility of oxygen, or, to cite a few other possibilities, as a result of tectonic activity, changes in the configuration of the ocean basins, and changes in ocean circulation. Moreover, it takes the atmosphere a few million years to adjust to any shift in the equilibrium abundance of oxygen, so departures from the equilibrium value are to be expected.

Subject to these fluctuations, atmospheric oxygen is stabilized, over the long term, by the requirement that the average rate of supply of oxygen to the deeper levels of the ocean be equal to the average rate of supply of organic material. As explained above, this means that the equilibrium oxygen content of the atmosphere is related to the average nutrient content of sea water by way of the carbon-to-nutrient ratio in plankton and the solubility of oxygen. In Chap. 7 we shall draw on this discussion to develop a tentative geologic history of atmospheric oxygen.

NITROGEN

Nitrogen, like oxygen, is released to the atmosphere and removed by biological processes. In the case of oxygen, the driving process is photosynthesis. The rate of photosynthesis, called productivity, has been measured in many areas of the globe, both on land and at sea, so fairly reliable estimates can be made of the rate integrated over the whole earth. This is not the case for the biological processes that drive the nitrogen cycle, so we can make only very rough estimates of the rate of turnover of material in the nitrogen cycle.

Discussion of the geochemical budget of nitrogen is complicated by the number of *valence states* that are possible for nitrogen. In the atmosphere, for example, nitrogen occurs as ammonia, NH_3, molecular nitrogen, N_2, nitrous oxide, N_2O, nitric oxide, NO, nitrogen dioxide, NO_2, and nitric acid, HNO_3. We will not examine all of the possible transformations between the different compounds of nitrogen, but will consider just the compounds and the processes that might influence the atmospheric abundance of molecular nitrogen. As a unit of mass we shall here use the *gram atom* (1 gm atom $N = 14$ gm of nitrogen).

We shall first examine the budget of atmospheric nitrogen, concluding that the only reservoir large enough to compete with the atmospheric reservoir consists of organic matter incorporated in sedimentary rocks. Atmospheric nitrogen is therefore controlled by the same processes as atmospheric oxygen, the processes of burial of organic matter in sediments and of weathering of sedimentary rocks. We shall then examine the nitrogen budget of the ocean in an attempt to understand why the nitrate concentration is far below saturation. We shall suggest that nitrate, on the average, is maintained in constant ratio to phosphate by the activities of nitrogen-fixing organisms.

Nitrogen Reservoirs

Nitrogen does not escape from the atmosphere into space (see Chap. 4), so the total amount of nitrogen in the atmosphere, the oceans, the crust, and the upper mantle is conserved. Changes in the atmospheric nitrogen content, if they occur, involve the transfer of nitrogen between these reservoirs. Let us estimate the amount of nitrogen presently stored in various forms in the near-surface layers of the earth (see Fig. 3-7).

The atmosphere contains 3.9×10^{21} gm of molecular nitrogen, or 2.8×10^{20} gm atom N. Other nitrogen compounds in the atmosphere may be neglected (see Chap. 1). There are 1.5×10^{18} gm atom N in the form of molecular nitrogen dissolved in the ocean (Bowen, 1966, p. 57). This quantity is controlled by the solubility of nitrogen in sea water. Since the reservoir is small compared with the atmospheric reservoir and can hardly vary greatly in size, we may neglect it.

About 6.5×10^{17} gm of nitrogen are incorporated in organic material on land and in the sea, mostly in the form of the amino radical, NH_2 (Bowen, 1966, p. 57). Only about 6% of this nitrogen is in living organisms. Dissolved organic matter in sea water contains about 3.4×10^{17} gm of nitrogen (Vaccaro, 1965, p. 403). The rest is in the form of soil humus. The total organic reservoir therefore contains 4.6×10^{16} gm atom N.

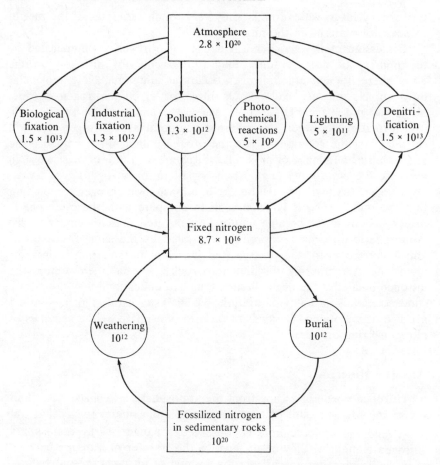

Fig. 3-7. Reservoirs (in gm atom N) and processes (rates in gm atom N yr^{-1}) affecting the nitrogen content of the atmosphere.

The oceans contain 5.7×10^{17} gm of nitrogen in the form of nitrate ions, NO_3^-, in solution (Vaccaro, 1965, p. 403). The nitrate content of soils as well as the ammonia and nitrite contents of soils and ocean are negligibly small by comparison. Adding the organic nitrogen and the nitrate nitrogen, we obtain a value of 12.2×10^{17} gm or 8.7×10^{16} gm atom N for the reservoir of *fixed nitrogen* at the surface of the earth. By fixed nitrogen we mean nitrogen in compounds containing an isolated N atom. Molecular nitrogen (N_2) and nitrous oxide (N_2O) are chemically inert and accumulate mainly in the atmosphere. Fixed nitrogen is reactive and is susceptible to incorporation in solid and liquid compounds.

Igneous rocks in the crust of the earth appear to contain a total of 1.6×10^{19} gm of nitrogen in the form of NH_4^+ ions incorporated into the crystal lattices of silicate minerals (Stevenson, 1962; Bowen, 1966, p. 57;

Eugster and Munoz, 1966; Eugster, 1972). We shall ignore the igneous rocks in our discussion of the nitrogen budget, as we did for oxygen, because of their small contribution, compared with sedimentary rocks, to the total rate of weathering.

Nitrogen is incorporated in sedimentary rocks as a constituent of the buried organic matter that we discussed in connection with atmospheric oxygen. We shall assume that the average carbon-to-nitrogen ratio (by numbers of atoms) in sedimentary organic matter is ten, although large variations in this ratio are observed (Arrhenius, 1950; Stevenson, 1962). This reservoir of what we shall call *fossilized* nitrogen therefore contains about 10^{20} gm atom N, approximately one-third of the amount in the atmospheric reservoir.

Burial of Organic Nitrogen and the Nitrogen Content of the Atmosphere

The fixed nitrogen reservoir is more than three orders of magnitude smaller than the atmospheric reservoir. We shall discuss below the possibility that this reservoir could grow large enough to affect the nitrogen content of the atmosphere. Ignoring this possibility for now, we see that only the fossil nitrogen reservoir is large enough to constitute a significant alternative to the atmosphere as a location for surface nitrogen. Changes in the nitrogen content of the atmosphere must therefore result from changes in the amount of organic nitrogen in sediments.

Using our estimated carbon to nitrogen ratio in sedimentary organic matter and results from our discussion of oxygen, we find that the rate of burial of organic nitrogen, which we take to be equal to the rate of removal of nitrogen from the fossilized nitrogen reservoir by weathering, is 10^{12} gm atom N yr^{-1}. Dividing this rate into the nitrogen content of the atmosphere, we find that the residence time of atmospheric nitrogen against the cycle of burial and weathering is 280 million years. The nitrogen content of the atmosphere must therefore vary much more slowly than the oxygen content, if it varies at all.

We have already discussed the factors that control the rate of burial and the rate of weathering of sedimentary organic matter. Since the carbon-to-nitrogen ratio in this material is determined largely by the composition of organisms, it should not vary markedly with either the oxygen or the nitrogen content of the atmosphere. It appears, therefore, that atmospheric nitrogen is controlled by the same set of circumstances as atmospheric oxygen. When conditions are such that the rate of burial of organic matter exceeds the rate of weathering, the oxygen content of the atmosphere increases and the nitrogen content of the atmosphere decreases. When weathering is faster than burial, oxygen decreases while nitrogen increases. The changes in the mass of oxygen in the atmosphere during periods of imbalance are roughly ten times as large as the changes in the mass of nitrogen.

We may speculate that before the development of life and the accumulation of organic matter in sedimentary rocks almost all of the surface nitrogen would have been in the atmosphere. If this were the case, the nitrogen content of the atmosphere would have been approximately 1.3 times its present value. The accumulation of sedimentary organic matter, more than 3.3 billion years ago (Schidlowski et al., 1975), would have been accompanied by a decline in atmospheric nitrogen. Fluctuations in the oxygen content of the atmosphere after the accumulation of the sedimentary organic reservoir would have been accompanied by much smaller fluctuations in the nitrogen content. An increase in atmospheric oxygen to a value five times the present value, for example, would be accompanied by a decrease in atmospheric nitrogen by 10%. It must be emphasized that this description of the factors controlling the nitrogen content of the atmosphere is speculative and subject to revision in the light of future research.

Sources and Sinks of Fixed Nitrogen

Material is removed from the fixed nitrogen reservoir when it is incorporated in sediments to become fossilized nitrogen. Over the long run, this sink is balanced by a source of fixed nitrogen provided by the weathering of sedimentary organic matter. The estimated rate of these processes is 10^{12} gm atom N yr^{-1}. Nitrogen is removed from the fixed nitrogen reservoir and restored to the atmosphere by the process of *denitrification*. Denitrification occurs mainly in oxygen-deficient environments, either in soils, sea-floor sediments, or in regions of the open ocean where the concentration of dissolved oxygen is very low (Goering et al., 1973; Hahn, 1974). As described in Chap. 1, organisms in these environments use nitrate and nitrite ions as sources of oxygen in order to derive energy from the oxidation of organic debris. Molecular nitrogen or nitrous oxide is produced as a result. The nitrous oxide escapes to the atmosphere where it is dissociated by sunlight, as we described in Chap. 2, to form molecular nitrogen and atomic oxygen. Thus, we may regard denitrification as a sink of fixed nitrogen and a source of atmospheric nitrogen, whether the product is nitrous oxide or molecular nitrogen. The global rate of denitrification has been estimated by Burns and Hardy (1975) as 1.5×10^{13} gm atom N yr^{-1}.

A number of processes restore the fixed nitrogen eliminated by denitrification. Of the photochemical reactions occurring in the atmosphere, the most important is probably the reaction

$$(3\text{-}1) \qquad\qquad N_2O + O(^1D) \rightarrow 2NO$$

discussed in Chap. 2 in connection with the stratosphere. It is not a globally significant source of fixed nitrogen. The role of lightning has been investigated by a number of authors (Hutchinson, 1944, 1954; Georgii, 1963; Commoner, 1970; Chameides et al., 1976). It too appears to be minor.

Man fixes nitrogen deliberately in the production of inorganic fertilizers and as a byproduct of industrial activity when oxygen and nitrogen combine in furnaces. According to Robinson and Robbins (1970a, b) the total pollution source of fixed nitrogen is 1.3×10^{12} gm atom N yr^{-1}, while industrial fixation of nitrogen occurs at about the same rate. We see, therefore, that man is currently fixing nitrogen at a rate that is not negligible compared to denitrification. Since nitrogen is also fixed biologically, we conclude that man's activities are increasing the fixed nitrogen reservoir at the expense of the atmosphere. It is only in recent decades that man has fixed nitrogen at a geologically significant rate, so we may suspect that the situation is temporary and that equilibrium will be restored in due course, either by an increase in the rate of denitrification or by a decrease in the rate of biological nitrogen fixation. Even if no adjustment were to occur in the natural rates it would take 100 million years to exhaust the atmospheric nitrogen reservoir at the present rate of consumption, assuming no other factors were to prove limiting.

There seems to be no way to arrive at a reliable estimate of the rate of biological nitrogen fixation. We shall simply assume that the nitrogen budget was in balance before man's intervention, and that biological processes fix nitrogen at a total rate of 1.5×10^{13} gm atom N yr^{-1} (cf. Burns and Hardy, 1975). Our results for the global nitrogen budget are illustrated in Fig. 3-7.

Fixed Nitrogen in the Ocean

It seems probable that the fixed nitrogen reservoir is presently increasing in size as a result of man's activities. We should, therefore, consider the factors that control the size of this reservoir and the possibility that it could grow large enough to affect the nitrogen content of the atmosphere. The fixed nitrogen reservoir at present consists of approximately equal amounts of humus and of nitrate ions dissolved in the sea. The humus component is identical to the reservoir of surface organic matter that we discussed in connection with the oxygen budget. Its size is determined by the cycle of photosynthesis followed by respiration and decay and depends, as we have already shown, on the rates of these processes and on the oxygen content of the atmosphere. For an oxygenated atmosphere, it is unlikely that the surface organic reservoir can grow large enough to affect the atmospheric reservoirs of oxygen and nitrogen.

On the other hand, the nitrate content of the ocean could be very much larger than it is without saturation occurring. Let us therefore

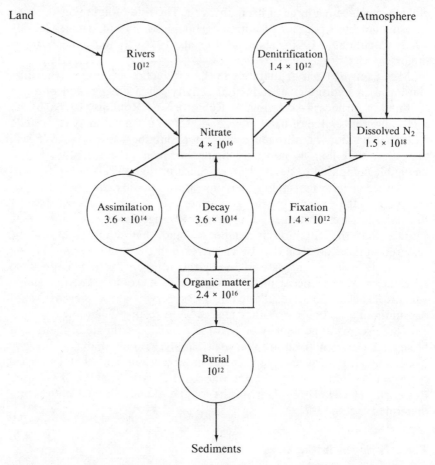

Fig. 3-8. Reservoirs (in gm atom N) and processes (rates in gm atom N yr⁻¹) affecting fixed nitrogen in the ocean.

explore the factors that limit the nitrate content of sea water. As the discussion will show, our estimates of components of the nitrate budget (Fig. 3-8) are very uncertain.

The rate of supply of fixed nitrogen to the sea by rivers is about 10^{12} gm atom N yr^{-1} (Hutchinson, 1954; Turekian, 1968, p. 101; Robinson and Robbins, 1970a, b). About half of this seems to be in organic matter and about half in the form of nitrate ions (Hutchinson, 1954). We have already estimated that fixed nitrogen is removed from the sea at about the same rate by burial of organic matter in sediments. We can derive the rate at which nitrogen is cycled between organic and inorganic forms from results presented earlier on the productivity of the sea and the average carbon to nitrogen ratio in plankton. Phytoplankton produce organic carbon at a rate of 2.4×10^{15} moles yr^{-1} (Bowen, 1966, p. 50) and incorporate, on average, 0.15 gm atom N for each mole of carbon. The turnover rate is therefore 3.6×10^{14} gm atom N yr^{-1}.

Denitrification in the open ocean has been discussed by Goering et al. (1973). They estimate that the rate of denitrification in the large bodies of oxygen-deficient water in the Eastern Tropical North Pacific is 10^{12} gm atom N yr^{-1}. Denitrification in anoxic basins such as the Black Sea and the Cariaco Trench is negligible by comparison.

Denitrification also takes place in anaerobic sediments. We may estimate this rate by calculating the rate of diffusion of nitrate ions into such sediments. We assume that the downward flux is given by $-D_s(dn/dx)$, where D_s is a diffusion coefficient with a magnitude of about 3×10^{-6} cm^2 sec^{-1} (Berner, 1971a), n is the concentration of nitrate ions, and x is depth in the sediment. Data summarized by Emery (1960, p. 108) for the Santa Barbara, Santa Monica, and Santa Catalina Basins off southern California show $n = 4.8 \times 10^{-7}$ gm cm^{-3} of nitrogen as nitrate at the surface of the sediments and nitrate concentrations that drop to undetectable amounts within the top few centimeters. Let us therefore take $dn/dx = 10^{-7}$ gm cm^{-4}. This gives a flux of nitrogen into the sediments of 3×10^{-13} gm cm^{-2} sec^{-1} or 10^{-5} gm cm^{-2} yr^{-1}. If we assume that this value applies to all of the area of the sea floor covered by blue-clay sediments $(5.6 \times 10^{17}$ cm$^2)$, we obtain a total rate of denitrification in sediments of 4×10^{11} gm atom N yr^{-1}, comparable to the rate in oxygen-deficient waters in the open ocean.

The fixation of nitrogen in the open ocean has been discussed by Goering et al. (1973). From data of Goering et al. (1966) they estimate that the blue-green algae, *Trichodesmium*, which appears to be most abundant in tropical waters, fixes nitrogen at a rate of about 4×10^{10} gm yr^{-1}. They mention a number of other organisms that may fix comparable or greater amounts of nitrogen, but quantitative information on these other organisms is lacking. We shall arbitrarily assume that atmospheric nitrogen is fixed at a large enough rate, 1.4×10^{12} gm atom N yr^{-1}, to bring the oceanic fixed nitrogen budget into balance.

Our budget of oceanic nitrogen is shown in Fig. 3-8. It is very tentative, of course, but it allows us to consider the processes that govern the nitrate content of the sea. The rate of river supply and the rate of burial of organic nitrogen are independent of the nitrate content of the sea, so they cannot be the controlling processes. As described above, the denitrification rate is probably proportional to the nitrate concentration in oxygen-deficient waters and in the waters overlying anaerobic sediments. We can therefore expect the denitrification rate to increase as the nitrate content of sea water increases, providing a measure of control. The extent of the anaerobic sediments and oxygen-deficient waters is important, however, and this depends on factors other than the nitrate content of sea water, as we have shown in our discussion of the oxygen budget.

In our discussion of the oxygen budget we noted, without comment, that the nitrogen-to-phosphorus ratio in sea water is approximately equal to the ratio in plankton. This near equality may be a coincidence, or it may reflect an accommodation by organisms over an extended period of evolution to the medium in which they live. It is possible, however, that the ratio is a direct reflection of some process controlling the nitrogen content of sea water. For example, control may be exercised by the nitrogen-fixing organisms, blue–green algae for example; they fix nitrogen only when they encounter water that does not contain enough nitrogen, relative to phosphorus, to meet their needs (Strickland, 1965, p. 544; Ryther and Dunstan, 1971; Hutchinson, 1973). These organisms utilize nitrate ions as a source of nitrogen in preference to molecular nitrogen as long as nitrate is present (Hutchinson, 1954). In water that has been stripped of nitrate but contains nutrient phosphorus, however, they fix nitrogen. Thus, these organisms could, in principle, maintain the observed ratio of nitrogen to phosphorus in sea water. If this explanation of the observed ratio is correct, then man's role as a nitrogen fixer will simply substitute for the nitrogen-fixing activities of bacteria and blue–green algae; little increase in the fixed nitrogen content of the sea will result. If man supplies nitrate to the sea faster than it is being removed by denitrification and burial, then biological fixation in the sea will cease altogether and the nitrate content of sea water will increase until denitrification can balance the input.

Because the ocean is nearly saturated with phosphate, this suggestion implies that the oceanic reservoir of fixed nitrogen can never have grown large enough to absorb a significant fraction of the nitrogen in the atmosphere. Neglect of the fixed nitrogen reservoir in our description of the controls on the abundance of atmospheric nitrogen is therefore justified.

CARBON DIOXIDE

Estimates of the rates of processes involved in the geochemical cycles of carbon dioxide are much more reliable than estimates of rates in the cycles of oxygen and nitrogen. Radioactive ^{14}C, produced in the atmosphere by cosmic rays, has been used as a tracer in determining the rates of transfer of carbon dioxide between the atmosphere and other reservoirs, providing a good understanding of the effects of short-term imbalances in the carbon dioxide budget. We shall discuss the short-term behavior first, including an analysis of the effects on atmospheric carbon dioxide of man's combustion of fossil fuel. In the short term, the carbon dioxide content of the atmosphere is governed by considerations of chemical equilibrium between atmosphere and ocean. What processes operate in the long term to control the carbon dioxide content of the atmosphere are still uncertain. We shall argue that chemical equilibrium considerations are not sufficient in the long term. Instead, the important factors are kinetic: CO_2 is continually added to the ocean–atmosphere system by volcanoes and hot springs. For long-term equilibrium, it must be removed equally fast by weathering of silicate minerals (at a rate that depends on the CO_2 partial pressure) followed by deposition in sediments as carbonate minerals.

Reservoirs and Exchange Rates

The atmosphere contains 5.6×10^{16} moles of carbon dioxide (see Chap. 1 and Fig. 3-9). Carbon dioxide is removed from the atmosphere by photosynthesis and restored by respiration and decay. These processes were discussed earlier in this chapter, in connection with the oxygen budget. The estimated rate of this circulation is 10^{16} moles yr^{-1}, so the average carbon dioxide molecule spends only 6 yr in the atmosphere before being incorporated in organic material. The reservoir of surface

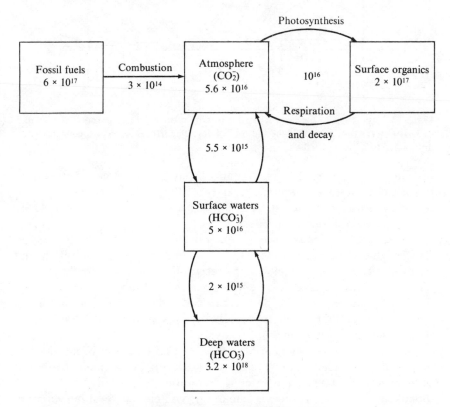

Fig. 3-9. Carbon dioxide budget for times less than a few thousand years (reservoirs in moles and rates in moles per year).

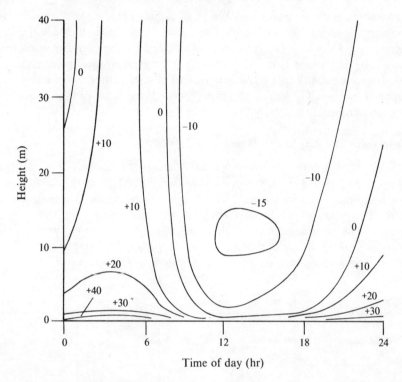

Fig. 3-10. Departures of the carbon dioxide mixing ratio in forest air from a mean value of 320 ppm as a function of height and time of day. Departures are expressed in parts per million. (After Bolin, 1970.)

organic matter, mostly humus, contains about 2×10^{17} moles of carbon, so it is larger than the atmospheric reservoir. Changes in atmospheric carbon dioxide can therefore result from imbalances in the rates of photosynthesis and of respiration and decay.

One source of imbalance is the diurnal cycle of light and dark. Photosynthesis proceeds only during daytime, whereas respiration and decay occur around the clock. Corresponding variation in the carbon dioxide partial pressure is observed near the ground, as shown in Fig. 3-10. The partial pressure decreases during the day as photosynthesis consumes carbon dioxide more rapidly than respiration and decay restore it. At night the partial pressure recovers.

Another source of imbalance is the seasonal variation in the rates of growth of green plants at higher latitudes. During the summer the rate of photosynthesis exceeds the rate of respiration and decay, and carbon dioxide declines. The reverse is true during winter. The resulting variation in atmospheric carbon dioxide is detectable throughout the troposphere, as shown in Fig. 3-11.

These periodic variations in carbon dioxide provide useful information on atmospheric mixing rates. Diurnal variations cannot affect the whole troposphere because tropospheric mixing times are of the order of weeks (see Chap. 2). Seasonal variations can affect the whole troposphere, but not the higher levels of the atmosphere because air spends about 60 yr in the troposphere, on average, before being carried into the stratosphere.

The ocean contains much more carbon dioxide than either the atmosphere or the surface organic reservoir, most of it in the form of bicarbonate ions, (HCO_3^-). Because of its importance to the carbon dioxide balance, it is worth describing the oceanic reservoir in some detail. Most of the surface of the sea is covered by a well-mixed layer of relatively warm water. This layer averages 70 m in thickness, and contains about 5×10^{16} moles of dissolved inorganic carbon (Broecker et al., 1971). Underlying the surface layer is the main thermocline, in which temperature decreases steadily with increasing depth. Such a temperature profile is stable in the ocean, as it is in the atmosphere, so mixing is severely inhibited in the region of the thermocline. According to Broecker et al. (1971), the thermocline averages 1000 m in thickness and contains 8×10^{17} moles of dissolved inorganic carbon. Most of the carbon

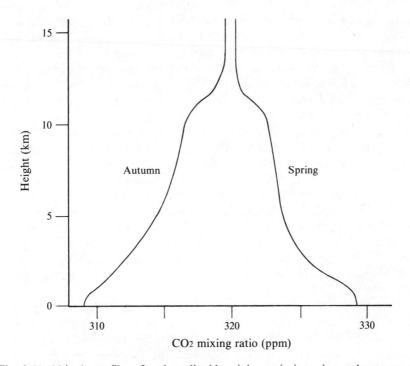

Fig. 3-11. Altitude profiles of carbon dioxide mixing ratio in spring and autumn at middle latitudes. (After Bolin, 1970.)

is in the cold, isothermal, deep ocean, which underlies the main thermocline. The deep ocean contains 2.4×10^{18} moles of dissolved inorganic carbon.

Variations occur in the carbon content of the deep water as a result of the circulation of the oceans. Deep water is formed in the North Atlantic, where it sinks and flows towards the Antarctic. Here it joins cold, dense water formed in the Weddell Sea and flows through the Indian and Pacific Oceans towards the Aleutian Arc. The deep current continuously loses water by upwelling to the surface, where a return flow balances the export of water from the deep-water, source areas in the North Atlantic and the Weddell Sea. As it flows towards the northern Pacific, the deep water is enriched in the nutrient elements, phosphorus, nitrogen, and silica, as well as in carbon, by the rain of organic debris falling into it from the illuminated surface waters where photosynthesis occurs. The Pacific therefore contains a higher concentration of the elements utilized by organisms than do the Indian or Atlantic Oceans.

This circulation implies that the Pacific deep water is older than the Atlantic deep water in the sense that it has been isolated from the atmosphere for a longer time. The age of the deep water is reflected in its ^{14}C to ^{12}C ratio relative to this ratio in the atmosphere. In Atlantic deep water this ratio is 0.90 of its atmospheric value, while in Pacific deep water it is 0.77 (Broecker et al., 1961). The ratio decreases with increasing age of the water because of decay of the radioactive ^{14}C.

From these figures we can estimate the average ages of carbon in the Atlantic and Pacific deep waters. The *decay constant* for ^{14}C is $\lambda = 1/8000 \ yr^{-1}$ (cf. Broecker et al., 1971), which means that the ratio of ^{14}C to ^{12}C evolves with time, t, in proportion to $\exp(-\lambda t)$. Equating this expression to 0.90 for the Atlantic we find an average age of 800 yr. For the Pacific the average age is 2100 yr. These ages are approximate because we have assumed that carbon supplied to the deep water has the atmospheric ratio of ^{14}C to ^{12}C, while part of the supply may, in fact, come from surface water with a lower value of the ratio (Broecker et al., 1971).

We will neglect the difference between the Atlantic and Pacific and adopt an average age of deep-water carbon of 1500 yr. By dividing this residence time into the carbon content of the reservoir, including the thermocline region in the deep-water reservoir, we obtain a figure for the rate of exchange of carbon between the deep water and the surface water. This rate is 2.1×10^{15} moles yr^{-1}.

The ^{14}C to ^{12}C ratio in surface water provides an indication of the rate of exchange between atmosphere and ocean. In the steady state, the rate at which carbon dioxide leaves the ocean for the atmosphere is equal to the rate at which it leaves the atmosphere for the ocean. But the carbon dioxide entering the ocean must carry more ^{14}C than that leaving in order to balance the decay of ^{14}C in the ocean (Broecker et al., 1971). We may write

(3-2) $$IC_A A - IC_s A = k_0 C_0 V \lambda$$

where I is the average CO_2 exchange rate per unit area and A is the area of the ocean surface; C_A, C_s, and C_0 are the ^{14}C to total C ratios in the atmosphere, the surface waters, and the whole ocean; k_0 is the average carbon concentration in the ocean, V is the volume of the ocean, and λ is the decay constant for ^{14}C. Solving for IA, we find

(3-3) $$IA = [(C_0/C_A)/(1 - C_s/C_A)]k_0 V \lambda$$

With $k_0 = 2.4$ moles m^{-3}, $C_s/C_A = 0.94$, $C_0/C_A = 0.82$, and $V = 1.35 \times 10^{18}$ m^3 (Broecker et al., 1971), we find an exchange rate, $IA = 5.5 \times 10^{15}$ moles yr^{-1}.

The elements of the carbon dioxide budget that we have described so far are shown in Fig. 3-9. With these results we can discuss the response of atmospheric carbon dioxide to changes in conditions having time scales of a few thousand years or less. Fluctuations over longer times involve additional factors, which we shall examine below.

Combustion of Fossil Fuels

Man has perturbed the carbon dioxide budget by extracting sedimentary organic carbon in the form of fossil fuels and burning this carbon to produce energy and, as a byproduct, carbon dioxide. This source is shown on the left of Fig. 3-9. Estimates of the carbon dioxide produced by combustion of different fuels in each decade since 1860 have been presented by Broecker et al. (1971). The cumulative production at the end of 1969 was 10^{16} moles, almost 20% of the carbon dioxide in the atmosphere. We wish to consider how this source affects the atmosphere.

Consider first exchange via photosynthesis and decay with the reservoir of surface organic material. Broecker et al. (1971) argue that the rates of exchange with this reservoir and the size of the reservoir are more or less independent of the partial pressure of carbon dioxide. Photosynthesis in the ocean is limited by the availability of the nutrient elements, phosphorus and nitrogen, and not by the supply of carbon. On land it is probable that productivity is also limited by factors other than carbon dioxide—nutrients, water, and illumination being among the most important. We shall accept their argument and ignore the reservoir of surface organics. It is, however, possible that the size of this reservoir has changed in recent decades, not because of combustion-induced changes in carbon dioxide, but because of man's agricultural activities. Thus it is possible that man has caused changes in atmospheric carbon dioxide other than those we seek to evaluate in this section.

Broecker et al. (1971) find that the rate of production of carbon dioxide by combustion of fossil fuels is growing at an average rate of

4.5% yr^{-1}. The average age of the added carbon dioxide is therefore about 20 yr. The time for the surface waters to come to equilibrium with the atmosphere is considerably shorter than this, so we may assume that equilibrium has been maintained between surface waters and atmosphere. The same authors have also analyzed the uptake of carbon dioxide by surface sea water. They find that the equilibrium partial pressure of carbon dioxide rises ten times as fast as the concentration of dissolved inorganic carbon in the water. Suppose, for example, that enough carbon dioxide is suddenly added to the atmosphere to increase the partial pressure by 11%. Equilibrium will be reestablished when only 1% of the additional carbon dioxide has been transferred to the ocean. With the exchange rates previously derived, this will take only about 1 yr. Because only one-tenth of the additional carbon dioxide must be transferred to the surface waters the equilibration time is ten times shorter than the residence time of carbon in the surface waters.

Penetration of the excess carbon dioxide into the main thermocline and into the deep sea is a much slower process, because of the stability of the thermocline. These reservoirs, therefore, cannot achieve equilibrium with the atmosphere in a time as short as 20 yr. Broecker and his associates have again analyzed the problem. They find that between 28 and 45% of the excess carbon dioxide has been taken up by the main thermocline and deep ocean combined. About 6% has been taken up by the surface waters, and between 50 and 65% of the addition is still in the atmosphere. These proportions remain unchanged as long as the growth rate of the carbon dioxide source is constant at 4.5%.

Independent observations confirm this theoretical result. Observations at Mauna Loa Observatory on Hawaii by Pales and Keeling (1965) show that the carbon dioxide concentration in the atmosphere increased at an average rate of 0.68 ppm yr^{-1} between March 1959 and June 1963. This is equal to the predicted rate of increase if 53% of the carbon dioxide produced by combustion in this period remained in the atmosphere.

Further confirmation is provided by the decrease in the ratio of ^{14}C to total C in the atmosphere caused by combustion of fossil fuels too old to contain any ^{14}C. The phenomenon is known as the *Suess effect*. Between 1850 and 1950 combustion produced an amount of carbon dioxide equal to 10% of the amount in the atmosphere. If all of this carbon dioxide had remained in the atmosphere, the ^{14}C-to-C ratio would have decreased by 10%. Radiocarbon measurements on tree rings of known age show that the reduction has amounted to only 2%. Evidently the atmosphere has exchanged carbon with a reservoir four times as large as the atmosphere itself. Broecker and his associates show that exchange with the reservoir of surface organics is not able to explain the effect. Exchange with living organisms is rapid, but the mass of carbon is too small; living organisms contain less carbon than the atmosphere (Bowen, 1966, p. 52). The humus reservoir is relatively large, but the rate of exchange is too small. The reduction in the Suess effect is mainly due to exchange of carbon between the atmosphere and the ocean. Broecker and his associates find that their model predicts an effect very close to that observed.

The solution of carbon dioxide in sea water increases the acidity of the water, thereby increasing its capacity to dissolve calcium carbonate. The reactions involved will be described in a later section. Much of the ocean is underlain by carbonate-rich sediments, so continued addition of carbon dioxide to the ocean–atmosphere system will lead, in due course, to solution of the calcium carbonate in sediments. Solution of calcium carbonate increases the capacity of the oceans for carbon dioxide, so once the process has started, a greater proportion of the excess carbon dioxide will be absorbed by the ocean and a smaller proportion will remain in the atmosphere. Solution of sedimentary carbonate will therefore decrease the rate of increase of atmospheric carbon dioxide.

The solubility of calcium carbonate in the ocean increases with depth. Three factors are responsible for this increase: solubility increases with increasing pressure, with decreasing temperature, and with increasing acidity (cf. Broecker, 1971). Acidity increases with depth because of the oxidation of carbon in organic debris. As a result, the deep waters are undersaturated with calcium carbonate while the upper layers of the sea are supersaturated. The phenomenon is illustrated in Fig. 3-12. Carbonate-containing sediments are restricted to the shallower parts of the sea floor, which are bathed with supersaturated water. Before solution of the sediments can commence, the acidity of these waters must be raised to the point where they become undersaturated. At present the carbon dioxide released by man is confined to the warm surface waters and the upper part of the main thermocline. As Fig. 3-12 shows, these are the most highly supersaturated waters in the sea. Broecker and his associates calculate that they will not become undersaturated until the excess carbon dioxide has achieved a value 15 times as large as its present value. Much smaller amounts of additional carbon dioxide are needed to bring the deeper waters to undersaturation, but penetration of the excess into these deeper layers is slow. Our knowledge of ocean circulation is not yet sufficient for us to predict whether solution will commence in the surface waters or near the transition from saturated water to undersaturated water. Broecker and his associates conclude that it will be a 100 yr or more, at present rates of increase of carbon dioxide production, before the solution process commences (cf. Whitfield, 1974).

Carbon Budget of the Ocean

Because of the size of the oceanic carbon reservoir, the carbon budget of the ocean merits closer examination. The budget illustrated in Fig. 3-13 is based on that of Pytkowicz (1967). Rates and reservoirs have

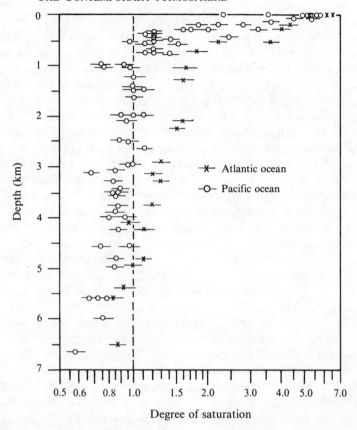

Fig. 3-12. The degree of supersaturation of calcite in the Atlantic and Pacific oceans. (From Li, Y.-H., Takahashi, T., and Broecker, W.S., Degree of saturation of $CaCO_3$ in the oceans. *J. Geophys. Res.* 74, 5507-5525, 1969. Copyrighted by American Geophysical Union.)

been derived from results of Garrels and Mackenzie (1971), Garrels and Perry (1974), Li et al. (1969), and Broecker et al. (1971).

Carbon is supplied to the surface layer of the ocean mostly in the form of bicarbonate ions in solution in river water. Much of this carbon is extracted from the atmosphere by the reactions that weather carbonate and silicate rock minerals. The weathering reactions cited by Pytkowicz (1967) may be taken as examples: for carbonate minerals

$$(3\text{-}4) \qquad CaCO_3 + H_2O + CO_2 \rightarrow Ca^{++} + 2HCO_3^-$$

and for silicate minerals

$$(3\text{-}5)$$
$$2NaAlSi_3O_8 + 2CO_2 + 3H_2O \rightarrow 2Na^+ + 2HCO_3^- + Al_2Si_2O_5(OH)_4 + 4SiO_2$$

albite   kaolinite

Bicarbonate ions are extracted from the water by organisms, which convert bicarbonate back to carbonate for incorporation in shells and hard parts, called carbonate tests. The reaction is the reverse of reaction (3-4), and carbon dioxide is released in the process. Some of this carbon dioxide returns to the atmosphere, while some of it is incorporated into organic matter by photosynthesis. Much of the organic matter decays in the surface ocean, restoring carbon dioxide to the water, but some

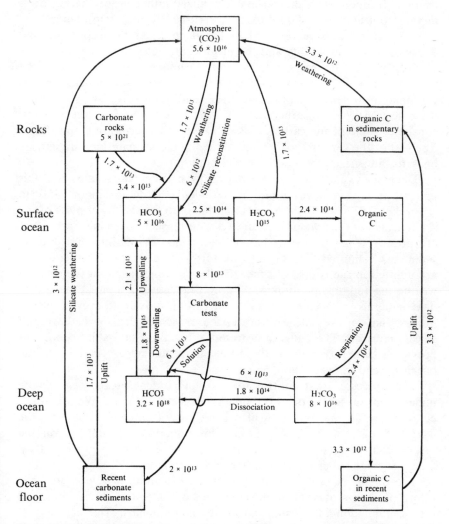

Fig. 3-13. The flow of carbon through the ocean. Reservoirs are expressed in moles and rates of transfer in moles per year. The rapid cycles that connect H_2CO_3 to HCO_3^-, atmospheric CO_2, and organic C are not shown explicitly; only the net flow of carbon is given. The rate of burial of organic carbon shown here is smaller than that in Figure 3-1. The difference arises in part because sulfide and ferrous iron are included with carbon in Figure 3-1. The difference also reflects real uncertainties in this rate.

organic debris settles into the deep ocean, where decay continues. The settling of organic debris constitutes a drain on the carbon dioxide of the surface water and a source of carbon dioxide in the deep water. A small fraction of the organic carbon produced by photosynthesis becomes incorporated into sea-floor sediments, providing a sink for oceanic carbon. We have discussed this fossilization of organic carbon earlier in the chapter. In due course the carbon is oxidized either in weathering or in metamorphism, and carbon dioxide is returned to the atmosphere.

The carbon dioxide released in the deep water by respiration and decay enhances the acidity and therefore contributes to the corrosiveness of deep water with respect to calcium carbonate. As discussed in the previous section, deep-sea water is undersaturated in carbonate ions, so carbonate tests settling from the surface tend to dissolve. The solution of these tests releases bicarbonate ions to the water and, in the process, consumes some of the carbon dioxide released by respiration and decay. The rest of the carbon dioxide and all of the bicarbonate ions are carried back to the surface by upwelling and eddy mixing.

Deep-sea waters are so corrosive to calcium carbonate that this mineral is almost never found in sediments at depths below 4500 m (Broecker, 1971). Only those tests that fall onto the shallower parts of the sea floor, where the water is saturated with respect to calcium carbonate, are preserved to become carbonate sediments. Formation of carbonate sediments is the most important sink for oceanic carbon. The sediments are uplifted in the course of time, to take part once again in the cycle of weathering and precipitation. Metamorphism converts some of the carbonate minerals to silicate minerals and carbon dioxide (Siever, 1968), a process that restores the carbon dioxide consumed in the weathering of silicate minerals. The budget illustrated in Fig. 3-13 shows this process as providing carbon dioxide to the atmosphere, but release may be to the ocean instead (Siever, 1968).

The processes that control the carbon content of the sea can be understood in terms of this budget (Broecker, 1971). We discussed controls on the rate of burial of organic carbon in connection with the oxygen budget of the atmosphere. This rate does not depend on the carbon content of the sea so it cannot be the controlling process. It is, in any event, smaller than the other terms in the budget.

The dominant processes are the supply of carbon in river water, the return of carbon dioxide from the surface ocean to the atmosphere, and the accumulation of calcium carbonate in sediments. The net rate of release of carbon dioxide from the surface of the ocean must be approximately equal to the rate of deposition of calcium carbonate because the source of the carbon dioxide is the carbonate precipitation reaction,

$$(3\text{-}6) \qquad Ca^{++} + 2HCO_3^- \rightarrow CaCO_3 + H_2O + CO_2$$

In an equilibrium ocean, in fact, the rate of release of carbon dioxide is equal to the rate of deposition of calcium carbonate minus the rate of burial of organic carbon.

The rate of deposition of calcium carbonate depends on the carbonate ion content of the sea. If the average concentration of carbonate ions is reduced, more of the sea floor is exposed to water corrosive to calcium carbonate and fewer carbonate tests survive to become incorporated in sediments. The rate of removal of carbon from the ocean therefore falls, and the carbon content of the sea tends to increase. An increase in the carbonate ion concentration, on the other hand, causes supersaturation with respect to calcium carbonate to extend to greater depths. The rate of accumulation of carbonate sediments therefore increases, along with the rate of release of carbon dioxide from the surface, and the carbon content of the ocean tends to fall.

We see, therefore, that the total carbon content of the ocean depends on kinetic factors involving the carbonate ion concentration and the input and output of carbon. In the next section, we shall show how the concentration of carbonate ions depends on the concentration of other carbon species in sea water and on the partial pressure of carbon dioxide in the atmosphere.

Carbonate Equilibria in Sea Water

According to Fig. 3-9, the average carbon dioxide molecule spends only 10 yr in the atmosphere before dissolving in the ocean and being replaced by a molecule coming out of solution. The exchange of carbon dioxide between atmosphere and ocean is so fast that the distribution of carbon between the two reservoirs must approach equilibrium in times of the order of a few thousand years, sufficiently long to permit mixing of carbon throughout the deep sea. Let us now examine the equilibrium between atmospheric carbon dioxide and the various carbon-containing species in sea water, using values of the equilibrium constants from Riley and Chester (1971).

As already indicated, the most abundant carbon species in sea water is the bicarbonate ion, HCO_3^-. From data presented in Table 1-3 we calculate that its average concentration in sea water is 2.3×10^{-3} moles l^{-1}. This concentration is related to the concentrations of hydrogen ions, H^+, and carbonate ions, $CO_3^=$, by the equilibrium reaction

$$(3\text{-}7) \qquad HCO_3^- \rightarrow H^+ + CO_3^=$$

The dissociation constant is

$$(3\text{-}8) \qquad K_2 = [H^+][CO_3^=]/[HCO_3^-] = 10^{-9.1}$$

The pH of sea water has an average value of about 8,[*] which means that $[H^+] = 10^{-8}$ moles l^{-1}. Substituting this value into Eq. (3-8) we see that the carbonate ion concentration is less than 10% of the bicarbonate ion concentration.

Water also undergoes dissociation—

(3-9) $H_2O \rightarrow H^+ + OH^-$

—with dissociation constant

(3-10) $K_W = [H^+][OH^-] = 10^{-14}$

Evidently $[OH^-] = 10^{-6}$ mole l^{-1}.

Of the charged species we have discussed so far, we see that bicarbonate ions are by far the most abundant. This is important because sea water must be electrically neutral, containing as many positively charged as negatively charged ions. The most abundant cations in the ocean are Na^+, K^+, Mg^{++}, and Ca^{++}, while Cl^-, $SO_4^=$, and HCO_3^- are the most abundant anions. The ions H^+, OH^-, and $CO_3^=$ are all negligible. The concentration of HCO_3^- is therefore fixed by the requirement of charge neutrality. It cannot change unless the concentration of one of the other abundant charge-carrying species changes also.

The source of change in both cation and anion abundances that is most closely related to the carbon cycle is the weathering and precipitation of calcium carbonate. If this were the only source of change, we could say that charge neutrality requires that any change in the bicarbonate concentration must be accompanied by a change one-half as great in the calcium ion concentration:

(3-11) $\Delta[HCO_3^-] = 2\Delta[Ca^{++}]$

We shall use this as a convenient and compact way of expressing the requirement of charge neutrality, remembering that $2\Delta[Ca^{++}]$ really refers to the net change in charge caused by changes in concentrations of all the ions other than bicarbonate.

The carbonic acid concentration is related to the bicarbonate concentration by the equilibrium reaction

(3-12) $H_2CO_3 \rightarrow H^+ + HCO_3^-$

with dissociation constant

(3-13) $K_1 = [H^+][HCO_3^-]/[H_2CO_3] = 10^{-6.0}$

Since $[H^+] = 10^{-8}$, we see that the bicarbonate concentration is 100 times the carbonic acid concentration. This result, combined with the ratio of carbonate to bicarbonate concentrations already derived, confirms that

[*] $pH = -\log_{10}[H^+]$

bicarbonate ions are by far the most abundant carbon-containing species in sea water.

Carbonic acid is formed by the solution in sea water of carbon dioxide:

(3-14) $$CO_2(gas) + H_2O(liquid) \rightarrow H_2CO_3(aqueous)$$

The solubility is given by

(3-15) $$K_p = [H_2CO_3]/P_{CO_2} = 10^{-1.5}$$

where P_{CO_2} is the partial pressure of CO_2 in atmospheres and $[H_2CO_3]$ is expressed in moles l^{-1}. Over time scales sufficiently long to permit complete mixing of the sea (more than a few thousand years) the carbonic acid concentration is simply proportional to the partial pressure of carbon dioxide in the atmosphere.

When combined with Eq. (3-13), this result implies that the hydrogen ion concentration in sea water is proportional to the carbon dioxide partial pressure, provided the bicarbonate ion concentration is fixed by requirements of charge neutrality. By combining Eqs. (3-8), (3-13), and (3-15) we obtain

(3-16) $$K_2/K_1 K_P = [CO_3^=]P_{CO_2}/[HCO_3^-]^2 = 10^{-1.6}$$

So, for fixed bicarbonate ion concentration, the carbonate ion concentration is inversely proportional to the carbon dioxide partial pressure.

The importance of the carbonate ion concentration has already been described. By determining whether carbonate particles settling from the surface will be dissolved in the deep sea or allowed to accumulate in sea-floor sediments it controls the rate of precipitation of calcium carbonate. The solution reaction is

(3-17) $$CaCO_3(solid) \rightarrow Ca^{++}(aqueous) + CO_3^=(aqueous)$$

with solubility product

(3-18) $$K_{sp} = [Ca^{++}][CO_3^=] = 10^{-6.2}$$

Solution will occur if the product of calcium and carbonate ion concentrations is less than K_{sp}. Precipitation will occur if the product is greater. Enhanced carbon dioxide partial pressure therefore reduces the rate at which calcium and carbon are removed from the sea.

Some reservations should be stated concerning the equilibrium relations we have presented in this section. For one thing, the equilibrium constants are functions of temperature and pressure, and these parameters vary through the ocean (cf. Riley and Chester, 1971). The values we have quoted are for a temperature of 25°C and a pressure of a few atmospheres, appropriate for warm surface waters. Our model is also oversimplified in that we have neglected borate ions, which interact with carbon dioxide as do carbonate ions (Broecker et al., 1971).

With these reservations in mind, let us check the theory by comparing the theoretical relationship between bicarbonate and hydrogen ion concentrations and the partial pressure of carbon dioxide with the known average values of these quantities. The product of Eq. (3-13) and (3-15) is

(3-19) $$[H^+][HCO_3^-]/P_{CO_2} = K_1 K_p = 10^{-7.5}$$

With $[H^+] = 10^{-8}$, $[HCO_3^-] = 10^{-2.6}$, and $P_{CO_2} = 10^{-3.5}$, the experimental value of this expression is $10^{-7.1}$. The agreement is satisfactory in view of the approximate nature of the theory and the variability of pH and bicarbonate ion concentration within the ocean.

It is possible to imagine various circumstances that would alter the equilibrium relationship between the amounts of carbon in the atmosphere and in the ocean, thereby changing the partial pressure of carbon dioxide. The most obvious of these circumstances is a change in ocean temperature, which changes the values of the equilibrium constants. Another possibility is a decrease in ocean volume resulting from growth of glaciers on the continents. Decreased volume causes increased concentration of the carbon-containing species in seawater. A third possibility is a change in the extent to which surface waters are depleted in carbon as a result of biological activity. A reduction in biological activity would lead to an increase in the carbon concentration in surface waters and thus to an increase in the partial pressure of carbon dioxide.

The effects of changes such as these in the balance between oceanic and atmospheric carbon have been explored by Eriksson (1963). He finds that resultant changes in the carbon dioxide partial pressure are not likely to be large. Over geological time scales, moreover, the apportionment of carbon between atmosphere and ocean is only one element of the geochemical cycle that controls the carbon dioxide content of the atmosphere. The central question, which we shall now consider, is what controls the total carbon content of atmosphere and ocean combined?

Controls on the Carbon Content of Atmosphere and Ocean

We shall begin by discussing the equilibrium system, in which none of the concentrations are changing with time, drawing on ideas presented in papers by Siever (1968) and Broecker (1971). Then, in order to gain a clearer understanding of the controlling processes, we shall present a qualitative description of the evolution of the system in response to a perturbation.

We have already described how carbon is removed from the combined system of ocean and atmosphere by incorporation into sediments, partly as organic carbon but mostly in the form of carbonate minerals. We can ignore the organic carbon; it is controlled by oxygen, not by carbon dioxide. Equation (3-6) is an example of a typical carbonate precipitation reaction. It shows that carbonate precipitation affects not only the carbon content of the ocean but also the cation content. For long-term stability, both carbon and cations must be added to the ocean–atmosphere system as fast as they are removed by carbonate precipitation.

In the equilibrium system, the only source of cations is the weathering of carbonate and silicate minerals. Equations (3-4) and (3-5) are examples of weathering reactions. For simplicity in the discussion that follows, we shall use the following as a schematic representation of a silicate weathering reaction:

$$(3\text{-}20) \qquad CaSiO_3 + 2CO_2 + H_2O \rightarrow Ca^{++} + 2HCO_3^- + SiO_2$$

If we now imagine that cations are removed from the ocean by carbonate precipitation as fast as they are added by weathering, we see that there is a drain on the carbon content of the ocean–atmosphere system corresponding to the overall conversion of silicate minerals to carbonate minerals. To maintain long-term stability, carbonate minerals must be converted to silicate minerals and carbon must be returned to the ocean–atmosphere system at a rate equal to the rate of silicate weathering.

The process that closes the cycle, permitting long-term stability, is called *silicate reconstitution.* A schematic silicate reconstitution reaction is

$$(3\text{-}21) \qquad CaCO_3 + SiO_2 \rightarrow CaSiO_3 + CO_2$$

Silicate reconstitution may occur when sediments containing a mixture of carbonates, quartz, and clay minerals are subjected to temperatures high enough to cause metamorphism (Siever, 1968). For ease of reference we may call the source of carbon dioxide provided by silicate reconstitution the volcanic source, although the release of carbon dioxide need not be localized in volcanoes.

The equilibrium, then, the volcanic source of carbon dioxide must be balanced by the removal of carbon from the combined ocean and atmosphere as a result of the weathering of silicate minerals followed by the precipitation of carbonate minerals. The volcanic source is hardly likely to depend on the carbon content of ocean and atmosphere, but the rate of weathering should increase with carbon dioxide partial pressure, and the partial pressure increases with the carbon content of ocean and atmosphere. It is the dependence of the weathering rate on carbon dioxide partial pressure that ultimately controls the carbon content of the ocean and atmosphere.

We can summarize the equilibrium situation as follows: For long-term stability the partial pressure of carbon dioxide must achieve a value that insures that the rate of silicate weathering is equal to the rate of silicate reconstitution (the volcanic source of carbon dioxide). The carbon dioxide partial pressure, in turn, determines the rate of weathering of carbonate minerals, the principal source of oceanic carbon and cations. For stability, the carbonate ion concentration in sea water must achieve a value that insures that carbon and cations leave the ocean by carbonate precipitation as fast as they are added. The carbon dioxide partial pressure and the carbonate ion concentration are therefore determined by kinetic considerations. Once they are established, the bicarbonate ion concentration and the pH of sea water are determined by the chemical equilibria that we discussed in the last section:

(3-22) $$[HCO_3^-] = \{P_{CO_2}[CO_3^=]K_1K_P/K_2\}^{\frac{1}{2}}$$

(3-23) $$[H^+] = \{K_1K_2K_PP_{CO_2}/[CO_3^=]\}^{\frac{1}{2}}$$

Thus the requirements that the rate of silicate weathering equal the rate of silicate reconstitution and the rate of carbonate precipitation equal the rate of carbonate-plus-silicate weathering ultimately determine the carbon contents of the atmosphere and the ocean as well as the pH of the ocean.

The long-term carbon budget is illustrated in Fig. 3-14. During variations with characteristic times longer than about 10^4 yr, the atmospheric reservoir remains in equilibrium with the oceanic reservoir and the

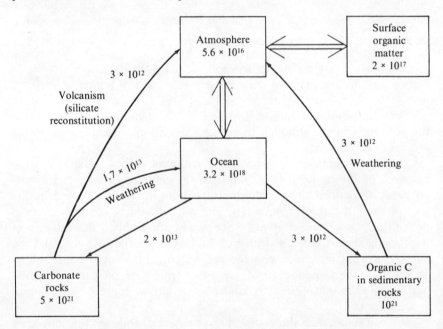

Fig. 3-14. Carbon budget for times longer than 10^4 years (reservoirs in moles and rates in moles per year). Double arrows indicate an equilibrium between reservoirs.

reservoir of surface organic matter. The size of the surface organic reservoir is controlled by the oxygen content of the atmosphere and is insensitive to the carbon dioxide partial pressure. Imbalances in the rate of weathering and precipitation take more than 10^5 yr to cause significant change in the combined atmosphere–ocean reservoir. The cycle of burial and weathering of reduced organic carbon is not involved in the determination of the carbon contents of ocean and atmosphere under conditions of equilibrium. This cycle is balanced unless the oxygen content of the atmosphere is changing. However, imbalances in this cycle can be a significant source of temporary change in the carbon budget.

In order to gain a clearer understanding of how the system responds to perturbations, let us resume our discussion of the effects of fossil-fuel combustion, concentrating now on the long–term behavior. As noted in our discussion of the oxygen budget, our present rate of fossil-fuel combustion will exhaust the fossil-fuel reservoir in 2000 yr (see Fig. 3-1 or 3-9). If the rate of consumption were to continue to increase by 4.5% yr^{-1} (doubling every 20 yr), however, the reservoir would be exhausted in less than 100 yr.

For simplicity, we shall assume that the entire fossil-fuel reservoir, 6×10^{17} moles, is suddenly converted into carbon dioxide. Since there are only 5.6×10^{16} moles of carbon dioxide in the atmosphere, the immediate effect of such a sudden addition would be to increase the partial pressure of carbon dioxide by a factor of eleven. We want to consider how the system responds to this impulse source of carbon dioxide and how it returns to its original state of secular equilibrium. Figure 3-15 summarizes the conclusions of the discussion that follows.

In a time less than 2000 yr (the mixing time of the ocean) the partial pressure of carbon dioxide and the carbonic acid concentration in sea water achieve equilibrium. Since the number of moles of carbonic acid in the sea is approximately equal to the number of moles of carbon dioxide in the atmosphere, and since the carbonic acid concentration in equilibrium with the atmosphere is proportional to the carbon dioxide partial pressure, this equilibrium is achieved when both the carbonic acid concentration and the carbon dioxide partial pressure are approximately five times their present values.

If we ignore the possibility of solution of sea floor sediments, the cation content of the sea and therefore the bicarbonate concentration (constrained by charge neutrality) cannot change significantly in a time as short as a few thousand years. The carbonate ion concentration is therefore depressed at this time by a factor of about five, while the hydrogen ion concentration is enhanced by a similar factor. The ocean therefore becomes more acidic, with pH falling from 8.0 to 7.3.

The reduction in carbonate ion concentration greatly reduces the rate of carbonate precipitation, while the increase in carbon dioxide partial

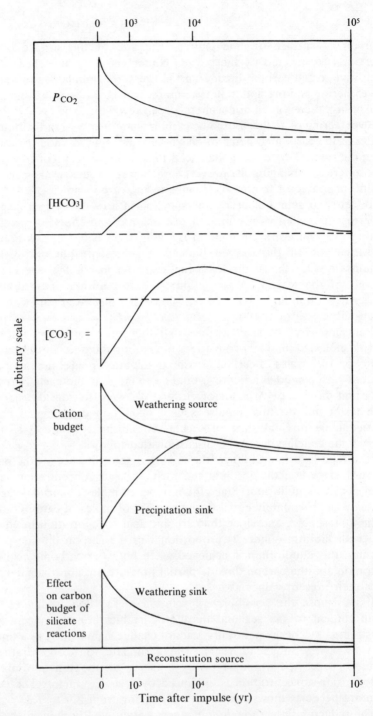

Fig. 3-15 Schematic response of carbon in the atmosphere and ocean to the sudden addition of carbon dioxide to the atmosphere.

pressure increases the rate of weathering. As a result, cations accumulate in the sea and still more carbon is added to the ocean as a result of the weathering of carbonate minerals. The cation content of the sea will continue to rise until most of the original influx of carbon dioxide has been converted into bicarbonate ions. Since the fossil fuel addition amounted to 20% of the initial bicarbonate content of the sea, and since weathering of carbonate minerals adds two bicarbonate ions to the sea for each carbon dioxide molecule extracted from the atmosphere, the bicarbonate concentration rises by approximately a factor of 1.4. At present weathering rates, this rise would take about 4×10^4 yr.

As oceanic bicarbonate grows at the expense of atmospheric carbon dioxide, however, the carbonate ion concentration in sea water increases, leading to an increase in the rate of precipitation of carbonate minerals. At the same time, the decrease in the carbon dioxide partial pressure reduces the rate of weathering. In time, therefore, the cation content of the sea ceases to rise, achieving a new equilibrium between the source of cations provided largely by the weathering of carbonate minerals and the sink provided by precipitation of carbonate minerals. This new equilibrium does not represent the final state of the system. Not only the cation content, but also the carbonate ion concentration and the carbon dioxide partial pressure are greater than they were before the original perturbation, and the system must ultimately return to its original state.

Silicate weathering provides the mechanism for this return. We assume that the volcanic source of carbon dioxide is not affected by any of the changes we have described. But the enhanced carbon dioxide level leads to an enhanced rate of silicate weathering. Therefore the carbon budget is unbalanced. Silicate weathering followed by carbonate precipitation removes carbon from the combined ocean and atmosphere faster than silicate reconstitution restores it. As a result, the carbon dioxide partial pressure is reduced and the carbonate ion concentration is increased. Now, carbonate precipitation exceeds carbonate weathering and the cation content of the sea begins to decrease. As the cation content decreases, so too do the bicarbonate and carbonate ion concentrations as well as the carbon dioxide partial pressure. Equilibrium is finally established when all parameters have returned to their preperturbation levels. The time scale for this return to equilibrium is the time for silicate weathering to consume the fossil fuel carbon dioxide. At present-day weathering rates this time is approximately 2×10^5 yr.

There are thus three distinct stages in the response of the atmosphere and ocean to a sudden addition of carbon dioxide. In the first stage, with a time scale of a few thousand years, the carbonic acid concentration approaches equilibrium with the partial pressure of carbon dioxide: the partial pressure drops by about a factor of two. The second stage represents the response of the cation content of the sea to imbalance between the rate of weathering and precipitation of carbonate minerals. As the cation and bicarbonate concentrations rise, the carbon dioxide partial pressure and the weathering rate fall, while the carbonate ion concentration and the precipitation rate rise. Equilibrium is achieved after a time of a few tens of thousands of years. During the third stage, with a characteristic time approximately ten times as long, the system returns to its initial state, driven by imbalance between the volcanic source of carbon dioxide and the rate of removal of carbon dioxide from the ocean and atmosphere by weathering of silicate minerals followed by precipitation of carbonate minerals.

Evidently the carbon dioxide content of the atmosphere is determined over time scales exceeding 10^5 yr by the requirement that the weathering of silicate minerals consume the carbon dioxide released to the atmosphere as a result of metamorphic and volcanic processes. The equilibrium carbon dioxide content of the atmosphere should be increased by an increase in volcanic activity, provided other factors remain unchanged. On the other hand, it should be reduced by an increase in the rate of weathering caused, for example, by an increase in the rate at which erosion exposes fresh rock to the atmosphere. There is therefore every reason to suppose that atmospheric carbon dioxide has varied over the course of geologic history, in response to changes in the rates of weathering and reconstitution of silicate minerals.

Climatic Effects of Changes in Carbon Dioxide Partial Pressure

There is a possibility that climate could be affected by changes in atmospheric carbon dioxide. Carbon dioxide affects the earth's thermal balance by absorbing some of the infrared radiation emitted by the earth's surface and returning it to the ground (a process called the *greenhouse effect*). Evaluation of the change in average temperature at the ground caused by changes in atmospheric carbon dioxide is difficult (Kellogg and Schneider, 1974) because of uncertainty concerning simultaneous changes in atmospheric water vapor, which also contributes to the greenhouse effect, and in cloudiness, snowcover, and atmospheric circulation. Here we shall simply describe some of the results that have been presented. There has been little attempt to apply these results to geologic history.

Manabe (1970) has presented results of a numerical study of the problem using a radiative–convective model and various assumptions concerning the cloudiness and humidity of the atmosphere (Manabe and Wetherald, 1967). If the relative humidity remains constant and if the average cloudiness of the atmosphere does not change, then he finds that a factor-of-two increase in the carbon dioxide concentration increases surface temperature by 2.3°C. A factor-of-two decrease causes a decrease of surface temperature by about the same amount.

Manabe has also presented vertical profiles of temperature calculated for various levels of carbon dioxide. He finds that increased carbon dioxide warms the troposphere uniformly, causes little temperature change at the tropopause, and cools the stratosphere quite significantly. Stratospheric temperature falls because the stratosphere loses energy to space by infrared radiation emitted by carbon dioxide as well as by water vapor and ozone.

The results of Rasool and Schneider (1971) show smaller increases in surface temperature resulting from increases in atmospheric carbon dioxide. They find that a factor-of-ten increase in the carbon dioxide partial pressure causes a temperature increase of less than 2.5°K, and further increases in partial pressure cause little change in temperature. It appears that the climatic effects of possible changes in atmospheric carbon dioxide are of little significance, but climate is not well understood, and there may be more complex interactions than have yet been explored between atmospheric composition, climate, organisms, and the oceans.

Effect of Life on Atmospheric Oxygen, Nitrogen, and Carbon Dioxide

In this chapter we have found that the long-term oxygen budget of the atmosphere is dominated by the consumption of oxygen in weathering, principally of organic carbon in sedimentary rocks, and a source corresponding to the burial of photosynthetically produced organic carbon in sea-floor sediments. The rate of consumption of oxygen in weathering may be largely independent of the oxygen partial pressure; weathering is therefore not the process that controls the oxygen content of the atmosphere. Atmospheric oxygen appears to be controlled by the source; excess oxygen leads to a reduced rate of burial of carbon and vice versa. We have argued that the equilibrium oxygen content of the atmosphere can be roughly calculated from the solubility of oxygen in sea water, the phosphorus content of the ocean, and the carbon-to-phosphorus ratio of plankton. Together with the oxygen partial pressure, these quantities determine the extent to which deep-sea water is anaerobic, promoting the preservation of carbon, or aerobic, inhibiting it.

We have found that the long-term nitrogen budget of the atmosphere is closely related to the oxygen budget. Burial of organic matter in sediments extracts nitrogen from the atmosphere. Weathering of organic matter in sedimentary rocks restores it. No direct control on the nitrogen content of the atmosphere has been identified; the apportionment of nitrogen between atmosphere and rocks appears to be governed by the oxygen cycle and the carbon to nitrogen ratio in sedimentary organic matter. The amount of nitrogen in the atmosphere is larger than the amount in sedimentary rocks and very much larger than the amount in the ocean. The oceanic reservoir could, in fact, contain much more nitrogen, in the form of nitrate ions, than it does. It appears that organisms keep the nitrate content of the sea low, maintaining a concentration that is related to the phosphorus concentration by the nutritional needs of the organisms themselves.

The long-term carbon dioxide budget of the atmosphere is dominated by a volcanic source and consumption in the weathering of silicate minerals followed by precipitation of carbonate minerals. The source does not depend on the carbon dioxide partial pressure, but the weathering sink does. The equilibrium level of carbon dioxide is therefore determined by the requirement that the weathering sink just balance the volcanic source. We do not, at present, know how to predict this equilibrium level for varying rates of volcanism and erosion.

We shall discuss, in Chap. 6, the composition of the atmosphere before the origin of terrestrial life. In anticipation of this discussion, let us here consider how the abundances of the major atmospheric constituents would change if there were no life. That the abundance of oxygen in our atmosphere is completely dependent on life, has already been pointed out. If there were no photosynthetic source, there would be essentially no oxygen in the atmosphere.

Most nitrogen is already in the atmosphere. Life has taken a fairly small part of earth's nitrogen and incorporated it into organic compounds in sedimentary rocks. If there were no life, there would be more nitrogen in the atmosphere, but not a lot more.

Life does not play an essential role in the long-term cycles of carbon dioxide, but it has important effects on the rates of controlling processes. Life probably accelerates the rate of weathering of silicate minerals, partly by facilitating the mechanical breakdown of rocks and partly by producing humic acids that are more corrosive than air. In the absence of life, therefore, a higher partial pressure of carbon dioxide would be needed to maintain a rate of silicate weathering equal to the rate of release of carbon dioxide from volcanoes. Life also manufactures the carbonate particles that accumulate in sediments, thereby removing carbon and cations from the ocean. In the absence of life, inorganic precipitation would presumably substitute for organic precipitation in order to maintain a balanced flow of carbon and cations through the ocean, but a higher carbonate ion concentration would probably be needed in order to maintain a sufficient rate of precipitation. We conclude, tentatively, that life has the effect of depressing the carbon contents of both the atmosphere and the ocean. Unfortunately, we do not know how large this depression is.

Chapter Four
Loss of Atmospheric Gases to Space

The processes that we have discussed in the preceding two chapters have been *reversible*, involving no change in the overall composition and mass of the earth. In the rock cycle, for example, gas is removed from the atmosphere during rock weathering and restored to the atmosphere during metamorphism and volcanism. In this chapter we shall examine processes that lead to *irreversible* changes in the composition of the earth. These processes involve the loss of atmospheric gases to space.

The theory of the loss of atmospheric gases to space is more highly developed than any other subject of importance to atmospheric evolution. The processes involved are physical and chemical, and they are susceptible to detailed quantitative treatment. Mathematical equations are therefore fairly abundant in this chapter. A nonmathematical summary of the most important results appears at the end of the chapter.

We shall describe a number of mechanisms that cause an atmosphere to lose matter to space. First we shall show that simple hydrodynamical considerations imply that an atmosphere must expand into a surrounding vacuum. The most notable example of the hydrodynamic escape of gas from an atmosphere is provided by the *solar wind*.

Under many circumstances it is simpler and more accurate to describe the expansion of an atmosphere by means of the kinetic theory of gases rather than hydrodynamic theory. When the rate of loss of gas is derived from considerations of the motions of individual molecules rather than the motion of a continuous fluid, the loss process is called *Jeans escape*. We shall derive an expression for the rate of loss of gas from an atmosphere as a result of Jeans escape. We shall then use this expression to examine the importance of Jeans escape for various gases in various atmospheres. The escape rate turns out to be negligible for most constituents of the terrestrial atmosphere, but significant for the lightest gases, hydrogen and helium.

Hydrogen, in fact, escapes into space almost as soon as it reaches the level of the atmosphere from which escape is possible. The rate of loss of hydrogen is therefore limited to the rate at which hydrogen and its compounds are transported upwards from lower levels. We shall show that the transport processes can be described with sufficient generality to

SOLAR WIND

Interplanetary space is filled with a fully ionized, electrically neutral plasma (composed of equal numbers of positively and negatively charged particles) flowing radially outwards from the sun. This flowing plasma, called the solar wind, results from the expansion into space of the hot upper levels of the solar atmosphere. Typical properties of the solar wind are summarized below (Snyder, 1967).

The solar wind contains a weak, variable magnetic field, which represents lines of force of the solar magnetic field that have been swept away from the sun by the outward-moving plasma.

Solar wind properties measured by Mariner II

PARAMETER	OVERALL AVERAGE	DAILY AVERAGES	UNIT
Velocity	504	338–750	$km \, sec^{-1}$
Flux	2.3	0.44–5.1	$10^8 \, cm^{-2} \, sec^{-1}$
Hydrogen number density	5.1	0.68–14.6	cm^{-3}
Temperature	1.9	0.46–3.9	$10^5 \, °K$
Helium fraction	5.5	2–20	%

enable us to estimate the hydrogen escape rate under a wide range of circumstances.

For helium, however, upward transport is not a limitation, and the rate of loss depends sensitively on the density profile of helium and on the temperature of the thermosphere. It is therefore impossible to determine accurately the rate of Jeans escape of helium averaged over a geologically significant period of time. There is, however, some evidence that Jeans escape by itself is not able to bring the atmospheric helium budget into balance.

An additional loss process that may be important is the *polar wind*. The polar wind is a hydrodynamic expansion into space of the ionospheric plasma at high altitudes and high latitudes. Estimates of the rate of this process are uncertain, but they suggest that it is not of evolutionary significance. There are other possibilities that may play a role on Mars and Venus, if not on Earth. A highly speculative process is *solar wind sweeping*, in which charged particles in a planetary atmosphere are swept away by the magnetic field embedded in the solar wind. Another process, important only on Mars, is *photochemical escape*, in which atoms are produced by chemical reactions with sufficient kinetic energy to escape from the gravitational field.

Our discussion in this chapter will be largely limited to the present atmospheres of the Earth and other planets. In later chapters we shall explore possible effects on atmospheric evolution of the loss of gases to space.

BREAKDOWN OF THE BAROMETRIC LAW

We shall begin by showing that an atmosphere must expand into the near vacuum of space. In Chap. 1 we found that the balance of pressure gradient and gravitational forces in a stationary atmosphere requires that pressure decrease with altitude according to the barometric law

(4-1) $$(1/p)\,\mathrm{d}p/\mathrm{d}r = -mg(r)/kT$$

where r is the distance from the center of the earth, p is pressure, m is the molecular mass of the gas, g is the acceleration due to gravity, k is Boltzmann's constant, and T is absolute temperature. Let us investigate the behavior of pressure, according to the barometric law, at altitudes so high that the variation of gravitational acceleration with geocentric distance cannot be neglected. For simplicity, we consider a hypothetical isothermal atmosphere containing only one constituent.

Using the expression $g(r) = g_s(r_s/r)^2$, where g_s is the acceleration due to gravity at the ground (at geocentric distance r_s), and integrating Eq. (4-1) from r_s to r we obtain

(4-2) $$p(r) = p_s \exp[-r_s(1 - r_s/r)/H_s]$$

where p_s is the pressure at the ground and

$$H_s = kT/mg_s$$

According to Eq. (4-2), the pressure approaches a constant value as r approaches infinity,

(4-3) $$p(\infty) = p_s \exp(-r_s/H)$$

This result cannot be correct, for it implies that the atmosphere is of infinite extent and of infinite mass, or, alternatively, that the pressure at the ground is proportional to the pressure of the interplanetary medium.

If we assume, as a first approximation, that the pressure of the interplanetary medium is zero, we deduce that the atmosphere is expanding into the surrounding void, and that the barometric law breaks down because of our neglect of inertial forces in its formulation. The discussion that follows is based on the work of Banks and Holzer (1969a).

DERIVATIVES IN FLUID DYNAMICS

Consider some property of a fluid, say temperature T, that is a function of position (x, y, z) and time t. A small change in this property is given by

$$dT = dt\, \partial T/\partial t + dx\, \partial T/\partial x + dy\, \partial T/\partial y + dz\, \partial T/\partial z$$

where the partial derivatives are evaluated with all other independent variables held constant. The rate of change with time in the property of a given element of the fluid is obtained by dividing this equation by dt

$$dT/dt = \partial T/\partial t + u\, \partial T/\partial x + v\, \partial T/\partial y + w\, \partial T/\partial z$$

where u, v, and w are the components of the velocity of the fluid. There are two contributions to the rate of change with time of the property. The first, represented by $\partial T/\partial t$, is called the *local derivative* with respect to time. It represents the rate at which the property would change if the element of fluid did not move. In the case of temperature, the local derivative could reflect the effects of heating or cooling of the fluid. The second contribution is represented by the terms proportional to velocity. Together these terms describe the changes that the element would experience as it moved through a fluid in which the property is a function of position but not of time. This contribution is called the *convective derivative* with respect to time.

The *total derivative*, which is the rate of change experienced by a moving fluid element, is the sum of the local derivative and the convective derivative (cf. Hess, 1959, p. 172).

Allowing for expansion with radial velocity v, the difference between gravitational and pressure-gradient forces is balanced by the acceleration of the gas,

(4-4) $$nm\, dv/dt = -\partial p/\partial r - nmg$$

where the total derivative is

(4-5) $$dv/dt = \partial v/\partial t + v\, \partial v/\partial r$$

and n is the number density of the gas. In the steady state we have

(4-6) $$nmv\, \partial v/\partial r = -\partial p/\partial r - nmg$$

With the ideal gas law, $p = nkT$, this becomes

(4-7) $$v\, \partial v/\partial r = -(kT/nm)\, \partial n/\partial r - g$$

where we are retaining the isothermal assumption.

With the gas in motion, we must invoke the steady-state continuity equation

$$\partial(r^2 nv)/\partial r = 0$$

or

(4-8) $(1/n)\,\partial n/\partial r = -(1/v)\,\partial v/\partial r - 2/r$

Substituting this in Eq. (4-7), we obtain

(4-9) $(1/v)(\partial v/\partial r)(v^2 - kT/m) = -g + 2kT/mr$

It is convenient to use a dimensionless number to represent the velocity of the gas. This number is $M = (mv^2/kT)^{\frac{1}{2}}$; it is approximately equal to the velocity divided by the speed of sound in the gas. We also introduce a reference altitude, r_0, and let $M_0 = M(r_0)$, $g_0 = g(r_0)$, and $H_0 = kT/mg_0$. The equation becomes

(4-10) $(1/M)(\partial M/\partial r)(M^2 - 1) = -(r_0/r)^2/H_0 + 2/r$

Integration yields

(4-11) $\frac{1}{2}M^2 - \frac{1}{2}M_0{}^2 - \ln(M/M_0) = -(r_0/H_0)(1 - r_0/r) + 2\ln(r/r_0)$

We cannot solve this transcendental equation explicitly for velocity as a function of radius, but we can examine its behavior in two limiting cases.

Suppose we assume, for a start, that the flow is everywhere *subsonic*, i.e., $M \ll 1$. Then we obtain

(4-12) $M(r) = M_0(r_0/r)^2 \exp[(r_0/H_0)(1 - r_0/r)]$

The flow velocity initially increases with height, but at high altitudes the exponential term approaches a constant and the velocity decreases as $1/r^2$. The continuity equation, Eq. (4-8), therefore implies that the density and pressure approach a constant value as r approaches infinity, so this solution does not correspond to our assumption that the pressure of the interplanetary medium is zero.

Indeed, if we use the continuity equation and our solution, Eq. (4-12), for $M(r)$ to calculate the density profile, we recover the barometric law

(4-13) $n(r) = n_0 \exp[-(r_0/H_0)(1 - r_0/r)]$

So, while there is an expansion of the atmosphere in this solution, with an outward flux of $n(r)v(r) = n_0 v_0 (r_0/r)^2$, as long as $M \ll 1$ the inertial term, $nm\,dv/dt$, in the equation of motion, Eq. (4-4), is negligible, and the density profile is not affected by the expansion.

If the pressure at infinity is to approach zero, the flow must be *supersonic* at high levels. Since the phenomenon we are examining is the acceleration of the atmosphere into the vacuum of space, there is a transition from subsonic flow at low levels to supersonic flow above. Let the reference level r_0 be the level at which this transition occurs. Then

$M(r_0) = M_0 = 1$. But when $M = 1$, the left-hand side of Eq. (4-10) is zero, and the right-hand side must also be zero. This gives $r_0/H_0 = 2$. In terms of g_s, the gravitational acceleration at the surface r_s, this becomes $r_0 = r_s^2/2H_s$, where $H_s = kT/mg_s$. For atomic hydrogen at 1500°K in the earth's atmosphere, we calculate $r_0 = 16,000$ km.

In the region of supersonic flow, $M \gg 1$, Eq. (4-11) becomes

$$(4-14) \qquad M^2 = 1 + 4\ln(r/r_0) - 4(1 - r_0/r)$$

and in the limit of large r

$$(4-15) \qquad M^2 = 4\ln(r/r_0) - 3$$

In this case, therefore, the expansion velocity continues to increase as the altitude increases. From the continuity equation, Eq. (4-8), we may confirm that supersonic flow yields a density at large r that goes to zero more rapidly than $1/r^2$, so this solution is consistent with our assumption of a finite atmosphere and zero pressure in the interplanetary medium.

We have presented, here, a highly simplified treatment of a complex problem that has been quantitatively examined by a number of authors, both for its relevance to planetary atmospheres and as the source of the solar wind (Parker, 1958, 1964, 1971; Banks and Holzer, 1968, 1969a, 1969b; Yeh, 1970; Holzer et al., 1971; Strobel and Weber, 1972). We have found that there is an outward flow of gas from the atmosphere into space unless the pressure of the interplanetary medium is large—larger than the pressure at infinity predicted on the basis of the barometric law. If the background pressure is zero, the flow must be supersonic at high altitudes. Banks and Holzer (1968, 1969a) have shown that the flow becomes supersonic even in the case of a small, finite, background pressure, but is converted back to subsonic flow by a shock that forms at very high altitudes. The situation is complicated by a mixture of gases (Strobel and Weber, 1972), but the essential features of the flow are not changed.

By assuming, in our treatment, that the temperature is independent of altitude, we have ignored all questions of the energy budget of the expanding gas. Quantitative predictions of the magnitude of the flow must include an examination of the sources of energy (Banks and Holzer, 1969b; Holzer et al., 1971; Raitt et al., 1975). We have also assumed that the gas behaves as a compressible fluid, and that pressure is proportional to density. This description of the gas is valid only as long as collisions between molecules are sufficiently frequent to maintain a random, isotropic distribution of molecular thermal velocities. This assumption must be invalid, however, at very high altitudes, where gas densities are so small that molecules can travel great distances, possibly even orbiting the earth, between collisions. Holzer et al. (1971) have found that the altitude profiles of density and velocity are little modified if the transition from *continuum* to *free molecular flow* occurs above the transition from sub-sonic to supersonic flow. If the transition to free molecular flow occurs at lower altitudes, however, the problem can be more easily studied with a model based on the kinetic theory of gases.

JEANS ESCAPE

As altitude increases and density decreases, there is a gradual transi-tion, extending over several scale heights, from conditions of continuum flow to conditions of free molecular flow. In order to simplify the problem, let us assume that this transition is sharp. Let us assume that there is a level in the atmosphere, called the *critical level* or *exobase*, above which collisions between molecules are so infrequent as to be negligible and below which collisions are sufficiently frequent to maintain a completely isotropic and random distribution of molecular velocities. At and below the exobase, therefore, the velocity distribution of the molecules of a given atmospheric constituent is the *Maxwellian dis-tribution*.

Since collisions are negligible above the exobase, the molecules in this region, called the *exosphere*, move along ballistic trajectories under the action of the earth's gravitational field. Some of the upward-moving molecules have velocities sufficiently great to carry them on hyperbolic trajectories away from the earth, out into space. The resulting flux of escaping molecules is the kinetic-theory equivalent of the hydrodynamic expansion of the atmosphere into space that we have just been describing. We wish to evaluate the escape flux.

If an upward-moving molecule is to escape the earth's gravitational field altogether, its kinetic energy must exceed its gravitational potential energy. Escape requires, therefore, that

$$(4\text{-}16) \qquad\qquad \tfrac{1}{2}mv^2 > mMg/r$$

where m is the mass and v the velocity of the molecule in question, M is the mass of the earth, G is the universal constant of gravitation, and r is the distance from the center of the earth. The minimum velocity for escape, called the *escape velocity*, is therefore

$$(4\text{-}17) \qquad\qquad v_e = \sqrt{[2g(r)r]}$$

where $g(r) = MG/r^2$ is the acceleration due to gravity. We note that the escape velocity is independent of the mass of the molecule concerned, and is independent, also, of the direction in which the molecule is moving. At the exobase, however, only upward-moving molecules can escape, because downward-moving molecules encounter lower levels of the at-mosphere.

At the exobase the molecules of a given atmospheric constituent have a Maxwellian distribution of velocities. That is, the number of molecules with velocities between v and $v+dv$ is

(4-18) $f(v)\,dv = n_c 4\pi^{-\frac{1}{2}}(m/2kT)^{\frac{3}{2}}v^2 \exp(-mv^2/2kT)\,dv$

where n_c is the number density of the constituent in question at the exobase, and T is the absolute temperature. The distribution is isotropic, so the number of molecules traveling at angles between θ and $\theta+d\theta$ from the vertical is $f(v)\,dv2\pi \sin\theta\,d\theta/4\pi$. The vertical flux of these molecules is obtained by multiplying this quantity by the vertical component of their velocity, $v\cos\theta$. The total vertical flux of molecules with velocity v is given by the integral over direction,

$$\tfrac{1}{2}\int_0^{p/2} f(v)\,dv\,v\cos\theta\sin\theta\,d\theta$$

where we consider only upward-moving molecules. The angular integration yields

$$\tfrac{1}{4}vf(v)\,dv$$

for the vertical flux of molecules with velocity v.

We obtain the escape flux at the critical level by integrating this expression over all velocities greater than the escape velocity:

(4-19) $F_c = n_c\pi^{-\frac{1}{2}}(m/2kT)^{\frac{3}{2}}\int_{v_e}^{\infty} v^3 \exp(-mv^2/2kT)\,dv$

$\qquad = n_c\pi^{-\frac{1}{2}}(m/2kT)^{\frac{3}{2}}(kT/m)\exp(-mv_e^2/2kT)(v_e^2+2kT/m)$

Expression (4-17) for the escape velocity can be used to rewrite the escape flux as

(4-20) $F_c = \tfrac{1}{2}n_c\pi^{-\frac{1}{2}}\exp(-r_c/H_c)U(1+r_c/H_c)$

where r_c is the geocentric distance of the exobase,

$$U = \sqrt{(2kT/m)}$$

is the most probable random velocity, and

$$H_c = kT/mg(r_c)$$

In order to evaluate the escape flux, we must locate the exobase. The concept of a sharp transition between the region of the atmosphere dominated by collisions and an overlying collisionless region is an idealization of a gradual transition, so the location of the exobase is to some extent arbitrary. Most commonly used is the definition of Jeans (1925). The exobase is the height from which a fraction $1/e$ of molecules traveling straight up with enough energy to escape will do so without further collisions. The probability that a molecule will collide while traveling a distance dr is $N'\sigma\,dr$, where N' is the number density of all atmospheric molecules, and σ is the average collision cross section. The probability of traveling from r_1 to r_2 without collisions is therefore

$$\exp[-\int_{r_1}^{r_2} N'(r)\sigma \, dr]$$

And since the probability of escaping from the exobase is e^{-1}, we have

$$\int_{r_c}^{\infty} N'(r)\sigma \, dr = 1$$

We have shown in our discussion of the barometric law (see Chap. 1) that

$$\int_{r}^{\infty} N'(h) \, dh = N'(r)H(r)$$

where H is the ambient atmospheric scale height, $kT/\bar{m}g(r)$, and $\bar{m}$ is the mean molecular mass. The breakdown of the barometric law that we described earlier in this chapter becomes significant only at heights well above the exobase.

A typical cross section for collisions between atmospheric molecules is $3 \times 10^{-15} \, cm^2$ (Mason and Marrero, 1970), so the exobase is at the altitude where the ambient atmospheric density is

(4-21) $$N'(r_c) = 3 \times 10^{14}/H(r_c) \, cm^{-3}$$

For the terrestrial atmosphere, this density is found in the 400–500 km altitude region, the precise height depending on the temperature of the thermosphere, which varies with time of day and solar activity.

Chamberlain (1963) has shown that the neglect of collisions occurring above the exobase does not lead to an overestimate of the escape flux. Particles removed from the escaping population by collisions above the exobase are replaced by escaping particles that suffer their last collision at altitudes below the exobase. A slight overestimate does, however, result from the assumption that the Maxwellian distribution is fully populated in the region from which escape occurs. In fact, the distribution is deficient in downward-moving molecules with velocities in excess of the escape velocity because these molecules do not return from their ballistic flights through the exosphere. This deficiency causes departures from the Maxwellian distribution for the escaping gas, even in the upward-moving particles. The effect has been extensively studied (Hays and Liu, 1965; Chamberlain and Campbell, 1967; Chamberlain, 1969; Brinkmann, 1970, 1971a; Chamberlain and Smith, 1971), and it appears that corrections to the expression for the escape flux derived above are generally smaller than 30%. We shall neglect this correction.

The arbitrary nature of the definition of the exobase is not a matter of concern (Jeans, 1925; Chamberlain, 1963). Escape occurs from an isothermal region of the atmosphere and from a level sufficiently high that all of the atmospheric gases have density profiles governed by diffusive equilibrium. Under these conditions, as shown in Chap. 1, the density of each constituent varies approximately exponentially with altitude, at a rate determined by the mass of the constituent and not by the other gases present.

From Eq. (4-13) we have

(4-22) $n_c = n_1 \exp[-(r_1/H_1)(1 - r_1/r_c)]$

where n_1 is the density and H_1 the scale height, $kT/mg(r_1)$, of the constituent in question at arbitrary reference level r_1. From the variation of gravitational acceleration with r we have $H_c/H_1 = (r_1/r_c)^2$. Using this relationship and substituting Eq. (4-22) into Eq. (4-20), we obtain

(4-23) $F_c = \frac{1}{2}n_1\pi^{-\frac{1}{2}}\exp(-r_1/H_1)U(1 + r_c/H_c)$

The escape flux can therefore be evaluated at any level, r_1, in the isothermal region of the thermosphere. The height of the exobase appears only in the slowly varying term on the right of this expression; the escape flux is therefore not sensitive to the precise value chosen for r_c (Hunten, 1973).

Most of our discussion of atmospheric evolution will be based on the kinetic theory description of escape rather than on the hydrodynamic theory. The kinetic theory formulation is appropriate for gases that are relatively tightly bound to their planets by gravity, and these are the gases that survive for geologically significant periods of time. It is for these gases that we need a reliable value of the escape flux. In situations where the escape is better described by the hydrodynamic approach, an approximate value for the escape flux will usually be sufficient. In what follows we shall therefore use the kinetic theory value for the escape flux.

CHARACTERISTIC TIMES FOR ESCAPE

It is convenient to assess the importance of escape as a process modifying atmospheric composition by evaluating *characteristic escape times* for the different atmospheric constituents. To calculate an escape time, we divide the escape flux into an appropriate column density (the number of molecules in a vertical column of unit cross section). If the escape time is short compared with the times that characterize competing processes, we conclude that escape will modify the column density. Different column densities and therefore different escape times are appropriate for different aspects of our problem, as we shall show.

For a start, let us divide the escape flux for a given constituent, given by Eq. (4-20), into the column density of that constituent present in the exosphere (that is, above the exobase). Since this column density is $n_c H_c$, the characteristic escape time that results is

(4-24) $\tau_e = H_c/\langle v_c \rangle$

where

$$(4\text{-}25) \qquad \langle v_c \rangle = F_c/n_c = \tfrac{1}{2}\pi^{-\frac{1}{2}} \exp(-r_c/H_c) U (1 + r_c/H_c)$$

is the *mean expansion velocity*. If all sources of supply of a given constituent to the exosphere were cut off, the exospheric density would be reduced by a factor of $1/e$ in a time τ_e as a result of Jeans escape.

In order to calculate τ_e, let us take a value for the exospheric temperature, $T = 1500°K$, which is higher than average, and let us place the exobase at a height of 500 km ($r_c = 6870$ km). At this height the acceleration due to gravity, $g(r_c)$, is 844 cm sec^{-2}. The results of the calculation are shown in Table 4-1. We see that the characteristic escape time for atomic hydrogen is only a few hours. This is shorter than the replenishment time associated with the diffusion of hydrogen upwards through the lower thermosphere, so we may conclude that Jeans escape affects the density profile of atomic hydrogen in the thermosphere. For molecular hydrogen the characteristic escape time is about 8 d. Patterson (1966) has shown that molecular hydrogen in the thermosphere is destroyed in a matter of seconds in the reaction with atomic oxygen

$$(4\text{-}26) \qquad O + H_2 \rightarrow OH + H$$

We may therefore conclude that escape has a negligible direct effect on molecular hydrogen densities.

For helium the escape time is approximately 100 yr, which is very much longer than the time associated with replenishment by upward diffusion from the lower atmosphere. For atomic oxygen the time for depletion of the exosphere by Jeans escape is longer than the age of the earth ($\sim 1.4 \times 10^{16}$ sec). We may conclude that Jeans escape of oxygen is

TABLE 4-1. Characteristic time for the reduction of exospheric density by Jeans escape

	H	H_2	He	O
Scale height, H (km)	1460	730	365	91
Most probable speed U (km sec^{-1})	4.96	3.52	2.48	1.24
Mean expansion velocity $\langle v_c \rangle$ (km sec^{-1})	7.26 (−2)[a]	8.6 (−4)	9.6 (−8)	6.2 (−32)
Exospheric escape time τ_e (sec)	2 (4)	8.5 (5)	3.8 (9)	1.47 (33)

[a] $7.3\,(-2) \equiv 7.3 \times 10^{-2}$.

far too slow to have had any effect on the evolution of the atmosphere. The same may be said for the escape of all molecules more massive than atomic oxygen.

The results of similar calculations for the moon and several of the planets are shown in Table 4-2. The characteristic times in this table suggest that Mercury and the moon could retain atmospheres of argon and heavier gases for times longer than the age of the solar system, but that lighter gases would be lost. Belton et al. (1967) have shown, however, that even an argon atmosphere on Mercury must be so tenuous that the exobase is at the ground. Because the inert gases are not able to dissipate energy by emission of infrared radiation, a greater abundance of argon or krypton would lead to a temperature at the exobase so high that Jeans escape would destroy the atmosphere in a geologically short period of time. The same is true of the moon, so we may conclude that Mercury and the moon can retain only very tenuous atmospheres consisting of argon and heavier gases. Hydrogen and helium escape at significant rates from the atmospheres of Mars and Venus, but, as on Earth, oxygen and heavier gases are retained.

Jupiter, on the other hand, is so cold and so massive that not even atomic hydrogen can escape. Spectroscopic analysis of Jupiter's atmosphere indicates that hydrogen, helium, carbon, and nitrogen are present in approximately the same ratios as in the solar system as a whole (McElroy, 1969a; Newburn and Gulkis, 1973; Hunten and Münch, 1973; Houck et al., 1975). It is therefore possible that Jupiter represents a sample of the primitive solar nebula that has changed very little since the solar system was formed.

The time for depletion of the exosphere is only one of the possible

TABLE 4-2. Characteristic escape times for the planets

	MOON[f]	MERCURY[f]	MARS	VENUS	JUPITER
$T°K$	300[a]	600[b]	365[c]	700[d]	155[e]
r_c km	1738	2439	3590[c]	6255[d]	69,500
$g(r_c)$ cm sec^{-2}	162	376	332	827	2620
$\tau_e(H)$ sec	3.55 (3)[g]	3.32 (3)	1.39 (4)	5.70 (5)	7.59 (617)
$\tau_e(He)$ sec	2.03 (4)	1.40 (5)	2.66 (8)	2.87 (16)	9.59 (2455)
$\tau_e(O)$ sec	2.25 (9)	7.32 (13)	1.02 (28)	7.61 (61)	1.20 (9818)
$\tau_e(A)$ sec	3.35 (20)	2.64 (32)	2.09 (68)	3.24 (153)	3.04 (24532)
$\tau_e(Kr)$ sec	3.97 (41)	1.10 (67)	4.34 (142)	1.07 (323)	9.99 (51509)

Data from Goody and Walker (1972), except as indicated below:
[a] Kaula (1968)
[b] Belton et al. (1967)
[c] McIlroy (1972)
[d] McElroy and Hunten (1969a)
[e] Strobel and Smith (1973)
[f] For the Moon and Mercury the critical level is taken as the solid surface.
[g] $3.55 (3) \equiv 3.55 \times 10^3$

characteristic times associated with Jeans escape. In order to assess the influence of Jeans escape on the bulk composition of the atmosphere, we must divide the escape flux for a given constituent into the column density of that constituent in the whole atmosphere, not just in the exosphere. The characteristic time, τ_a, that results is the time for Jeans escape to cause substantial changes in the amount of a given constituent present in the atmosphere, assuming that other sources and sinks of that constituent are absent.

Take hydrogen as an example. Observations of the H density and the temperature in the exosphere (Donahue, 1966; Joseph, 1967; Brinton and Mayr, 1971; Vidal-Madjar et al., 1973; Meier and Mange, 1973; Tinsley, 1974) have shown that the escape flux is about 10^8 atoms $cm^{-2} sec^{-1}$, independent of the temporal variations in exospheric temperature. (The reasons for this constancy of the flux will be discussed in the next section.) Because hydrogen is produced by photochemical reactions from water vapor, we should consider the effect of escape on the total water vapor content of the atmosphere. From Table 1-2, this column density is 10^{23} molecules cm^{-2}. The characteristic time for escape of atmospheric water vapor is therefore 2×10^{15} sec or 6×10^7 yr. This time is very much longer than the 0.3 yr residence time associated with the cycle of evaporation and precipitation (cf. Skinner, 1969). We conclude, therefore, that while escape is fast enough to affect hydrogen densities in the exosphere, it is too slow to affect the water vapor content of the atmosphere as a whole.

Alternatively, we may determine the time for escape to dissipate all of the water on the earth's surface by dividing the escape flux into the number of water molecules in a square centimeter column of ocean and atmosphere. This time is 6×10^{12} yr, longer than the age of the earth. At present, therefore, Jeans escape is not sufficiently rapid to affect the abundance of water on earth.

We see that for Earth, Venus, and Mars, the planets of greatest interest to us, Jeans escape is important only for the lightest gases, hydrogen and helium. We will discuss the escape of helium later in this chapter. Hydrogen escape is of greater importance for atmospheric evolution because it affects the oxidation state of the atmosphere and because it results in the loss of atmospheric water vapor. We must try to understand the processes that control the rate of escape of hydrogen, so that we may estimate what this rate may have been in the past.

THE LIMITING FLUX AND THE ESCAPE OF HYDROGEN

We have already noted that the time for escape to deplete the hydrogen in the exosphere is shorter than the time for transport from below to replenish it. Hydrogen escapes almost as soon as it reaches the exosphere. The escape flux of hydrogen is therefore limited by the rate at which exospheric hydrogen is replenished by transport upwards from the lower atmosphere. It is this transport problem that we must examine. Our treatment follows Hunten (1973).

We consider the motion of constituent i (hydrogen for example) through the ambient atmosphere, a. The difference in the vertical velocities of i and a is given by Chapman and Cowling's (1970) Eq. (14.1,2):

$$(4\text{-}27) \quad w_i - w_a = -\frac{b_i}{n_a}\left[\frac{1}{n_i}\frac{dn_i}{dz}+\frac{m_i g}{kT}+\left(1+\frac{\alpha_i}{1+f_i}\right)\frac{1}{T}\frac{dT}{dz}\right]-\frac{K}{f_i}\frac{df_i}{dz}$$

The first term in this equation represents molecular diffusion; $b_i/(n_i+n_a)$ is the diffusion coefficient D_i, n is the number density, and α_i is the thermal diffusion factor. The second term represents eddy mixing; f_i is the mixing ratio, defined as n_i/n_a, and K is the eddy mixing coefficient.

Since we are interested in a light gas moving through a stationary background gas, we may set $w_a = 0$. The vertical flux of i is therefore

$$(4\text{-}28) \quad \phi_i = n_i w_i = -b_i f_i[(1/n_i)\,dn_i/dz+1/\mathcal{H}_i]-Kn_a\,df_i/dz$$

where we have introduced the equilibrium density scale height

$$(4\text{-}29) \quad \mathcal{H}_i = \{m_i g/kT+[1+\alpha_i/(1+f_i)](1/T)\,dT/dz\}^{-1}$$

This is the scale height that the density of i would have if it were in diffusive equilibrium; if $(1/n_i)\,dn_i/dz = -1/\mathcal{H}_i$, the diffusive flux of i is zero.

Now $n_i = f_i n_a$, so

$$(4\text{-}30) \quad (1/n_i)\,dn_i/dz = (1/f_i)\,df_i/dz + (1/n_a)\,dn_a/dz$$

Corresponding to Eq. (4-27) is an equation for the velocity of a relative to i:

$$(4\text{-}31) \quad w_a - w_i = -\frac{b_i}{n_i}\left[\frac{1}{n_a}\frac{dn_a}{dz}+\frac{m_a g}{kT}+\left(1-\frac{\alpha_i f_i}{1+f_i}\right)\frac{1}{T}\frac{dT}{dz}\right]+\frac{K}{f_i}\frac{df_i}{dz}$$

$$= -\frac{b_i}{n_i}\left(\frac{1}{n_a}\frac{dn_a}{dz}+\frac{1}{\mathcal{H}_a}\right)+\frac{K}{f_i}\frac{df_i}{dz}$$

Addition of Eqs. (4-27) and (4-31) yields the equation of hydrostatic balance for this gas mixture:

$$(4\text{-}32) \quad d(n_i+n_a)/dz = -(n_i m_i + n_a m_a)g/kT+[(n_i+n_a)/T]\,dT/dz$$

Setting $w_a = 0$ in Eq. (4-31) we obtain

$$(4\text{-}33) \quad (1/n_a)\,dn_a/dz = -1/\mathcal{H}_a + \phi_i/b_i - (Kn_a/b_i)\,df_i/dz$$

From Eq. (4-30), therefore,

$$(4\text{-}34) \quad (1/n_i)\,dn_i/dz = (1/f_i)\,df_i/dz - 1/\mathcal{H}_a + \phi_i/b_i - (Kn_a/b_i)\,df_i/dz$$

Substitution in Eq. (4-28) yields

(4-35) $\qquad \phi_i = -b_i\dfrac{df_i}{dz} - f_i\phi_i - (1+f_i)Kn_a\dfrac{df_i}{dz} + b_if_i\left(\dfrac{1}{\mathscr{H}_a} - \dfrac{1}{\mathscr{H}_i}\right)$

Solving for ϕ_i we obtain

(4-36) $\qquad \phi_i = \phi_l - [Kn_a + b_i/(1+f_i)]\,df_i/dz$

$\qquad\qquad\quad = \phi_l - (K + D_i)n_a\,df_i/dz$

where

(4-37) $\quad \phi_l = [b_if_i/(1+f_i)](1/\mathscr{H}_a - 1/\mathscr{H}_i)$

$\qquad\quad = [b_if_i/(1+f_i)][(m_a - m_i)g/kT - (\alpha_i/T)\,dT/dz]$

The quantity ϕ_l is called the *limiting flux,* for reasons we shall present below. Equation (4-36) shows that the flux is equal to the limiting flux when the mixing ratio is independent of altitude. The flux in this situation is entirely due to molecular diffusion because eddy mixing leads to transport only when the mixing ratio varies. Equation (4-37) shows that the diffusive flux is caused by the difference in the gravitational force on the background gas and the diffusing gas. The upward flux of a light constituent in a well-mixed atmosphere may therefore be considered a consequence of buoyancy.

If the mixing ratio is constant, the limiting flux is also constant in an isothermal atmosphere with constant gravitational acceleration. In a real atmosphere, it is nearly constant. It might seem strange that diffusion can sustain as large a flux at low altitudes, where the diffusion coefficient is small, as at high altitudes. The diffusion coefficient and the diffusion velocity are inversely proportional to ambient density; at low altitudes they are both small. But the diffusion flux is the product of velocity and density and, for constant mixing ratio, the density of the diffusing constituent is proportional to the ambient density. Ambient density therefore cancels in the expression for diffusive flux, and a constant flux results.

Consider the case of a light gas, say hydrogen, immersed in a heavy background gas such as nitrogen. Equation (4-37) shows that the limiting flux in this case is positive, and therefore directed upwards. Suppose that the flux, ϕ_i, is equal to the limiting flux. Equation (4-36) shows that the mixing ratio, then, is independent of altitude. This result is independent of the magnitude of the eddy mixing coefficient K. If the upward flux is equal to the limiting flux, the mixing ratio is constant regardless of whether K is very large or very small.

If the upward flux is less than the limiting flux, the mixing ratio of a light constituent increases with altitude. This is a result of the gravitational separation which we described in Chap. 1. The rate of increase may be imperceptibly small if K is large, as is the case in the terrestrial atmosphere below the homopause. If the upward flux exceeds the limiting flux, the mixing ratio must decrease with altitude. This is a circumstance that cannot be extended to great heights, however, without the mixing ratio becoming negative because $(K + D_i)n_a$ is either constant or decreasing with altitude, depending on the relative magnitudes of K and D_i.

The way in which the mixing ratio profile varies with flux can best be illustrated by solving the differential equation for the mixing ratio as a function of altitude. We consider an isothermal atmosphere with constant gravity, for which $\mathcal{H}_a$, $\mathcal{H}_i$, and b are all independent of altitude. We assume that K is also independent of altitude. We consider a minor atmospheric constituent, for which $f_i \ll 1$; both hydrogen and helium in the terrestrial atmosphere satisfy this condition. We further assume that the constituent is neither produced nor destroyed by photochemical reactions within the atmosphere. The flux, ϕ_i, is therefore independent of altitude.

The equation to be solved is derived from Eqs. (4-36) and (4-37):

(4-38) $df_i/dz = [b_i f_i(1/\mathcal{H}_a - 1/\mathcal{H}_i) - \phi_i]/(K + D_i)n_a$

If we let $z = 0$ at the level where $K = D_i$, we have

(4-39) $(K + D_i)n_a = b[1 + \exp(-z/\mathcal{H}_a)]$

since $n_a \propto \exp(-z/\mathcal{H}_a)$ and $D_i = b/n_a$ for $n_a \gg n_i$. The solution of Eq. (4-38) is

(4-40) $$\frac{f_i}{f_0} = \left(1 - \frac{\phi_i}{\phi_{l0}}\right)\left\{1 + e^{z/\mathcal{H}_a}\right\}^{1 - \mathcal{H}_a/\mathcal{H}_i} + \frac{\phi_i}{\phi_{l0}}$$

where f_0 is the value of f_i for large negative z, and

$$\phi_{l0} = f_0 b_i(1/\mathcal{H}_a - 1/\mathcal{H}_i)$$

is the corresponding value of the limiting flux.

The behavior of f_i is illustrated in Fig. 4-1 for a light gas i, with $\mathcal{H}_i > \mathcal{H}_a$. When $z \ll -\mathcal{H}_a$, corresponding to $K \gg D_i$, f_i is constant and equal to its low altitude value, f_0. This region of the atmosphere is the homosphere. For $z \gg \mathcal{H}_a$, corresponding to $D_i \gg K$, we find

(4-41) $f_i/f_0 = (1 - \phi_i/\phi_{l0}) \exp(z/\mathcal{H}_a - z/\mathcal{H}_i) + \phi_i/\phi_{l0}$

(4-42) $n_i = f_i n_a = f_i n_0 \exp(-z/\mathcal{H}_a)$

$= f_0 n_0[(1 - \phi_i/\phi_{l0}) \exp(-z/\mathcal{H}_i) + (\phi_i/\phi_{l0}) \exp(-z/\mathcal{H}_a)]$

where n_0 is the value of n_a at $z = 0$, the level where $K = D_i$.

Since $\mathcal{H}_i > \mathcal{H}_a$, the first term dominates at high altitudes, and we recover the diffusive equilibrium solution for n_i. The altitude at which the constant-mixing-ratio solution gives way to the diffusive-equilibrium solution depends on the value of ϕ_i/ϕ_{l0}, as shown in Fig. 4-1. If ϕ_i is equal to the low altitude limiting flux, ϕ_{l0}, the first term in Eq. (4-40) is zero, and

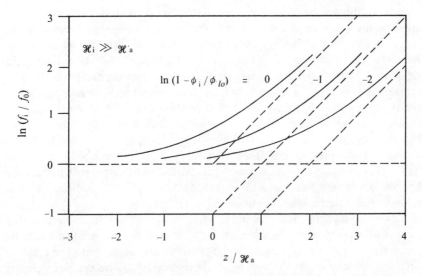

Fig. 4-1. Density profiles of a light minor gas in the vicinity of the homopause (z=0), as given by equation (4–40). The dashed lines represent the asymptotic solution ($z \gg \mathcal{H}_a$) of equation (4–41). A decrease of $1 - \phi_i / \phi_{lo}$ by a factor of e raises the effective transition from the well-mixed distribution to the diffusive equilibrium distribution by one scale height ($\mathcal{H}_a$).

the mixing ratio is independent of altitude. The diffusive-equilibrium solution disappears.

Suppose, now, that $\phi_i > \phi_{lo}$, so that the first term in Eq. (4-40) is negative. For $\mathcal{H}_a < \mathcal{H}_i$ this term must dominate at sufficiently high altitudes because of its exponential increase with altitude. Negative mixing ratios and negative densities are the result. Since these are not possible, this solution is invalid. The upward flux of a light, nonreactive constituent cannot exceed the limiting flux in the steady state. Although we have derived this important result for an isothermal atmosphere, it is equally true for the real atmosphere, that the upward flux cannot exceed the limiting flux by a significant amount for a significant period of time.

Now let us consider the relevance of the limiting flux to the escape of hydrogen from the terrestrial atmosphere. Hydrogen is transported upwards through the atmosphere by eddy mixing and by diffusion until it reaches the exosphere, from which it can escape into space. If we neglect, for the time being, any photochemical reactions of hydrogen, the flux at any level is equal to the escape flux. The question to be considered, therefore, is how the escape flux compares with the limiting flux.

Two types of possible behavior have been identified by Hunten (1973). In the first, the escape flux is very much less than the limiting flux. The constituent in question has a constant mixing ratio up to the homopause and is approximately in diffusive equilibrium above that. The density at the exobase is therefore determined, and the escape flux is equal to the product of this density and the mean expansion velocity for Jeans escape given by Eq. (4-25). This describes the situation of helium in the terrestrial atmosphere (Kockarts, 1973), as we shall show below. In the second type of behavior, applicable to hydrogen on earth, the limit is imposed by upward transport not by Jeans escape. The escape flux is nearly equal to the limiting flux. The transition from constant mixing ratio to diffusive equilibrium occurs at altitudes well above the homopause. The density at the exobase and the Jeans escape rate depend on the altitude of this transition. This altitude therefore adjusts so that the escape flux is equal to the upwards flux carried by transport. If the escape flux is too large, the transition altitude rises and the density at the exobase falls, thereby reducing the escape flux.

Let us now verify that it is indeed transport and not Jeans escape that limits the rate of loss of hydrogen from the earth. We shall do this by comparing the mean expansion velocity for Jeans escape with the limiting diffusive velocity at an altitude near the homopause. The smaller velocity will identify the limiting process. If the mean expansion velocity is very much smaller than the diffusive velocity, the hydrogen is in diffusive equilibrium above the homopause. Because of the large scale height of hydrogen, the density in this situation varies little between the homopause and the exobase, so the smaller velocity corresponds to the smaller flux. If the mean expansion velocity is much larger than the diffusive velocity, the hydrogen density must decrease markedly between the homopause and the exobase in order that the escape flux not exceed the limiting flux. This decrease corresponds to the case of diffusion-limited escape.

The limiting diffusive velocity is given by $\phi_l/n_i \sim b_i/n_a \mathcal{H}_a$ from Eq. (4-37), with $f_i \ll 1$ and $m_i \ll m_a$. Values of b_i for various gas mixtures appear in Table 4-3. Using atmospheric data from Table 1-1, at 100 km we calculate a limiting diffusive velocity of about 4 cm sec^{-1}. From Table

TABLE 4-3. The binary diffusion parameter for various gases as a function of temperature and at representative temperatures.

GASES	$b_i(T)$ (cm^{-1} sec^{-1})	T (°K)	b_i (cm^{-1} sec^{-1})
H in air	$6.5 \times 10^{17} T^{0.7}$	208	2.73×10^{19}
H_2 in air	$2.67 \times 10^{17} T^{0.750}$	208	1.46×10^{19}
H_2O in air	$0.137 \times 10^{17} T^{1.072}$	250	0.51×10^{19}
He in air	$2.77 \times 10^{17} T^{0.729}$	208	1.36×10^{19}
H in CO_2	$8.4 \times 10^{17} T^{0.6}$	180	1.89×10^{19}
H_2 in CO_2	$2.23 \times 10^{17} T^{0.750}$	180	1.10×10^{19}

After Hunten (1973); data from Colegrove et al. (1966), Mason and Marrero (1970).

TABLE 4-4.   The limiting flux of H at 100 and 120 km.

Height (km)	$T(°K)$	M_a(amu)	$\dfrac{(m_a - m_i)g}{kT}$ (km^{-1})	$\dfrac{\alpha_i}{T}\dfrac{dT}{dz}$ (km^{-1})	b_i(cm^{-1} sec^{-1})	$\dfrac{\phi_l}{f_i}$ (cm^{-2} sec^{-1})
100	209.2	28.54	1.52(−6)[a]	−5.09(−8)	2.74(19)	4.30(13)
120	324.0	26.21	8.90(−7)	−1.01(−7)	3.72(19)	3.68(13)

[a] $1.52(-6) \equiv 1.52 \times 10^{-6}$

4-1 the mean expansion velocity for hydrogen at 1500°K is 7.26×10^3 cm sec^{-1}. The limiting process is therefore upward transport, and the escape flux is equal to the limiting flux.

The mean expansion velocity is sensitive to the temperature of the exobase while the diffusion velocity is not. Only at temperatures below about 500°K, however, does upward transport cease to limit the loss of hydrogen from the terrestrial atmosphere. The temperature of the thermosphere almost never falls this low.

The fact that hydrogen loss from the earth is limited by upward transport and not by Jeans escape leads to a great simplification of the escape problem. In order to calculate the escape rate, we do not need to use the Jeans formula Eq. (4-23), which depends on the density in the thermosphere and which is very sensitive to the variable and imperfectly known temperature of the exobase. Instead we can equate the escape flux to the limiting flux evaluated at low levels where the hydrogen mixing ratio is known. The limiting flux, Eq. (4-37), is not sensitive to atmospheric parameters and can be determined with fair accuracy. The temperature gradient term is usually small, and the temperature itself nearly cancels the temperature dependence of the collision coefficient, b_i. As an illustration of this insensitivity, we show, in Table 4-4, the values of the limiting flux for hydrogen calculated at 100 and 120 km with atmospheric parameters from Table 1-1. For the thermal diffusion factor α, we use −0.25 (cf. Hunten, 1973).

In the next section we shall compare the predictions of the transport theory we have just presented with observations of the rate of escape of hydrogen from the upper atmosphere. But first we must consider how the theory should be modified to allow for the photochemical reactions that produce the atomic hydrogen that escapes.

SOURCE OF ESCAPING HYDROGEN

We have implicitly assumed in this discussion that hydrogen is carried upwards in the thermosphere only in the atomic form; that is, that the densities of hydrogen compounds are negligible compared with the density of atomic hydrogen. A theoretical study by Hunten and Strobel (1973) indicates, however, that H_2 is several times as abundant as H at altitudes near 100 km. All other hydrogen compounds are negligible. The abundance of H_2 does not significantly alter our findings. In the thermosphere the reaction with O, reaction (4-26), followed by

$$(4\text{-}43) \qquad\qquad OH + O \rightarrow O_2 + H$$

converts H_2 to H, which subsequently escapes. Destruction of H_2 is balanced by diffusive flow through the base of the thermosphere at a flux very close to the limiting flux for H_2. Let us evaluate Eq. (4-37) for the limiting flux of H_2 at 100 km, using parameters from Tables 1-1 and 4-3 and $\alpha = -0.30$ (cf. Hunten, 1973). We find

$$(4\text{-}44) \qquad\qquad \phi_l(H_2) = 2.23 \times 10^{13} f(H_2)$$

For a mixture of H and H_2 at 100 km we have an escape flux of hydrogen atoms equal to the sum of the limiting flux of H and twice the limiting flux of H_2:

$$(4\text{-}45) \quad F_c = \phi_l(H) + 2\phi_l(H_2) = 4.30 \times 10^{13} f(H) + 2 \times 2.23 \times 10^{13} f(H_2)$$

where $f(H)$ and $f(H_2)$ are the respective mixing ratios at 100 km. The mixing ratio of hydrogen atoms in all forms is $f_t = f(H) + 2f(H_2)$. For constant f_t, we see that the escape flux varies by only a factor of two as the proportions of H and H_2 at 100 km are varied. A factor-of-two uncertainty is not a matter of great concern.

With f_t defined to include both H and H_2, Hunten and Strobel (1973) find

$$(4\text{-}46) \qquad\qquad F_c = 2.5 \times 10^{13} f_t \, cm^{-2} \, sec^{-1}$$

and this is the expression we shall adopt. Measurements of the density of hydrogen and the temperature at the exobase have shown that the escape flux is about $10^8 \, cm^{-2} \, sec^{-1}$ (Donahue, 1966; Joseph, 1967; Brinton and Mayr, 1971; Vidal-Madjar et al., 1973; Meier and Mange, 1973; Hunten and Strobel, 1973; Tinsley, 1974). Therefore, f_t is 4×10^{-6} at the bottom of the thermosphere. How does this value compare with the mixing ratios of hydrogen compounds at lower altitudes?

In the stratosphere, hydrogen exists principally in the form of water vapor, which has a mixing ratio of 3×10^{-6} by volume (Mastenbrook, 1968; 1971; Goldman et al., 1973). Possible additional contributions to stratospheric hydrogen in the form of CH_4 and H_2 have been discussed by Hunten and Strobel (1973). They derive a total mixing ratio for hydrogen atoms in all forms of 9×10^{-6}, but argue that this estimate may be too high as a result of contamination of the high-altitude measurements. The data are uncertain, but it appears that the total hydrogen mixing ratio is somewhat higher in the stratosphere than in the lower thermosphere (cf. Liu and Donahue, 1974a,b,c).

Photochemical reactions in the stratosphere and mesosphere convert the water and other hydrogen compounds into atomic and molecular hydrogen, but these reactions conserve hydrogen atoms. At all levels in the atmosphere, therefore, the upward flux of hydrogen in the steady state is equal to the escape flux, 10^8 atoms $cm^{-2} sec^{-1}$. Our previous discussion of the transport equation has shown that this is just the flux carried by molecular diffusion in an atmosphere in which the mixing ratio of hydrogen is constant. It is, therefore, at first sight surprising that the hydrogen mixing ratio declines between the stratosphere and lower thermosphere.

An explanation of this decline has been offered by Hunten and Strobel (1973). Above about 80 km hydrogen is predominantly in molecular or atomic form, and the limiting diffusive flux of either of these species carries almost the same number of hydrogen atoms. At lower altitudes, however, the dominant compound is water vapor, which has a diffusion coefficient only one-sixth as large as that of atomic hydrogen. The relatively large mass of the water molecule causes a further reduction in the flux that diffusion of water vapor molecules can carry [cf. Eq. (4-37)]. Below 80 km, therefore, molecular diffusion alone cannot support the required upward flux of hydrogen, and some of this flux must be carried by eddy mixing. Upward transport by eddy mixing requires that the mixing ratio decrease with altitude; the magnitude of the decrease depends on the values of the eddy mixing coefficient in the middle atmosphere. A numerical treatment by Hunten and Strobel (1973) has shown that data on the variation of mixing ratio with altitude are consistent with reasonable estimates of eddy mixing coefficients.

We may summarize the results of this analysis by saying that the escape of hydrogen is limited by the upward transport of hydrogen and its compounds by eddy mixing below about 80 km and by diffusion above this altitude. If eddy mixing were considerably more rapid, particularly in the stratosphere, the mixing ratio of hydrogen at the base of the diffusion region could approach its value at the tropopause. This would increase the escape flux of hydrogen by a factor of only about two.

We have been able to analyze the escape of hydrogen without explicit consideration of the photochemical reactions that convert water vapor to atomic hydrogen. This does not imply that these reactions are unimportant. The formation of atomic hydrogen is initiated by photodissociation of water vapor—

$$(4\text{-}47) \qquad\qquad H_2O + h\nu \rightarrow OH + H$$

—a process that occurs at a rate of about $2 \times 10^9 cm^{-2} sec^{-1}$ (Brinkmann, 1969). When this figure is compared with the escape flux, $10^8 cm^{-2} sec^{-1}$, we find that almost all of the hydrogen atoms produced by photo-dissociation must recombine to form water. Three facts have made it possible to analyze the escape problem without reference to these recombination reactions. The first is that the flux of hydrogen in all forms carried by eddy mixing does not depend on the compounds in which hydrogen occurs. The second is that the limiting flux of hydrogen in the diffusion region is insensitive to whether the hydrogen is atomic or molecular, and these are the most abundant hydrogen species in the lower thermosphere. The third is that molecular hydrogen is rapidly converted into atomic hydrogen above the base of the thermosphere.

A very complex set of physical and chemical processes therefore has a very simple result. The rate of escape of hydrogen is proportional to the concentration of hydrogen compounds in the stratosphere. The proportionality constant is known and is insensitive to atmospheric temperatures, eddy mixing rates, or the details of photochemistry. It is, therefore, a fairly easy matter to predict what escape rates would have been in the past. The escape rate would have been about the same in the past as it is today, unless the concentration of hydrogen compounds in the stratosphere has changed. Since the most abundant stratospheric hydrogen compound is water vapor, one possible source of change is a change in the temperature of the tropopause. Higher tropopause temperatures would permit larger hydrogen escape fluxes.

There is a limit, however, to the escape rate that can be achieved by increasing the water content of the upper atmosphere. This limit is imposed by the flux of solar photons that are able to dissociate water vapor. Hydrogen atoms cannot escape faster than they are produced by the destruction of water. The limit is about 10^{13} atoms $cm^{-2} sec^{-1}$ (Hunten, 1973). We are a long way from this limit today, because of the aridity of the upper atmosphere, but the limit may have been reached in some primitive atmospheres.

We shall return to the subject of primitive atmospheres in Chap. 6. Let us now apply the theory of hydrogen escape to the atmospheres of Mars and Venus.

JEANS ESCAPE OF HYDROGEN FROM MARS AND VENUS

The study of atmospheric evolution is principally theoretical. A theory, if it is to carry weight, must be subjected to as many tests as possible. Aspects of any theory applicable to the terrestrial atmosphere can be tested by seeing how well they account for what we know about the atmospheres of other planets. One such aspect is the theory of the escape of gases from planetary atmospheres. In this section we test the theory just developed for Earth by applying it to Venus and Mars, the only other planets for which the rate of escape of hydrogen has been measured.

The atmospheres of Venus and Mars are predominantly carbon dioxide (see Chap. 1), and the most abundant hydrogen compound at homopause levels in these atmospheres is probably H_2 (Hunten and McElroy, 1970). We shall therefore use the limiting flux for H_2 diffusing through CO_2 as an indication of what the escape flux might be.

Molecular hydrogen on Venus and Mars is converted to atomic hydrogen at heights above the homopause by the ionospheric reactions

$$(4\text{-}48) \qquad\qquad CO_2^+ + H_2 \rightarrow CO_2H^+ + H$$

followed by

$$(4\text{-}49) \qquad\qquad CO_2H^+ + e \rightarrow CO_2 + H$$

The rate of Jeans escape of hydrogen from these planets can therefore be estimated from data on the H density and temperature at the exobase. We want to know whether the escape flux is equal to the limiting flux and, if it is not, we want to know why.

The diffusion coefficient of H_2 in CO_2 appears in Table 4-3. The approximate expression for the limiting flux—

$$(4\text{-}50) \qquad\qquad \phi_l \sim b_i f_i / H_a$$

—is adequate for present purposes. This expression is a slowly varying function of temperature and thus of altitude. It therefore matters little at what altitude we choose to evaluate it.

On Mars let us use $T = 180°K$, corresponding to a height of about 100 km (McElroy and McConnell, 1971a). We calculate $b_i = 1.10 \times 10^{19}$ cm^{-1} sec^{-1} and $H_a = 9.6$ km, yielding

$$(4\text{-}51) \qquad\qquad \phi_l \sim 1.1 \times 10^{13} f_t$$

where f_t is still used to denote the total mixing ratio of hydrogen atoms in all compounds. The observed escape flux on Mars is 2×10^8 atoms cm^{-2} sec^{-1} (Anderson and Hord, 1971), so we find a value of $f_t = 1.8 \times 10^{-5}$. Eddy mixing coefficients are believed to be large on Mars (McElroy and McConnell, 1971a; McElroy and Donahue, 1972), so this value of the mixing ratio should apply also at the tropopause. The mixing ratio of water in the troposphere is larger than this (Hunten and McElroy, 1970), but the tropopause value is limited to about 3×10^{-7} by the saturated vapor pressure of water at the temperature of the tropopause. Hunten and McElroy have shown that the most abundant form of hydrogen in the Martian stratosphere is H_2; their detailed calculations yield densities close to those inferred here (McElroy and Hunten, 1969b; Hunten and McElroy, 1970). We conclude that the theory works on Mars.

The situation on Venus appears to be quite different from that on either Earth or Mars (Hunten, 1973). Let us again use $T = 180°K$ to evaluate the limiting flux. We calculate $H_a = 3.7$ km and $\phi_l = 2.9 \times 10^{13} f_t$. The observed mixing ratio of water vapor above the clouds on Venus is in the range $0.6–1.0 \times 10^{-6}$ (Fink et al., 1972), and temperatures in Venus's atmosphere are too high for condensation to limit the water vapor mixing ratio at higher altitudes (McElroy and Hunten, 1969a). Hydrogen may also be present in other compounds, including H_2 and HCl. Hunten (1973) has reviewed the evidence and found that the mixing ratio of H atoms can hardly be less than 10^{-6}, with 10^{-5} being a more probable value. Using the lower value and the expression for the limiting flux, we calculate an escape flux from Venus of 3×10^7 cm^{-2} sec^{-1}.

The escape flux actually observed by Mariner 5 on the day side of the planet was very much smaller, only 6×10^5 cm^{-2} sec^{-1} (Wallace, 1969). The explanation of this discrepancy may lie in our uncertain knowledge of the exospheric temperature on Venus. It is possible that this temperature is so low that Jeans escape rather than diffusion is the limiting process (Hunten, 1973; Kumar and Hunten, 1974). According to Kumar and Hunten, the discrepancy can be removed without violating other knowledge about the upper atmosphere of Venus if the temperature at the exobase is about 350°K. Hydrogen on Venus may therefore resemble helium on the earth in that the escape flux is much less than the limiting flux and is not simply related to the mixing ratio in the lower atmosphere.

ESCAPE OF HELIUM

As an illustration of what happens when escape is not limited by diffusion we shall now examine the escape of helium from the earth.

Characteristic times for the depletion of exospheric hydrogen are shorter than the time for transport of hydrogen upwards from the lower atmosphere. The escape flux is therefore limited by the transport properties of the atmosphere. The escape of helium is altogether different, however. Characteristic times for the escape of helium are much longer than characteristic transport times, so escape does not affect the profile of helium density vs. altitude. Alternatively, we may apply the criterion we used to establish that diffusion is limiting for hydrogen by comparing the mean expansion velocity for Jeans escape with the limiting diffusion velocity at the homopause. From Table 4-1, the mean expansion velocity of helium is 10^{-2} cm sec^{-1}; the limiting diffusion velocity at 100 km is about $2 \, \text{cm sec}^{-1}$. Since the density of helium at the exobase is less than the density at the homopause, the Jeans escape flux is less than the limiting flux by at least two orders of magnitude. To calculate the escape flux, we must therefore calculate the density at the exobase. In calculating the density profile, however, we can neglect the upward flux.

With $\phi_i \ll \phi_{l0}$ our solution for the mixing ratio, Eq. (4-40), becomes

$$(4\text{-}52) \qquad f_i/f_0 = [1 + \exp(z/\mathcal{H}_a)]^{1-\mathcal{H}_a/\mathcal{H}_i}$$

where $z = 0$ at the homopause (where $K = D_i$), and f_0 is the mixing ratio well below the homopause. At altitudes well above the homopause, the density profile, derived from Eq. (4-42), is

$$(4\text{-}53) \qquad n_i = f_0 n_0 \exp(-z/\mathcal{H}_i)$$

where n_0 is the ambient density at the homopause. This is just the diffusive equilibrium solution we would obtain if the mixing ratio were constant below the homopause and if mixing had no effect at all above the homopause. The full solution, of course, departs from this simple behavior in the neighborhood of the homopause; at $z = 0$ we have $f_i/f_0 = 2^{1-\mathcal{H}_a/\mathcal{H}_i}$, but we can calculate the density at high altitudes without reference to these departures. We shall therefore calculate the density at the exobase by assuming a constant mixing ratio up to the homopause and diffusive equilibrium above.

The procedure may be reversed to determine the value of the eddy mixing coefficient, K, in the vicinity of the homopause. For this purpose, argon density data are most useful. Measurements of argon density profiles in the thermosphere are extrapolated downwards, assuming that argon is in diffusive equilibrium. These extrapolated profiles intersect the profile of constant mixing ratio at an average height of 101 km (von Zahn, 1970). According to the derivation we have just presented, this is the height at which the eddy mixing coefficient equals the diffusion coefficient, at least in an isothermal atmosphere. The diffusion coefficient of argon in air at 101 km is $3.66 \times 10^5 \, \text{cm}^2 \, \text{sec}^{-1}$ (Colegrove et al., 1966). We shall use this value for K, which we assume to be independent of altitude.

For helium in air, the diffusion coefficient is $2.77 \times 10^{17} T^{0.729}/n_a \, \text{cm}^2 \, \text{sec}^{-1}$ (see Table 4-3). This is equal to K at a height of 94 km (see Table 1-1), the homopause for helium. We approximate the helium distribution by a well-mixed profile below the homopause and a diffusive-equilibrium profile above. Although this procedure does not correctly describe the density profile in the region of the homopause, it yields the correct helium density at high altitudes, and it is the density at the critical level that we need to know.

The density profile of helium in the thermosphere is significantly influenced by the increase in temperature between 100 km and 200 km.

We may allow for this temperature variation without greatly complicating the problem by fitting an algebraic expression to the temperature profile in the thermosphere. A convenient expression was suggested by Bates (1959):

$$(4\text{-}54) \qquad T(z) = T_\infty(1 - ae^{-\tau\zeta})$$

In this expression the temperature approaches a constant value of T_∞ at high altitudes. The parameters are

$$(4\text{-}55) \qquad \tau = dT/dz\big|_{z_0}[T_\infty - T(z_0)]^{-1}$$

and

$$(4\text{-}56) \qquad a = 1 - T(z_0)/T_\infty$$

while ζ is the *geopotential height* above reference height z_0, which we take as the homopause:

$$(4\text{-}57) \qquad \zeta = \int_{z_0}^z [g(h)/g(z_0)]\, dh = [(r_s + z_0)/(r_s + z)](z - z_0)$$

where r_s is the radius of the earth.

In diffusive equilibrium, the equation for the density profile is obtained from Eq. (4-27) by equating $w_i - w_a$ and K to zero and taking $f_i \ll 1$,

$$(4\text{-}58) \qquad (1/n)\, dn/dz + mg/kT + (1 + \alpha)(1/T)\, dT/dz = 0$$

We allow for the variation of g with altitude by changing the independent variable from z to ζ:

$$(4\text{-}59) \qquad (1/n)\, dn/d\zeta + mg(z_0)/kT + (1 + \alpha)(1/T)\, dT/d\zeta = 0$$

The solution of this equation, with T given by Eq. (4-54), is

$$(4\text{-}60) \qquad n(z) = n'(z_0)[(1 - a)/(1 - ae^{-\tau\zeta})]^{1+\alpha+\gamma}e^{-\tau\gamma\zeta}$$

(Stein and Walker, 1965; Walker, 1965, 1975a), where

$$(4\text{-}61) \qquad \gamma = mg_0/\tau kT_\infty$$

and $n'(z_0)$ is the reference density given by $fN(z_0)$, where f is the mixing ratio of helium in the lower atmosphere, and $N(z_0)$ is the ambient density at the homopause. As already noted, this solution is valid only some distance above the homopause, so $n'(z_0)$ is not the helium density at the homopause.

At high altitudes, where $\tau\zeta \gg 1$, Eq. (4-60) reduces to

$$(4\text{-}62) \qquad n(z) = n'(z_0)(T_0/T_\infty)^{1+\alpha+\gamma} \exp(-mg_0\zeta/kT_\infty)$$

At the exobase, in particular, we have

$$(4\text{-}63) \qquad n_c = n'(z_0)(T_0/T_\infty)^{1+\alpha+\gamma} \exp[(mg_0/kT_\infty)(r_0/r_c)(r_c - r_0)]$$

where r_c is the radius of the exobase, $r_c = r_s + z_c$, $r_0 = r_s + z_0$, and we have used Eq. (4-57) for the geopotential height.

Substituting Eq. (4-63) into the expression for the escape flux Eq. (4-20), we obtain

(4-64) $F_c = \frac{1}{2}\pi^{-\frac{1}{2}} n'(z_0)(T_0/T_\infty)^{1+\alpha+\gamma} \exp(-mg_0 r_0/kT_\infty) U(r_c/H_c + 1)$

Note that r_c does not appear in the exponential term, so the critical flux is insensitive to the height of the exobase.

The reference density at the homopause, $n'(z_0) = fN(z_0)$, where $N(z_0) = b/K$. For helium we have already found that $z_0 = 94$ km, $N(z_0) = 3.4 \times 10^{13}$ cm^{-3}, and $T(z_0) = 185°$K (Table 1-1). In the expression for F_c, the most important temporal variation is that in the exospheric temperature, T_∞, in the exponential term. In evaluating the other terms we use $T_\infty = 1500°$K, $dT/dz|_{z_0} = 10°$K km^{-1}, and $\alpha = -0.38$ (Kockarts, 1972). Then

(4-65) $F_c = 5.5 \times 10^{18} \exp(-30200/T_\infty)f$

where $f = 5.24 \times 10^{-6}$ (Kockarts, 1972) is the mixing ratio of helium in the lower atmosphere. The escape flux is proportional to the mixing ratio, which means that the mixing ratio would decrease exponentially as a result of Jeans escape if there were no sources and no other losses of atmospheric helium. The characteristic time for this exponential decay is given by the escape flux divided into the number of helium atoms in a square centimeter column of the atmosphere, $2.09 \times 10^{25}f$. This time is

(4-66) $\tau_a = 3.8 \times 10^6 \exp(30200/T_\infty)$

For an average exospheric temperature $T_\infty = 1500°$K, we find $\tau_a = 2 \times 10^{15}$ sec or 6×10^7 yr. This time is short compared with the age of the earth, so we expect the atmospheric helium content to have been significantly affected by Jeans escape.

In fact, if there were no source of atmospheric helium, we would expect any initial supply to have been long since exhausted. There is a source, however, provided by the release of helium from the crust, where it is produced by the decay of radioactive elements. The magnitude of this source has been estimated to be about 2×10^6 cm^{-2} sec^{-1} (MacDonald, 1963; 1964; Turekian, 1964; Axford, 1968; Craig and Clarke, 1970). By dividing this flux into the column density of helium in the atmosphere, 1.1×10^{20} cm^{-2}, we derive a residence time for atmospheric helium of only 2×10^6 yr, much smaller than the characteristic escape time derived above for an exospheric temperature of 1500°K. This result implies that the rate of Jeans escape at 1500°K is much smaller than the crustal source of helium. Since 1500°K is well above the average temperature of the exosphere, there appears to be a problem with the helium budget of the atmosphere.

The problem is not clearcut, however, because of the variability of exospheric temperature with time and the sensitive dependence of Jeans escape on temperature. MacDonald (1963, 1964) has evaluated the escape flux averaged over an entire 11-yr cycle of solar activity, using satellite data on exospheric temperature. He finds an average escape flux of $6 \times 10^4 \, cm^{-2} \, sec^{-1}$, a factor of 30 less than the source. It is still possible, nevertheless, that the bulk of the escape occurs during infrequent periods of unusually high temperature (Spitzer, 1949; Hunten, 1973). Hunten has pointed out that if the temperature were to exceed 2000°K, diffusion would become the limiting process and the escape flux would be equal to the limiting flux, about $10^8 \, cm^{-2} \, sec^{-1}$. To provide an average loss rate of $2 \times 10^6 \, cm^{-2} \, sec^{-1}$, these hot episodes would therefore have to occupy about 2% of the time. Because of the million-year residence time of atmospheric helium, the hot episodes could be quite infrequent; perhaps we have not yet had an opportunity to observe one. According to this suggestion, helium accumulates slowly between hot episodes and drops very sharply during hot episodes.

An alternative possibility is that there is a loss process for helium in addition to Jeans escape. Mechanisms other than Jeans escape have been proposed for the escape of gases from planetary atmospheres (Cole, 1966, Axford, 1968; Michel, 1971; Sheldon and Kern, 1972; Torr et al., 1974; Liu and Donahue, 1974b,c). Most of these are speculative and of undetermined evolutionary significance. There is one other process that we must consider, however. This is *photochemical escape*. It occurs only on Mars, but there it may substantially affect the evolution of the atmosphere.

PHOTOCHEMICAL ESCAPE FROM THE ATMOSPHERE OF MARS

Even heavy atoms, such as carbon, nitrogen, and oxygen, can escape from the upper atmosphere of Mars because the gravitational acceleration is small (Brinkmann, 1971b; McElroy, 1972). The velocity of escape from Mars is only $5 \, km \, sec^{-1}$. The corresponding escape energies are 1.49 eV, 1.74 eV, and 1.99 eV for carbon, nitrogen, and oxygen atoms, respectively. The exospheric temperature of about 350°K is too low for Jeans escape of these species to be important, but atoms with kinetic energies in excess of these limits are produced by photochemical reactions in the ionosphere (McElroy, 1972). On Earth and Venus, on the other hand, the escape velocities are twice as large, the escape energies are four times as large, and photochemical escape is not possible.

Rates of photochemical escape can be derived from considerations of

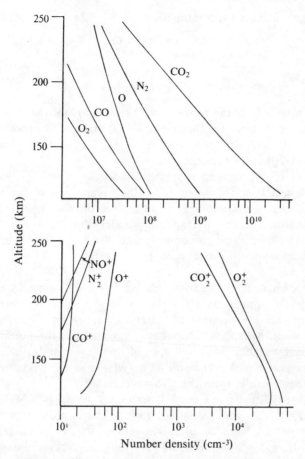

Fig. 4-2. Theoretical densities of neutrals and ions in the upper atmosphere of Mars. (From McElroy, M.B., *Science,* Vol. 175, pp. 443-445, 28 January 1972. Copyright 1972 by the American Association for the Advancement of Science.)

ionospheric photochemistry. McElroy's models of the ion and neutral densities of the upper atmosphere of Mars are shown in Fig. 4-2. The base of the exosphere is at about 200 km. Atoms produced above this level escape if they are moving upwards with sufficient energy.

Photoionization of the most abundant constituent, CO_2, is the principal source of ions:

$$(4\text{-}67) \qquad CO_2 + h\nu \rightarrow CO_2^+ + e$$

The photoionization coefficient for unattenuated solar radiation on Mars is $3.4 \times 10^{-7} \, sec^{-1}$ (McElroy, 1967, 1968), so CO_2^+ is produced at a rate of $2.2 \times 10^8 \, cm^{-2} \, sec^{-1}$ in the sunlit exosphere. These ions are removed by reaction with atomic oxygen,

$$(4\text{-}68) \qquad CO_2^+ + O \rightarrow O_2^+ + CO$$

followed by dissociative recombination

$$(4\text{-}69) \qquad\qquad O_2^+ + e \rightarrow O + O$$

or by

$$(4\text{-}70) \qquad\qquad CO_2^+ + e \rightarrow CO + O$$

The reaction with O is the more rapid below 250 km; the reaction with electrons above that height. Reactions (4-69) and (4-70) produce oxygen atoms with sufficient energy to escape. The CO molecules produced by (4-68) and (4-70) cannot escape.

Since half of the oxygen atoms produced by these reactions travel upwards while half travel downwards, the escape flux is approximately equal to the rate of photoionization of carbon dioxide in the exosphere. McElroy's value for the ionospheric model in Fig. 4-2 is 1.2×10^8 escaping atoms $cm^{-2} sec^{-1}$. Since only half the exosphere is exposed to sunlight, the escape flux averaged over the surface of the planet is $6 \times 10^7 cm^{-2} sec^{-1}$.

The flux of hydrogen lost from the Martian atmosphere as a result of Jeans escape is approximately twice as large. Since the hydrogen is produced by photodissociation of water vapor, we find that the oxygen left behind when hydrogen escapes is removed from the atmosphere by photochemical escape. Jeans escape of hydrogen and photochemical escape of oxygen are balanced so that no change in the oxidation state of the atmosphere results from the two processes.

McElroy (1972) has explained why this is so (cf. McElroy and Donahue, 1972; Liu and Donahue, 1976). The oxygen escape rate depends only on the rate of photoionization of carbon dioxide in the exosphere. It will remain constant so long as carbon dioxide is the predominant atmospheric constituent and the solar ultraviolet flux does not change. The hydrogen escape rate, on the other hand, is proportional to the mixing ratio of H_2 in the lower atmosphere, as we explained earlier in this chapter. Molecular hydrogen is formed from the products of water vapor photodissociation by reactions in the troposphere and at the surface. The integrated rate of photodissociation is $10^9 cm^{-2} sec^{-1}$ (Hunten and McElroy, 1970). About 90% of the photodissociations are followed by reformation of water. The remaining 10% lead to the formation of H_2, which is converted to H by reactions with $O(^1D)$ and CO_2^+ at higher altitudes. The atomic hydrogen escapes.

The fraction of the water vapor photodissociations that yields H_2 depends on the partial pressure of oxygen. If this partial pressure were too low, more H_2 would be formed, the H_2 mixing ratio would increase, and the rate of escape of H would increase also. The oxygen left behind by the escaping hydrogen would increase the partial pressure of oxygen. If, on the other hand, the partial pressure of oxygen were too high, the H_2

mixing ratio would be depressed, the escape flux of hydrogen would be less than twice the photochemical escape flux of oxygen, and the partial pressure of oxygen would fall. The time constant for the approach to equilibrium is the residence time of H_2, about 10^3 yr (Hunten and McElroy, 1970).

We see, therefore, that the rate of loss of hydrogen and thus of water from Mars is controlled by the rate of photochemical escape of oxygen in just such a way as to maintain the oxidation state of the atmosphere. If we assume that the flux of solar ultraviolet radiation has not varied much with time, then the rate of escape of oxygen and the rate of loss of water have not varied much either, over the lifetime of the planet. The total amount of water that has escaped from Mars is therefore about 10^{25} molecules per square centimeter of Martian surface. There are at present only 5×10^{19} molecules cm^{-2} of water in the Martian atmosphere (Hunten and McElroy, 1970), but there may be substantial amounts of water in subsurface permafrost layers (Leighton and Murray, 1966; Smoluchowski, 1968; Sagan et al., 1973).

Carbon atoms can also escape from the upper atmosphere of Mars. McElroy calculates that the dissociative recombination of exospheric carbon monoxide ions—

$$(4\text{-}71) \qquad\qquad CO^+ + e \rightarrow C + O$$

—contributes an escape flux of C equal to 1.5×10^5 cm^{-2} sec^{-1}. Dissociation of CO by photoelectrons and by solar radiation contributes about 3×10^5 cm^{-2} sec^{-1} and dissociation of CO_2 contributes an equal amount. The total of these fairly well-determined escape fluxes is 7.5×10^5 cm^{-2} sec^{-1}, large enough to have removed almost one half of the carbon now present in the atmosphere over the lifetime of the planet.

There are other possible loss processes for carbon. McElroy estimates an upper limit for the escape flux resulting from dissociative recombination of carbon dioxide ions—

$$(4\text{-}72) \qquad\qquad CO_2^+ + e \rightarrow C + O_2$$

—of 4×10^5 cm^{-2} sec^{-1}, and an upper limit for the rate of loss of carbon dioxide ions by solar wind sweeping of 5×10^5 cm^{-2} sec^{-1}. The escape flux of carbon therefore lies between 7.5×10^5 cm^{-2} sec^{-1} and 1.65×10^6 cm^{-2} sec^{-1}. This flux is too small to affect the oxidation state of the atmosphere, but it is large enough to have removed between 10^{23} cm^{-2} and 2.3×10^{23} cm^{-2} atoms of carbon from the atmosphere over the age of the planet. There are at present only 2×10^{23} cm^{-2} molecules of carbon dioxide in the atmosphere, but there may be much larger amounts in the polar caps (Sagan et al., 1973). The escape flux of carbon, like the escape flux of oxygen, is independent of the carbon dioxide content of the atmosphere as long as carbon dioxide is the principle constituent.

Because nitrogen is a minor atmospheric constituent, the escape rate is proportional to the nitrogen concentration at ionospheric levels. This concentration is proportional to the mixing ratio of nitrogen in the lower atmosphere and depends on the value of the eddy mixing coefficient, which determines the height of the homopause. For an eddy mixing coefficient of 5×10^6 cm^2 sec^{-1}, which corresponds to a homopause height of 90 km, McElroy (1972) finds that dissociative recombination of N_2^+—

$$(4\text{-}73) \qquad N_2^+ + e \to N + N$$

—produces a nitrogen atom escape flux of $3 \times 10^7 f$ cm^{-2} sec^{-1}, where f is the N_2 mixing ratio in the lower atmosphere. Atoms produced by dissociative recombination of NO^+ do not have enough energy to escape.

Brinkmann (1971b) has derived the escape flux resulting from absorption of solar radiation by exospheric nitrogen molecules followed by predissociation into atoms having sufficient energy to escape:

$$(4\text{-}74) \qquad N_2 + h\nu \to N_2^* \to N + N$$

His value is $10^7 f$ cm^{-2} sec^{-1}. The combined flux due to these two fairly well-evaluated mechanisms is $4 \times 10^7 f$. There is some evidence for eddy diffusion coefficients on Mars as large as 5×10^8 cm^2 sec^{-1}, corresponding to a homopause height of 140 km (McElroy and McConnell, 1971a; McElroy and Donahue, 1972). This much eddy mixing would reduce the nitrogen density in the exosphere and thus the escape flux by a factor of 4.2.

McElroy's estimate for the nitrogen loss as a result of solar wind sweeping is $4 \times 10^7 f$ cm^{-2} sec^{-1} for the low homopause model. Additional loss may result from dissociative collisions between N_2 and photoelectrons. Altogether, therefore, the nitrogen escape flux may lie between $10^7 f$ and $8 \times 10^7 f$ cm^{-2} sec^{-1}, depending on the height of the homopause and the importance of solar wind sweeping. Since there are 2×10^{23} molecules cm^{-2} in the atmosphere of Mars, the residence time of nitrogen, given by $2 \times 10^{23} f$ divided by the escape flux, is between 8×10^7 and 6×10^8 yr. We may therefore conclude that escape has had a significant effect on the nitrogen content of the Martian atmosphere as well as on the water and carbon dioxide contents.

SUMMARY

This chapter has been devoted to the derivation of the rates of loss of gases from planetary atmospheres. Application of these rates to topics in atmospheric evolution has been deferred to later chapters. Here we summarize the major findings.

Jeans escape is the most important loss process for the terrestrial atmosphere. Two situations can be distinguished; they are exemplified by hydrogen and helium. The rate of loss of hydrogen is limited by upward diffusion from the lower atmosphere. The loss rate is equal to the limiting flux, which is insensitive to atmospheric structure and can be estimated with fair confidence both for the present and for hypothetical primitive atmospheres. The helium escape rate, on the other hand, is too small for upward transport to be a limiting factor. The escape rate therefore depends sensitively on the height of the homopause and the temperature profile in the thermosphere. Most helium escape may occur during infrequent periods of unusually high exospheric temperature.

Photochemical escape from the atmosphere of Mars appears to be well established. The rates of loss of carbon and oxygen should have remained constant as long as the flux of ionizing radiation from the sun has remained constant. The rate of Jeans escape of hydrogen is controlled by the rate of photochemical escape of oxygen. The rate of loss of nitrogen is proportional to the nitrogen mixing ratio and depends on the height of the homopause. For all these gases, the total losses over the age of the planet are substantial compared with the amounts now in the atmosphere.

Part Three
THE ANCIENT ATMOSPHERE

Chapter Five
Origin of the Atmosphere

The first four chapters of this book have been devoted to the processes that control the composition and mass of the present atmosphere. The same processes have presumably controlled atmospheric composition and mass during much of earth history, so an understanding of them is surely essential to an understanding of atmospheric evolution. In the remaining chapters we shall try to develop an account of the history of the atmosphere that is based on plausible extrapolation, backwards in time, of the processes described in the previous chapters.

The subject of atmospheric history is conveniently divided into three topics. First there is the origin of the atmosphere, which we take up in this chapter. This topic includes the origin of the ocean, which would, of course, be part of the atmosphere if the surface temperature were higher. Next, if we assume that some time elapsed between the origin of the atmosphere and the origin of life, there is the question of what the atmosphere was like in the absence of life: This topic is considered in Chap. 6. We have already shown, in Chap. 2 and 3, the extent to which biological processes influence the composition of the atmosphere. We can therefore expect the development of life to have had a profound effect on the history of the atmosphere. The interaction of biological and atmospheric evolution is explored in Chap. 7, where we present a tentative, qualitative account of the geological history of the atmosphere.

We shall begin the present chapter with an account of the evidence that the terrestrial atmosphere was formed by the release of volatile compounds from the solid planet. The origin of the atmosphere can

179

therefore be described by the history of the release of gas and the composition of the gases released. The composition depends, however, on the chemical state of the upper layers of the earth at the time when the atmosphere originated. Most important is the question of whether the metallic iron that is now in the core was originally distributed throughout the body of the earth. We shall present geochemical and geophysical evidence that it never was, and go on to describe how the earth may have formed inhomogeneously, with the core and mantle segregated, before release of gases to the atmosphere began.

We shall then turn to the history of the release of volatiles. There are several arguments to suggest that most of the release was concentrated very early in earth's history, close to the time of origin of the earth; during most of earth's history, the addition of new material to the atmosphere—and the ocean—can probably be ignored. Based on this evidence, we develop a tentative account of the growth of the mass of *surface volatiles*—atmosphere, ocean, and volatiles incorporated in sedimentary rocks—to approximately their present masses in less than 1 billion years or so after earth's formation.

We estimate the total mass of material released and its composition by adding to the volatile material that now constitutes the ocean and the atmosphere an estimate of the volatiles that are now locked up in sedimentary rocks and an estimate of the material that may have been lost from the atmosphere to space. We can now compare the rate of release of volatiles required by our model of rapid initial release with the rate of release of gases by modern volcanoes (volcanic gases consist largely of recycled surface volatiles). We shall show that our model is consistent with a much greater rate of tectonic activity in early earth history.

Finally, we examine possible sources of atmospheric gases other than the solid earth. These sources include accretion of material from the solar wind and accretion of interstellar matter. We find that these other sources are negligibly small.

SECONDARY ORIGIN OF THE ATMOSPHERE

Here we shall review the evidence that the earth has a *secondary* atmosphere, produced by the release of gas from the interior, rather than a *primordial* atmosphere, a remnant of the gaseous nebula from which the solar system condensed. (The origin of the solar system is described later in this chapter.)

In Table 5-1 we compare estimates of the abundances of various elements in the whole earth with estimates of their abundances in the

TABLE 5-1. **The abundances of some elements in the earth and in the solar system**

	ATOMIC NUMBER	WHOLE EARTH (ATOMS/10,000 ATOMS SI) (a)	SOLAR SYSTEM (ATOMS/10,000 ATOMS SI) (b)	DEFICIENCY FACTOR [log (b/a)]
H	1	84	3.5×10^8	6.6
He	2	3.5×10^{-7}	3.5×10^7	14
C	6	71	80,000	4.0
N	7	0.21	160,000	5.9
O	8	35,000	220,000	0.8
F	9	2.7	90	1.5
Ne	10	1.2×10^{-6}	50,000	10.6
Na	11	460	462	0
Mg	12	8900	8870	0
Al	13	940	882	0
Si	14	10,000	10,000	0
P	15	100	130	0.1
S	16	1000	3500	0.5
Cl	17	32	170	0.7
A	18	5.9×10^{-4}	1200	6.3
Kr	36	6×10^{-8}	0.87	7.2
Xe	54	5×10^{-9}	0.015	6.5

From Mason (1958). Copyright 1958 by John Wiley and Sons, Inc., New York. Used by permission of the publisher.

solar system. The abundances are expressed relative to silicon, but the choice of this particular element for normalization is arbitrary. It is the ratios of abundances that are of interest to us. The table shows that there is a group of elements that are relatively as abundant on earth as they are in the solar system. These elements, which include Si, Mg, and Al, are called *refractory* elements because they enter into compounds which volatilize only at high temperatures. There are other elements, including hydrogen and carbon, which either are gaseous at relatively low temperatures or form compounds that are gaseous. These are the *volatile* elements. They are depleted on earth relative to the solar system. The depletion is most notable in the case of the *inert gases*, He, Ne, A, Kr, and Xe, which condense at very low temperatures indeed and which form very few chemical compounds.

Since we know of no process that could selectively remove inert gases from the vicinity of the earth, the depletion of the inert gases suggests that gases were not retained to a significant extent by the earth when it formed (Moulton, 1905; Aston, 1924; Russell and Menzel, 1933; Brown, 1952; Sagan, 1967; Meadows, 1973). The evidence suggests that the earth consists mainly of materials that were solid or liquid at the temperatures that prevailed around the earth at the time of its formation. The present atmosphere, therefore, is not a remnant of the gaseous nebula from which the sun and the planets condensed. Instead, it has been formed by the release of gases from materials that were originally incorporated into the earth in the form of nongaseous chemical compounds.

We may use the inert gas abundance data to set an upper limit on the mass of a primordial terrestrial atmosphere of solar composition (Fanale, 1971a; Walker, 1976a). Neon is a convenient gas to use since it does not escape from the atmosphere, is not produced by radioactive decay, and is not incorporated to any significant extent into rocks (Canalas et al., 1968; Fanale and Cannon, 1971a; Phinney, 1972). The mass of neon in the atmosphere is 6.48×10^{16} gm (Verniani, 1966). This is an upper limit on the mass of neon retained on earth in a hypothetical primordial atmosphere of solar composition. Using the table of solar system abundances compiled by Cameron (1968; cf. Ross and Aller, 1976), we calculate corresponding upper limits on the amounts of other elements that could have been retained in the primordial atmosphere. The results in Table 5-2 show that the mass of the primordial atmosphere must have been less than 1% of the mass of the present atmosphere (5.1×10^{21} gm). We conclude that nearly all of the atmosphere (and the ocean) has been released from the solid earth.

This conclusion is strengthened by evidence that the inert gases on earth today are not the remnants of a primordial atmosphere. Studies of meteorites have revealed two components of their inert gas complements (Pepin and Signer, 1965). One, called "*solar*," exhibits relative concentrations that are characteristic of the sun; it is thought to consist of trapped solar wind particles. The other, called "*planetary*," exhibits a much less rapid decrease of concentration with increasing atomic mass than does the solar component. The planetary component is believed to be gas adsorbed from the solar nebula by a process that favors the heavier species. The terrestrial complement of inert gases is "planetary" rather than "solar" (when allowance is made for the xenon content of shales), indicating that inert gases were incorporated into the earth adsorbed onto solids and not in a primordial atmosphere of solar composition.

With the secondary origin of the atmosphere established, we turn to a consideration of the gases released from the solid earth. The composition

TABLE 5-2. Upper limits on the masses of the most abundant elements in a primordial atmosphere

	H	He	C	N	O	S
10^{16} gm	3500	1100	22	4.6	51	2.2

Form Walker (1976a). Copyright 1976 by John Wiley and Sons, Inc., New York. Used by permission of the publisher.

of these gases would have depended on the composition of the upper layers of the earth (the region now occupied by the upper mantle and the crust). In particular, the oxidation state of the gases would have depended on the oxidation state of the upper mantle region (Holland, 1962, 1964).

For present purposes, the oxidation state of the upper mantle is sufficiently characterized by the presence or absence of metallic (or free) iron. The upper mantle today contains no free iron, which is all segregated into the core. As a consequence, the gases released from modern volcanoes are only weakly reducing, containing much more water vapor than hydrogen and much more carbon dioxide than carbon monoxide. Release of such gases early in earth history would have produced a weakly reducing primitive atmosphere. If, however, the metallic iron of the core was distributed more or less uniformly throughout the earth at the time degassing occurred, the gases released would have contained abundant hydrogen (Holland, 1962, 1964), and the original atmosphere would have been strongly reducing.

Holland (1964) has searched for direct evidence of change in the oxidation state of the upper mantle as reflected in the ratio of ferrous iron to ferric iron in ancient basalts and diabases (rocks that originate in the upper mantle). He finds no indication of change in at least the last 1.8 billion years. But we are interested in much earlier times, for which a suitable rock record is not available. Other arguments must therefore be invoked. These are described in the next section, where we shall conclude that the earth had an inhomogeneous origin without free iron in the upper mantle.

HOMOGENEOUS ORIGIN OF THE EARTH

Metallic iron is denser than silicate minerals and must therefore settle to the center of the earth unless the materials of the earth have considerable strength. An earth that was initially homogeneous must also have been cold, therefore, because hot rocks can flow. Such an earth presumably would have warmed up gradually, as a result of heating by radioactive elements, until the interior became soft enough for iron to begin to settle. The gravitational energy released by settling iron would have caused additional heating, leading to more melting and more rapid settling. It is likely that the segregation of the core would have been a catastrophic event, melting the earth entirely and destroying all surface rocks.

Hanks and Anderson (1969) have argued in this way that the segregation of the core must have preceded the emplacement of the oldest known surface rock, with an age of 3.8 billion years (Moorbath et al., 1972, 1973). By calculating the evolution of internal temperature under various assumptions concerning the radioactive heat source, and by requiring that substantial melting occur within the first billion years, they have been able to set a lower limit on the initial heat content of the earth. This initial heat was provided by gravitational energy released during the accretion of the solids from which the earth was formed and also by adiabatic compression of the interior under the weight of material accreted later. From a consideration of the total available gravitational energy and the lower limit on the initial heat content of the earth, they have been able to set an upper limit on the time during which the earth grew to substantially its present size. This upper limit is 200,000 yr. Most of the energy that led to melting and permitted segregation of core and mantle came from accretion and not from radioactive heating. From this Hanks and Anderson conclude that the earth may well have melted during the course of its growth rather than subsequently. There may never have been a homogeneous earth.

Ringwood (1959, 1966) has suggested further arguments against an initially homogeneous earth (Turekian and Clark, 1969; Clark et al., 1972). The most compelling concerns the high observed abundance of nickel in ultramafic rocks and basalts, which originate in the mantle (see Chap. 1). Nickel would have been almost entirely removed from these rocks if they had ever been in contact with the iron–nickel alloy that now forms the core. A second argument concerns the oxidation state of the upper mantle as reflected in the Fe_2O_3/FeO ratio in ultramafic rocks and basalts. This ratio is large enough to suggest that the rocks of the upper mantle were never intimately mixed with metallic iron.

From these various lines of evidence we conclude that the earth did not form homogeneously, and that there never was metallic iron in the upper mantle. In the absence of metallic iron, the gases released from the upper layers of the earth would have contained little hydrogen, and the primitive atmosphere would not have been strongly reducing. Before considering the origin of the atmosphere further, however, let us describe how the earth may have formed inhomogeneously.

INHOMOGENEOUS ACCRETION OF THE EARTH

In this section we present the main features of a theory of the origin of the solar system and the earth that is based largely on papers by Turekian and Clark (1969, 1975), Clark et al. (1972), Grossman (1972), Cameron and Pine (1973), Cameron (1973), Lewis (1972, 1974), Grossman and Larimer (1974), and Arrhenius et al. (1974). The solar system originated, about 4.6 billion years ago, as a cloud of gas, dust, and ice of a few solar

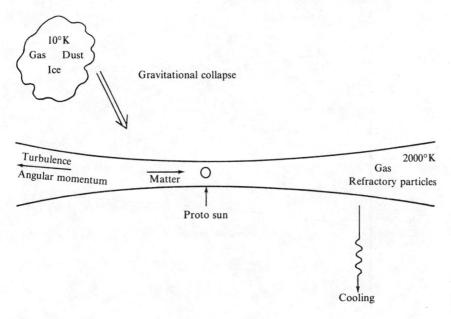

Fig. 5-1. Contraction of the primitive solar nebula.

masses at a temperature of about $10°K$ collapsing under its own gravitational attraction. Compression caused the temperature to rise to several thousand degrees, vaporizing all but the most refractory compounds, while conservation of angular momentum flattened the cloud into a disk (Fig. 5-1).

The properties of this disk have been calculated by Cameron and Pine (1973). Their results are summarized in Fig. 5-2. The thickness of the disk was governed by its own gravitational attraction and therefore was smaller near the center, where matter was more abundant, than near the edges. Much of the disk was optically opaque and in a state of turbulent convection.

The gravitational field of the disk caused the refractory solids to settle to the midplane, where they began to accumulate into larger bodies. Accretion was facilitated by the turbulence and also by viscous drag, which caused particles of different size to move at different speeds relative to the gas. At this time, temperatures in the inner regions of the nebula were sufficiently high to prevent the condensation and accretion of any but the most refractory materials. The nuclei of the inner planets therefore consisted largely of corundum (Al_2O_3), perovskite ($CaTiO_3$), gehlenite ($Ca_2Al_2SiO_7$), and metallic iron (Grossman and Larimer, 1974).

The turbulence in the nebula transported angular momentum outward and matter inward, so the sun grew while gas pressures decreased. Simultaneously, the nebula cooled by emission of radiation to space,

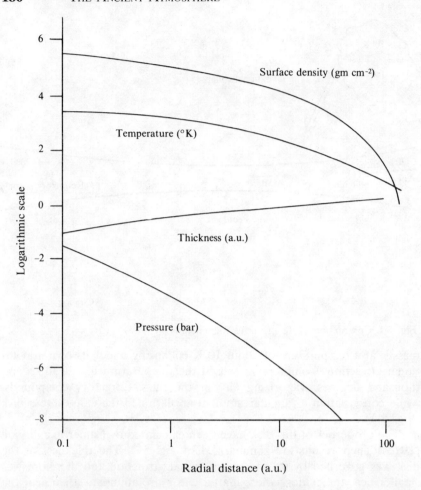

Fig. 5-2. Conditions in the nebular disk according to Cameron and Pine (1973). Pressure and temperature are shown for the central plane of the disk. Surface density is the mass of matter in a column of unit cross section at right angles to the plane of the disk. (From Walker, 1976a. Copyright 1976 by John Wiley & Sons, New York. Used by permission of the publisher.)

allowing less refractory minerals to condense and accrete. This behaviour is sketched in Fig. 5-3.

The rate of growth of the planetary nuclei increased rapidly with their masses and the strengths of their gravitational fields. The largest nuclei grew at the expense of their smaller neighbors. In the intermediate stages of planetary growth it is likely that the rate of accretion was large enough to melt the planets (Clark et al., 1972) or possibly even to volatilize some of the less refractory silicates (Cameron, 1973). The molten planets

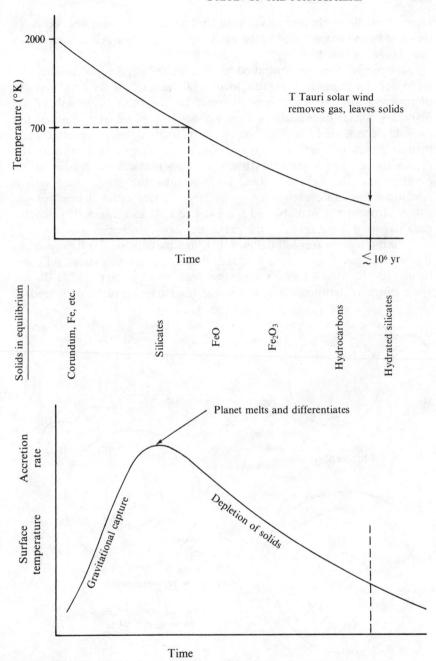

Fig. 5-3. Cooling and condensation in the nebula and accretion of the primitive planet. (From Walker, 1976b. Copyright 1976 by the Astronomical Society of Southern Africa. Used by permission of the publisher.)

segregated efficiently into nickel–iron cores and silicate mantles. Volatiles that had been incorporated in the earlier stages were largely driven out of the planets at this time.

Meanwhile, the sun continued to grow at the expense of the nebula, while nebular temperatures continued to fall, permitting the condensation of progressively less–refractory compounds. The rates of growth of the planetary nuclei decreased as they gathered up more and more of the available condensed material and as the nebular density decreased. The planets therefore began to cool and solidify, driving to their surfaces the radioactive elements and other trace elements that are not readily incorporated into the silicate lattice. By this time the inner solar system contained hot, differentiated, volatile-free planets, solid debris not yet gathered up by the planets, and a rapidly cooling and dissipating nebular gas. Stages in the growth of the earth are shown in Fig. 5-4.

When temperatures fell below 750°K any metallic iron still exposed to the nebular gas began to be oxidized, first to the ferrous state and then, below 400°K, to the ferric state (Grossman and Larimer, 1974). In the same range of temperatures, the stable high temperature compound of

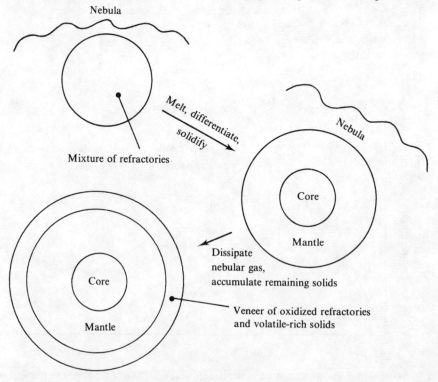

Fig. 5-4. Stages in the growth of the earth. (From Walker, 1976b. Copyright by the Astronomical Society of Southern Africa. Used by permission of the publisher.)

OXIDATION OF IRON IN THE SOLAR NEBULA

The important equilibrium reactions are the following:

$$CO + 3H_2 \rightleftharpoons CH_4 + H_2O$$

$$\underset{\text{Enstatite}}{2MgSiO_3} + 2Fe + 2H_2O \rightleftharpoons \underset{\text{Fayalite}}{Fe_2SiO_4} + \underset{\text{Forsterite}}{Mg_2SiO_4} + 2H_2$$

The free energies of the participants in these reactions are such that both reactions proceed leftwards at high temperatures (above about 800°K) and rightwards at low temperatures. As temperature decreases, therefore, the conversion of CO to CH_4 increases the ratio of H_2O to H_2 in the nebula, promoting the oxidation of iron in the second reaction. The second reaction is driven further to the right by the temperature dependence of its own equilibrium constant (L. Grossman, private communication, 1975).

carbon, carbon monoxide, began to react with hydrogen to form methane and other less volatile hydrocarbons (Lewis, 1972). Below 350°K magnesium silicates reacted with water vapor to form hydrated silicates. Lower temperatures were probably not achieved in the inner solar system before the nebula was dispersed.

The cooling of the nebula was interrupted when the sun had accumulated enough mass to become a star. During its approach to the main sequence it passed through a *T Tauri phase* during which it ejected into space several tenths of a solar mass of material in the form of a very strong solar wind (Hayashi, 1961; Ezer and Cameron, 1963, 1971). This wind carried the remaining nebular gas out of the inner solar system and also dissipated any primordial atmospheres that had accumulated around the inner planets (Cameron, 1973).

The protoplanets were left to gather up the solid debris left in their regions of the solar system from the condensation that had occurred during cooling of the nebula. This debris included refractory material that had condensed early but had not been accumulated right away. The refractory component provided the source of the nickel and iron in the upper mantle. The iron had been oxidized as a result of exposure to cold nebular gas. The debris also included volatile-rich material, resembling Type 1 carbonaceous chondrites, that had condensed when the nebula was quite cool. This volatile-rich material was the source of the atmosphere and ocean.

At the time of its dissipation, the nebula was, of course, cooler in its outer parts than in its inner parts. Planets closer to the sun are therefore composed, on average, of more refractory materials than planets further from the sun. Lewis (1974) has shown how this theory can explain the

METEORITES

Meteorites originate within the solar system; they may well be small asteroids. Radiometric dating has shown that only about 10^8 years elapsed between the time of formation of the chemical elements and the time of formation of the meteorites. Most meteorites condensed and cooled about 4.6×10^9 yr ago. Taken together, these two results indicate that meteorites originated at the same time as the solar system, 4.6 billion years ago.

Measurements of nuclides produced in meteorites by cosmic ray impacts indicate that most meteorites have been exposed to space for times much less than the age of the solar system. They may have been shielded in the interiors of larger asteroids which have been broken apart by collision comparatively recently.

There are three main groups of meteorites, *stones, irons,* and *stony irons.* Stones are by far the most abundant. Most stones contain small spheroids, about 1 mm in diameter (*chondrules*), with a distinctive mineralogy. These meteorites are called *chondrites.* Chondrites are remarkably uniform in chemical composition. Their mineralogy resembles that of basalt. The *achondrites* are a small group of stones that lack chondrules.

Another small group of stones is the *carbonaceous chondrites.* Unlike other meteorites, they contain substantial amounts of water, carbon, sulfur, and other volatile elements in various chemical compounds. They are classified into three groups on the basis of their volatile contents. Type I carbonaceous chondrites are the richest in volatiles; they lack chondrules (cf. Wood, 1967).

CARBONACEOUS CHONDRITES	% C	% S	% H_2O
Type I	2.7–5.0	5.2–6.7	18–22
Type II	1.1–2.8	2.3–3.7	8–17
Type III	0.2–0.6	1.8–2.4	0.1–1.5

observed differences in the mean densities of the inner planets and some of the satellites of the outer planets. Additional considerations involved in the formation of the outer planets themselves have been discussed by Cameron (1973).

According to the theory of inhomogeneous accretion, therefore, the earth differentiated into core and mantle before it ever acquired the material that now comprises the upper mantle, crust, ocean, and atmosphere. The iron in the material that formed the surface layers was oxidized by the time it arrived on earth. Volatiles were incorporated in the form of low temperature condensates resembling carbonaceous chondrites, carbon dioxide in the form of the carbon in hydrocarbons, and water as the hydrogen in hydrocarbons and possibly also in hydrated silicate minerals. The inert gases were added largely as adsorbed material in the late-forming condensate. Their abundance ratios, therefore, are "planetary." This is the model of earth formation that we shall adopt. Now we can examine its implications for atmospheric evolution.

First, let us consider the time scale for the evolution of the solar nebula and accretion of the inner planets. Theoretical arguments indicate that the duration of the whole process was geologically very short—less than 10^7 yr, possibly less by several orders of magnitude (Clark et al., 1972; Cameron, 1973; Hills, 1973). The time that is of most importance for the atmosphere, however, is the time for the release of volatiles from the late-forming and late-accreting veneer. Degassing of the volatile-rich veneer may have resulted from impact heating, and may have occurred as fast as material was accreted (Fanale, 1971a, b). This would have led to an essentially instantaneous origin of the atmosphere. The volatile-rich material may, however, have arrived in small enough pieces to escape degassing on impact. Small pieces are, indeed, implied by the oxidation of refractory material exposed to the cold solar nebula. Degassing in this case could have resulted from radioactive heating and more normal tectonic processes.

DEGASSING HISTORY

If the earth had been initially cold and had warmed up only gradually, it could have been tectonically inactive during the earliest portions of its history. With such an origin, it is possible that the rate of degassing could have increased with time. However, we have specifically rejected a cold origin for the earth in favor of the inhomogeneous accretion model described above. We prefer an initially hot earth, with radioactive heat sources more abundant in the past, and with additional energy provided by the last stages of the accretional process; it then seems clear that the earth must have been tectonically more active early in its history than it is now, and that the rate of degassing must have declined with time.

In this section we shall try to decide whether the decline was rapid or slow: Did the atmosphere and ocean reach approximately their present masses very early in earth's history, or have they grown steadily throughout geologic time? We shall present various arguments in favor of the first hypothesis, concluding that degassing was largely confined to the first billion years of earth's history. We shall not, however, be able to set a lower limit on the duration of degassing. There does not seem to be any way to distinguish between degassing that was essentially instantaneous, with a time scale of a few million years or less, and rapid degassing, with a time scale of a hundred million years or so. The later evolution of the atmosphere does not seem to depend on whether degassing was instantaneous or rapid.

First let us describe evidence that the upper layers of the earth were homogenized at a very early date in spite of their initially heterogeneous origin. The evidence is based on the concentrations of uranium and lead in rocks and on the isotopic composition of the lead. Uranium decays radioactively to produce the lead isotopes ^{206}Pb and ^{207}Pb, but not the isotope ^{204}Pb (Eicher, 1968). With the passage of time, therefore, the relative abundances of the different lead isotopes in a rock change in a predictable way at a rate that depends on the ratio of uranium to lead in the rock. Extensive measurements of lead isotopes show that nearly all rocks have had nearly the same uranium-to-lead ratio for nearly all of geological time. Lead isotope data would clearly distinguish any rock that had, at some time in the past, had a uranium-to-lead ratio markedly different from its present value for a time as short as 100 million years (Armstrong, 1968). Such rocks are not found.

This result is surprising, because the inhomogeneous accretion theory predicts that the ratio of uranium to lead would have varied quite markedly in the materials of the primitive earth. Uranium forms refractory compounds that would have accumulated along with other refractories early in the course of earth accretion. When the earth melted and differentiated the uranium would have been included in the minerals that were the last to crystallize. It would therefore have been concentrated at the surface of the cooling earth, where it would have been covered by the last stages of accretion. Lead, on the other hand, would not have condensed early from the solar nebula. Instead it would have been incorporated into the earth with other volatile elements in the late-accreting, volatile-rich veneer.

Initially, therefore, terrestrial lead was concentrated in the volatile component of the surface veneer, while uranium was concentrated at greater depth, at a level that had once been the surface of the molten earth. Suppose that tectonic processes drove the uranium gradually to the surface over an extended period of geological history. Then the uranium-to-lead ratio in surface materials would have risen gradually from an initially low value. Modern lead isotope data would reveal such a gradual increase in the uranium-to-lead ratio if it had occurred. Apparently it did not occur.

The likeliest explanation of the observed uniformity of uranium-to-lead ratios is that the uranium was driven to the surface very early in earth's history before much radioactive decay had taken place. The upper limit on the time during which uranium and lead could have remained apart without producing detectable anomalies in lead isotope data is probably 100 million years (Armstrong, 1968). The evidence therefore suggests that tectonic activity was sufficiently rapid to thoroughly mix uranium and lead within the first 100 million years of earth's history. Thorough mixing of uranium and lead would presumably have been associated with thorough degassing of the surface volatiles, so it appears that the atmosphere may have originated in a time not longer than 100 million years.

The case for early degassing does not rest on this evidence alone. We shall now argue that the rate of degassing today is negligibly small. In this discussion, degassing should be understood to refer to the release of *juvenile* volatiles that have never been in the atmosphere before, and not to volcanic release of recycled surface volatiles.

Direct evidence on the present rate of degassing on Earth is hard to find, but we can, with some assumptions, show that degassing on Venus today is negligibly slow. This we do by arguing that the rate of degassing cannot greatly exceed the rate at which the atmosphere is losing water by photolysis followed by escape of hydrogen to space (Walker et al., 1970). The escape rate, which we discussed in Chap. 4, is much too small today to have any effect on the evolution of the atmosphere. Therefore, the degassing rate is also negligibly small. Venus and Earth are very similar in mass, mean density, and position in the solar system. Therefore, we might reasonably assume that they have broadly similar bulk compositions and that their degassing histories have been similar. If degassing is negligible on Venus today, it seems likely that it is also negligible on Earth.

In our exploration of the degassing history of the earth, we have presented indirect evidence for rapid degassing in the first 100 million years of its history and indirect evidence that the present degassing rate is negligibly small. Let us now describe some direct evidence concerning the growth of the atmosphere.

In discussing the geological evidence, we shall take it for granted that the atmosphere and the ocean accumulated at the same time. Because the conversion of igneous rocks into sedimentary rocks requires a supply of volatiles, we can probably assume that growth of the mass of sedimentary rocks also accompanied growth of the atmosphere (Siever, 1968; Garrels and Mackenzie, 1971; Li, 1972). At the same time, it is not unreasonable that degassing should have accompanied the differentiation of the crust and the upper mantle from the disorderly mixture of refractory and volatile materials that originally constituted the upper layers of the earth. So we shall try to learn something about the growth of the atmosphere from evidence of the growth of the ocean, the sedimentary shell of the earth, and the crust itself.

First, let us consider the growth of the crust. Substantial areas of

Archean crust (older than 2.5 billion years) have now been found in the interiors of most of the continents (cf. Windley, 1976). They are not always exposed at the surface, of course, frequently being overlain by younger sedimentary rocks. Evidently, the area of Archean crust was large, but what can we say about its thickness?

The structure of the oldest rocks suggests that the pattern of tectonic activity in the Archean was markedly different from that in the Phanerozoic (Glikson, 1970; Anhaeusser, 1972; Sutton and Watson, 1974). There appears to have been much more vertical movement and much less horizontal movement. The lithosphere, which provides the rigid plates of modern plate tectonics, was probably thinner in the Archean than it is now because initial heat and more rapid decay of radioactive minerals maintained higher temperatures in the interior of the earth. Archean tectonics therefore provide information on the thickness of the lithosphere, but probably not on the thickness of the crust.

On the other hand, some of the rock types that have been found in the Archean crust appear to have been formed at substantial pressures. The evidence has been interpreted to indicate a crustal thickness of 40 km in some locations (cf. Windley, 1976), comparable to the thickness of the modern continental crust. Tentatively, we conclude that the Archean crust was approximately as thick as the Phanerozoic crust and, by 3 billion years ago, the total volume of crust was not very much smaller than the volume of the crust today. A very rough lower limit might be one-quarter of the present volume. It might have been larger. Evidence of crustal growth does not contradict our model of degassing concentrated early in earth's history, but neither does it provide strong support.

Now let us consider the growth of the ocean. In the interiors of many of the continents are large stable areas that have suffered little deformation since well back in the Precambrian. These stable regions are called platforms or cratons (see Chap. 1). Undeformed, shallow-water sediments ranging in age up to at least 1.7 billion years are found on the cratons (Armstrong, 1968). These sediments indicate that the surface of the sea and the surface of the cratons have been at approximately the same level throughout this period. The thickness of crust underlying the ancient cratons shows little variation with position or age, which indicates that the crustal structure of the cratons has remained relatively constant. If we assume that oceanic crustal structure has not changed much either, then isostasy insures that the elevation of the craton surface relative to the floor of the ocean basins has been fairly constant. Thus the average depth of the oceans has not changed.

On an earth of fixed surface area, constant continental thickness and constant ocean depth are possible if neither ocean volume nor continental volume have changed with time, or if one of these volumes has grown at the expense of the other. Since the differentiation of mantle material would cause both continent and ocean to grow together rather than one at the expense of the other, the second alternative is not attractive. Therefore, ocean volume has been constant for the last 2 billion years. Chase and Perry (1972) derive the same result from oxygen isotope data, although their argument has been contested by Becker (1973; Chase and Perry, 1973). If the ocean has not grown significantly in the last 2 billion years, then degassing was largely complete by this time. The history of ocean growth therefore provides strong support for our model of early degassing, but it does not tell us how early this degassing was.

Let us consider, now, the evidence provided by sedimentary rocks. The argument here is simple. There must have been substantial amounts of ocean and atmosphere at the time of deposition of the oldest sedimentary rocks (Donn et al., 1965). The oldest, well-preserved sedimentary rocks that we know are the Onverwacht group of South Africa, with an age of 3.2 billion years (Schopf, 1972), but metamorphic rocks of the Isua formation in Greenland appear to have been deposited originally as sediments (Moorbath et al., 1973). The Isua rocks are 3.7 billion years old. We conclude that substantial degassing occurred within the first billion years of earth's history, but the sedimentary evidence does not permit a more quantitative statement to be made.

In summary, our model of degassing concentrated early in earth's history is supported by geological evidence on the growth of the crust and the ocean and the deposition of sedimentary rocks.

EXCHANGE OF MATERIAL BETWEEN CRUST AND UPPER MANTLE

None of the above should be taken to imply that volatiles are not now being released from the upper mantle to the surface of the earth. In this section we shall discuss the evidence that such release is indeed occurring. Then we shall show how continuing release of volatiles from the upper mantle can be reconciled with early accumulation of the atmosphere and ocean (which we call early degassing) by continuing exchange of material, including volatiles, between the crust and upper mantle.

First there is the evidence of the supply of radiogenic inert gases, ^{40}A and ^4He, to the atmosphere (Turekian, 1964). The helium content of the atmosphere would long ago have fallen to very small values as a result of Jeans escape from the upper atmosphere if there were not a continual supply of helium produced by the radioactive decay of uranium (see Chap. 4). Since much of the earth's uranium is in the crust, however, and since helium diffuses readily out of crustal rocks (Turekian, 1964), helium provides little information on the release of gases from the mantle. Argon is much more useful.

In the atmosphere, ^{40}A comprises 99.6% of all argon, while the corresponding figure for the solar system is only 0.01% (Nier, 1950; Cameron, 1968). The explanation of this difference is that ^{40}A is produced by the radioactive decay of ^{40}K, which is much more abundant on earth, relative to argon, than in the solar system. Our knowledge of the potassium content of the interior of the earth is inadequate (Armstrong, 1968; Hanks and Anderson, 1969), but it appears that the crust does not contain enough K to have been the source of more than 20% of the radiogenic argon that is now in the atmosphere (Turekian, 1964). Most atmospheric argon has therefore been released from the mantle. The ^{40}A content of the mantle has grown gradually, as a result of the decay of ^{40}K, from essentially zero when the earth was formed. Release of volatiles from the mantle very early in earth's history could not, therefore, have provided the argon in the atmosphere because there would not have been enough argon to release. Even with a high estimate of the potassium content of the mantle it would have taken 140 million years to produce the argon now in the atmosphere. Release of argon from the upper mantle was therefore not restricted to the first 100 million years of earth's history (Fanale, 1971a). It has probably continued at a significant rate throughout geological time (Turekian, 1964).

There is direct evidence, in fact, for the release of inert gases from the mantle today (Dymond and Hogan, 1973; Fisher, 1974, 1976a, b). The basalts that comprise the new crust formed at midocean ridges are derived from the upper mantle (see Chap. 1). Most of these rocks contain traces of inert gases in relative abundances that resemble those in sea water. The inert gases in these rocks have evidently been derived from the surrounding water, probably by diffusion through the crystal lattice. In the rapidly chilled, glassy margins of some of these rocks, however, inert gases are found with different relative abundances. They appear to be a mixture of a component with the "planetary" composition of meteorites and a component with the terrestrial composition (depleted in xenon). These gases may have originated in the upper mantle and have been preserved from sea-water contamination by the glassy structure of the rocks in which they are found.

There is as yet no evidence as to whether the rate of release of nonradiogenic inert gases from the mantle is significant or not, and the composition of the inert gases in the glassy margins is not sufficiently distinctive to imply that they are primordial gases that have never before been in the atmosphere. These observations therefore do not contradict the model we are developing, of rapid initial growth of the atmosphere giving way to a steady-state cycling of material between the crust and upper mantle.

Before describing this model, let us consider evidence for the release of volatiles other than the inert gases from the upper mantle. Since

---

KIMBERLITE PIPES AND PHENOCRYSTS

Kimberlite pipe. Kimberlite pipes are vertical, carrot-shaped bodies with diameters as large as 700 m. They are composed principally of peridotite, an igneous rock rich in MgO, FeO, and CaO, and poor in SiO_2, but they contain many inclusions of surrounding rocks. They appear to have been intruded from great depths, not as molten silicate minerals, but as a mixture of solids and fluids, principally water and carbon dioxide.

Phenocryst. Some igneous rocks contain distinct crystals of certain minerals imbedded in a finer-grained matrix. Such rocks are called porphyries. The larger crystals are called phenocrysts; the fine-grained material is the groundmass.

---

basaltic magmas originate in the upper mantle, it is entirely reasonable to suppose that the volatiles associated with them have come from the upper mantle also. There is a considerable amount of direct evidence for a high volatile content in the upper mantle: Roedder (1965) has described inclusions of liquid carbon dioxide at high pressure in nodules and *phenocrysts* contained in basalts derived from the upper mantle; diamonds also contain inclusions of water, carbon dioxide, and nitrogen (Mitchell and Crocket, 1971); the structure and mineralogy of *kimberlite pipes* imply a high content of volatiles in the source material, extending to depths of at least 300 km (Dawson, 1971); and consideration of the stability relations of carbon indicate that diamonds form at depths of at least 200 km in an environment where the partial pressure of carbon dioxide is approximately equal to the confining pressure of the overlying rock (Kennedy and Nordlie, 1968). Green (1972) has reviewed the evidence for high carbon dioxide concentrations in the upper mantle, and additional evidence is cited by MacGregor and Basu (1974).

If volatiles are abundant in the upper mantle, there is no reason to suppose that they are not released to the atmosphere whenever volcanic activity brings mantle-derived material to the surface (Sylvester-Bradley, 1972). It does not follow, however, that volatiles are still accumulating at the surface of the earth. They may be returned to the mantle as fast as they are released (Meadows, 1973; Arrhenius et al., 1974).

Armstrong (1968) has presented a model for the exchange of material between crust and upper mantle based largely on data concerning strontium and lead isotope distributions, but consistent with other geochemical evidence as well (Armstrong, 1971; Armstrong and Cooper, 1971; Armstrong and Hein, 1973). According to this model, sialic material (see Chap. 1) is eroded from the continents and accumulates as sediments on the ocean floor. In the course of sea-floor spreading, these sediments and other sialic material are dragged down into the mantle in subduction zones (see Chap. 1). There they melt and are returned to the crust as juvenile-appearing volcanic rocks. Before this happens, however, there is enough mixing with mantle material to significantly affect the isotopic composition of both crust and mantle.

Most of the volatiles contained in the subducted sediments (principally water and carbon dioxide in the form of carbonate minerals) are presumably returned to the surface with the sialic volcanics, but the volatiles, like the strontium and lead isotopes, should have ample opportunity to mix with mantle volatiles before this happens. Subduction of ocean sediments therefore provides a mechanism to recycle volatiles from the crust to the upper mantle, thereby maintaining the volatile content of the mantle.

The rate of exchange derived by Armstrong (1968) is fast enough to reprocess the entire volume of the crust in 1.5 billion years, and he argues that this rate was perhaps four times as fast in the Precambrian (Armstrong, 1968; Armstrong and Hein, 1972). The upper mantle is turned over by convection in a shorter time, several hundred million years, according to Armstrong (1968). These characteristic times are short enough to permit crust–mantle interchange to maintain an approximate equilibrium between the mantle reservoir of volatiles and the crustal reservoir, which includes the atmosphere, ocean, and volatiles incorporated in sedimentary rocks. If the mantle is temporarily depleted in volatiles, it retains more of the surface volatiles supplied to it in subduction zones; a temporary excess of mantle volatiles results in reduced retention.

Now we can understand how the atmosphere and ocean have grown, rapidly at first and then more slowly. At first, all of earth's volatiles were in the solid planet. Tectonic activity caused gases to be released, but there were no surface volatiles to be returned to the mantle by subduction of wet, weathered sediments. The surface volatile reservoir therefore grew at the expense of the mantle reservoir. Recycling of surface volatiles to the mantle increased, however, as the atmosphere, the ocean, and the sedimentary rock mass grew. Gradually, the partitioning of volatiles between the mantle and the surface approached an equilibrium and the rate of growth of the atmosphere declined. Release of volatiles from the upper mantle has continued through geologic history, but, except during the initial period of adjustment, it has been balanced by return of volatiles to the mantle in subduction zones.

This description of the exchange of material between crust and mantle is consistent with all of the evidence concerning the history of degassing that we have presented. The duration of the initial period of adjustment remains uncertain, however. There is strong geological evidence that it did not last longer than 1 or 2 billion years. The lead isotope evidence suggests that it may have lasted less than 100 million years. Degassing and crustal differentiation may even have occurred essentially instantaneously during the accumulation of the surface layers of the earth. Provided it is short, however, the time scale for degassing has little effect on the subsequent history of the atmosphere.

INVENTORY OF SURFACE VOLATILES

We have examined the history of the accumulation of volatiles in the atmosphere, ocean, and crust of the earth, but we have not yet considered the total mass or the composition of the gases released to the atmosphere during this period of growth. In this section we shall derive these quantities by estimating the masses of different volatile constituents in the surface reservoir. We shall have to make some allowance for water that may have been lost as a result of photolysis followed by escape of hydrogen to space and reaction of oxygen with surface rocks. We shall set an upper limit on the magnitude of this correction by estimating how much oxygen may have been absorbed into the upper layers of the earth.

By dividing the mass of volatiles released by one billion years we can set an approximate lower limit on the rate of degassing during early earth history. We shall show that this rate is greater than estimates of the modern rate of release of volcanic gases. Tectonic activity, volcanism, and crust–mantle interchange were probably more rapid early in earth's history than they are today. Finally, we shall estimate the oxidation state of the primitive volcanic gases by reference to the composition of modern volcanic gases and the expectation of chemical equilibrium between the gases and the upper mantle from which they originated.

Degassing introduced water and carbon dioxide to the primitive atmosphere and initiated weathering of the primitive, basaltic crust. Sedimentary rocks accumulated at the expense of igneous rocks, consuming volatiles in the process. In time, the primitive degassing came to an end and the mass of sedimentary rocks achieved its present equilibrium value (Garrels and Mackenzie, 1971; Li, 1972). We assume that subsequent volcanism and crust–mantle interchange has not changed the surface volatile inventory.

The composition of the atmosphere does not reflect, directly, the composition of the gases released from the interior by degassing. While relatively unreactive species such as nitrogen and the inert gases have largely accumulated in the atmosphere, other constituents, such as water and carbon dioxide have accumulated in other reservoirs. Degassed water now resides largely in the oceans; the atmospheric water vapor content is determined by the saturated vapor pressure of water at average atmospheric temperatures. Degassed carbon dioxide, on the other hand, is now incorporated as carbonate minerals in sedimentary rocks. The carbon dioxide content of the atmosphere is governed by the kinetic considerations described in Chap. 3.

The oxygen in the atmosphere is a special case. The reducing character of volcanic gases makes it most unlikely that free oxygen has been released from the solid earth. Our oxygen has been produced from water and carbon dioxide by biological and photochemical processes, which we discussed in Chap. 3.

In order to determine the average composition of the atmospheric source gas we must therefore estimate the amounts of the major volatile constituents, excluding oxygen, of the oceans and sedimentary rocks as well as of the atmosphere. These estimates are presented in Table 5-3. Reduced carbon, C, must be added to the total amount of carbon dioxide, because it is organic carbon, produced from carbon dioxide by photosynthetic organisms.

A small correction to these values must be made for the volatile contents of the igneous rocks weathered to produce sedimentary rocks. For the average composition of igneous rocks we use the compilation of Ronov (1968), and for the total mass of sedimentary rocks, approximately equal to the mass of igneous rocks that have been weathered, we take the figure 2×10^{24} gm (Ronov and Yaroshevskiy, 1967, 1969). The resulting corrections are shown in Table 5-4. The remaining amounts, which were not contributed by igneous rocks, are called *excess volatiles* (Rubey, 1951). Our values for the excess volatiles are close to those of Nicholls (1965, 1967), which were based on Poldervaart's (1955) compilation of crustal structure and composition. Very similar estimates have been derived from considerations of geochemical mass balance by Li (1972).

The excess volatiles provide an initial estimate of the gases released to

TABLE 5-3. Volatiles in atmosphere, oceans, and sedimentary rocks (gm)

	H_2O	C	CO_2	Cl	N	S^f
Atmosphere[a]	1.7 (19)[c]	—	2.45 (18)	—	3.87 (21)	—
Oceans[b]	1.4 (24)	2.7 (18)	1.38 (20)	2.62 (22)	2.18 (19)	1.22 (21)
Fresh water and ground water[c]	4.1 (22)	—	—	—	—	—
Sedimentary shell[d]	1.5 (23)[h]	1.0 (22)	2.42 (23)	5.0 (21)	1.0 (21)[g]	4.0 (21)
Total	1.6 (24)	1.0 (22)	2.4 (23)	3.1 (22)	4.9 (21)	5.2 (21)

[a] Verniani (1966); [b]Turekian (1968); [c]Skinner (1969); [d]Ronov (1968); [e]1.7 (19) $\equiv$ 1.7 $\times$ 10^{19};
[f]Sulfur occurs in various oxidation states;
[g] Computed from the value for C and C/N = 10 (Sverdrup et al., 1942, p. 1010). Dashes indicate negligible amounts.
[h] Unbound water estimated equal to bound water.

TABLE 5-4. Contribution from weathering of igneous rocks (gm)

	H_2O	C	CO_2	Cl	N	S
Total volatiles	1.6 (24)[a]	1.0 (22)	2.4 (23)	3.1 (22)	4.9 (21)	5.2 (21)
From igneous rocks	3.1 (22)	3.4 (21)	1.7 (22)	1.0 (21)	4.0 (19)[b]	8.0 (20)
Excess volatiles	1.6 (24)	6.6 (21)	2.3 (23)	3.0 (22)	4.9 (21)	4.4 (21)

[a] $1.6 (24) \equiv 1.6 \times 10^{24}$; [b] Bowen (1966)

the atmosphere during degassing. Water vapor predominated, with carbon dioxide next in abundance (the carbon which appears separately in Table 5-4 was released as carbon dioxide). Chlorine, which now exists largely as chloride ion in the ocean, was less abundant than carbon dioxide by more than an order of magnitude. Nitrogen and sulfur were even less abundant than chlorine. These conclusions would not be changed by making allowance for water that may have been lost as a result of escape of hydrogen, but we should try to set a limit on this loss nevertheless.

The present rate of loss of hydrogen corresponds to the destruction of 4.8×10^{11} gm yr^{-1} of water (see Chap. 4). If this rate had remained unchanged throughout the lifetime of the earth, only a negligible 2×10^{21} gm of water would have been lost. We may conclude that the loss today is too small to have any effect on the water budget of the earth, but we need other evidence concerning possible loss in the past.

Other evidence is provided by the oxygen that is left behind when hydrogen escapes (Holland, 1964; Abelson, 1966; Brinkmann, 1969; Arrhenius et al., 1974; Walker, 1976a). To set an upper limit on the hydrogen loss, let us assume that the surface layers of the earth originally contained CH_2, FeO, no Fe_2O_3, and enough water of hydration to provide the oceans and all of the oxygen now present in carbonate rocks and in ferric iron. The reactions we consider are

(5-1) $$CH_2 + 2H_2O \rightarrow CO_2 + 3H_2$$
(5-2) $$2FeO + H_2O \rightarrow Fe_2O_3 + H_2$$

where the hydrogen is assumed to escape. Other constituents, such as sulfur, are so much less abundant than carbon and iron that their oxidation need not be considered. The oxygen produced by dissociation of one-tenth of an oceanic mass of water would be sufficient to produce all of the Fe_2O_3 in the top 50 km of the mantle by oxidation of FeO (Clark et al., 1972). Dissociation of one-sixth of an ocean would produce all the oxygen now combined with carbon in crustal carbonate rocks. We do not need to be precise, so let us be generous and take one-third of the mass of the ocean as an upper limit on the water lost over the age of the earth. Brinkmann (1969) used an upper limit that was smaller than this by a factor of ten.

We conclude that the excess volatiles, presented in Table 5-4, provide an adequate description of the gases originally released to the atmosphere. In order to allow for the escape of hydrogen, we could augment the mass of water by no more than 33%. The actual loss of water has probably been much less.

RATE OF RELEASE OF VOLCANIC GASES

Let us now estimate the rate of release of volatiles by modern volcanoes in order to compare this rate with the rate that would have been required in order to provide the excess volatiles in a time less than about a billion years. We shall limit our estimates to carbon dioxide, for which the arguments that yield a modern rate of release are most secure.

We reiterate, first, that volatiles are consumed when igneous rocks are weathered to form sedimentary rocks (Siever, 1968; Garrels and Mackenzie, 1971, p. 243). In a steady-state system, these volatiles must be restored as fast as they are consumed. It is not hard to understand how such an equilibrium can be maintained in light of the model of sedimentary recycling that we have described above as well as in Chaps. 1 and 3.

Igneous rocks and volcanic gases are created simultaneously from sialic material carried down into the mantle in subduction zones. The igneous rocks are weathered to form sedimentary rocks, volatiles being consumed in the process. Corresponding to this source of sedimentary rocks is a sink in which sedimentary rocks are carried into the mantle and converted back into igneous rocks and volatiles. Material is cycled continuously between the reservoirs of igneous rocks, sedimentary rocks, and volatiles; but an equilibrium state is possible, in which the masses of the reservoirs do not change with time.

More extensive discussions of the cycles of sedimentary rocks have been given by Siever (1968), Garrels and Mackenzie (1971), and Garrels and Perry (1974). These discussions include examples of the chemical reactions involved in the weathering and reconstitution of silicate minerals. They also point out that only a small proportion of sedimentary rocks pass through the igneous rock reservoir. Most new sediments are formed from the erosion of old sediments, not of igneous rocks (Garrels and Mackenzie, 1972). The conversion of old sediments into new sediments does not, on average, lead to consumption or release of volatiles. The new sediments are lifted above sea level in time, becoming subject once more to erosion.

The overall system of atmosphere, ocean, crust, and upper mantle is illustrated schematically in Fig. 5-5. The notion of equilibrium in this overall system enables us to estimate rates of release of volcanic gases from data on the rate of weathering of igneous rocks.

We shall use an estimate by Garrels and Mackenzie (1971, p. 124) of the rate of consumption of carbon dioxide in chemical weathering. From a consideration of typical weathering reactions, they find that weathering of silicate minerals releases two SiO_2 molecules and one bicarbonate ion to solution for every molecule of carbon dioxide consumed. The weathering of carbonate minerals, on the other hand, releases two bicarbonate ions and one calcium or magnesium ion to solution for every carbon dioxide molecule consumed. This carbon dioxide is restored to the atmosphere when the carbonate minerals precipitate, once again, on the floor of the ocean.

From data on the silica, calcium, and magnesium contents of river waters on the various continents, Garrels and Mackenzie calculate that silicate weathering accounts for 19% of all of the carbon dioxide consumed in chemical weathering. Rivers carry about 19×10^{14} gm of bicarbonate into the sea each year (Garrels and Mackenzie, 1971, p. 108), about half of which is derived from atmospheric carbon dioxide. We calculate, therefore, that silicate weathering consumes 1.4×10^{14} gm of carbon dioxide per year. In equilibrium, this carbon dioxide must be restored to the atmosphere by volcanism.

This rate can now be compared with the net rate of release—volcanic emanations less volatiles returned to the mantle—that must have existed during the period of growth of the atmosphere. If the excess volatiles accumulated in a time less than a billion years, the net rate of addition of carbon dioxide to the crust must have exceeded 2.5×10^{14} gm yr^{-1}. The volcanic release rate must have been larger than this by an amount equal to the rate of return of volatiles to the mantle. The numbers are very uncertain, but they suggest that the rate of release of volcanic gases has declined since the beginning of earth's history. If degassing occurred in a time much less than a billion years, then the decline has been substantial. Since the intensity of radioactive heating within the earth has declined with time and since initial heat has gradually been dissipated, it is entirely reasonable that the rate of volcanic activity should have declined.

It is unfortunate that we can do little better than guess at the rate of release of volcanic gases to the primitive atmosphere, since before the evolution of life volcanoes may have been the major sources of the trace gases that influenced atmospheric properties. We shall consider the composition of volcanic gases in detail in Chap. 6. It is appropriate to consider their oxidation state (ratio of H_2 to H_2O) here, however, since this depends on the oxidation state of the upper mantle, a subject we

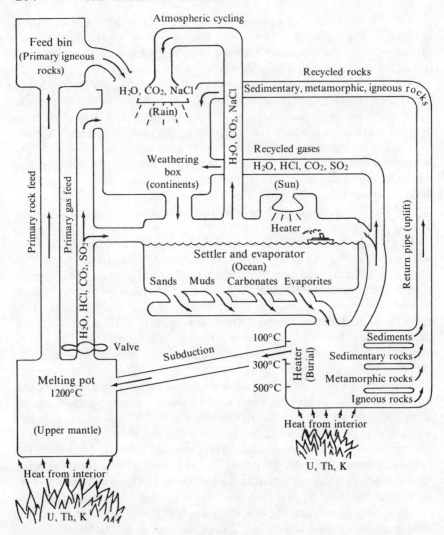

Fig. 5-5. The crust-ocean factory. (Reprinted from *Evolution of Sedimentary Rocks* by Robert M. Garrels and Fred T. Mackenzie. Copyright 1971 by W.W. Norton and Company, Inc., New York, N.Y. Used by permission of the publisher.)

discussed earlier in this chapter in the context of the inhomogeneous accretion model for the origin of the earth.

According to the inhomogeneous accretion model, there has never been free iron in the upper mantle; the upper mantle has always been approximately as oxidized as it is today. Since volcanic gases should achieve chemical equilibrium with the molten rocks in which they origi-

TABLE 5-5. Observed and predicted oxidation state of Hawaiian volcanic gases

	AVERAGE HAWAIIAN VOLCANIC GASES (volume %)		PREDICTED RATIOS
H_2O	79.31	$\dfrac{P_{H_2O}}{P_{H_2}} = 137$	105
H_2	0.58		
CO_2	11.61	$\dfrac{P_{CO_2}}{P_{CO}} = 31$	37
CO	0.37		
SO_2	6.48		
S_2	0.24		
H_2S	—		
N_2	1.29		
Cl_2	0.05		
A	0.04		

From Holland (1964). Copyright 1964 by the Institute for Space Studies. Used by permission of the Institute.

nate, we can assume, therefore, that these gases have always had about the same oxidation state as they have today. Evidence of chemical equilibrium in volcanic gases has been presented by Holland (1962, 1964), who has compared the oxidation state of Hawaiian volcanic gases with theoretical predictions for gases in equilibrium with basalt melts of the kind that might exist in the upper mantle source regions of these volcanoes. There is good agreement between the measured ratios of H_2O to H_2 and CO_2 to CO and the predicted equilibrium ratios, as shown in Table 5-5 (cf. Cruikshank et al., 1973; Fanale, 1971b; Heald et al., 1963).

Accepting chemical equilibrium, we can use the data in Table 5-5 to estimate rates of volcanic release of hydrogen and carbon monoxide. These estimates are very uncertain, however, because measured volcanic gases vary markedly in composition (White and Waring, 1963), and theory predicts that the equilibrium oxidation state of the gas is sensitive to the temperature of the molten material (Nordlie, 1968, 1972).

According to the data of Table 5-5, volcanoes emit 8×10^{-4} gm of H_2 with every gram of H_2O and 2×10^{-2} gm of CO with every gram of CO_2. If we adopt 10^{14} gm yr^{-1} as the modern rate of release of CO_2 and accept the H_2O to CO_2 ratio of Table 5-5, we derive modern rates of volcanic release of 2×10^{12} gm yr^{-1} of CO, 3×10^{14} gm yr^{-1} of H_2O, and 2×10^{11} gm yr^{-1} of H_2 (about the same as the rate of loss of hydrogen by Jeans escape). The rates would have been larger early in earth's history, but we do not know how much larger.

EXTRATERRESTRIAL SOURCES OF ATMOSPHERIC GASES

Our discussion in this chapter so far has been largely restricted to the origin of the atmosphere by the release of gases from the solid earth. It remains to show that possible extraterrestrial contributions to the atmosphere are negligibly small. Extraterrestrial sources we shall consider include accretion of solar wind ions and accretion of neutral atoms originating in the interstellar medium.

First, we again call attention to the limit imposed by the abundance of neon on earth on the total contribution to the atmosphere of any source with the composition of the sun (the average composition of the solar system). Earlier in this chapter we argued that a primordial atmosphere retained by the earth from the solar nebula must have had a negligibly small mass in order not to have left the earth with too much neon. The same limit applies also to atmospheric contributions from any of the extraterrestrial sources we have suggested, unless these sources have neon abundances that are relatively much lower than the neon abundance in the solar system. This argument applies to the total contribution from extraterrestrial sources over the age of the earth. Let us now consider evidence on the present-day magnitude of extraterrestrial sources.

The atmospheric helium budget provides an indication that extraterrestrial sources are negligible today. We described in Chap. 4 the difficulty of accounting for the loss from the atmosphere of ^{4}He produced in the crust of the earth by radioactive decay. This difficulty would be compounded by any additional source of ^{4}He. But since the ^{4}He budget is not well understood, this argument is not very strong.

Possibly stronger evidence is provided by the budget of the rare isotope of helium, ^{3}He (Johnson and Axford, 1969; Holzer and Axford, 1971). This isotope has an abundance relative to ^{4}He of 1.25×10^{-6} in the terrestrial atmosphere and 3×10^{-4} in the solar system (Cameron, 1968). Because of its lighter mass, its loss from the atmosphere is more likely to be controlled by Jeans escape than is the loss of ^{4}He. Johnson and Axford calculate that the average rate of Jeans escape of ^{3}He is $6 \text{ cm}^{-2} \text{ sec}^{-1}$. The average loss due to any other escape mechanism that is not sensitive to mass, including diffusion-limited Jeans escape at times of unusually high exospheric temperature, can be estimated from the ^{4}He budget and the atmospheric abundances of the two isotopes. A value of about $1 \text{ cm}^{-2} \text{ sec}^{-1}$ results, giving a total rate of loss of ^{3}He of $7 \text{ cm}^{-2} \text{ sec}^{-1}$.

The source of this ^{3}He is not clear. Craig and Clarke (1970) have presented evidence for a crustal source of $5.5 \text{ cm}^{-2} \text{ sec}^{-1}$, but a subsequent paper (Craig and Weiss, 1971) indicates that this is probably an overestimate. Johnson and Axford (1969) have estimated negligibly small values for the sources due to cosmic rays and meteoric dust (Lupton,

1973). They argue that the source is accretion of solar wind ions or neutral interstellar atoms (Holzer and Axford, 1971). It is sufficient, for our purposes, to know that the extraterrestrial source of ^{3}He is less than about 7 atoms $cm^{-2} sec^{-1}$. We now want to use this information to set limits on the extraterrestrial sources of ^{4}He and H. To do this, we need information on the composition of the sources.

Let us consider the solar wind source first. Ions of the solar wind are prevented by the earth's magnetic field from interacting with the atmosphere at low latitudes. They may, however, be guided to high latitudes, where they can appear as precipitating auroral particles. Whalen et al. (1971) have measured the He^{++}-to-H^+ ratio in auroral particles and found that it is equal to the value in the solar wind. Singly charged helium, He^+, is absent from the auroral primaries as it is from the solar wind. These observations provide evidence that solar wind ions can enter the atmosphere without significant change in their relative abundances.

The abundance of ^{3}He in the solar wind relative to protons is about 10^{-5} (Bame et al., 1968), so a ^{3}He influx of 7 $cm^{-2} sec^{-1}$ corresponds to a proton influx of $7 \times 10^5 cm^{-2} sec^{-1}$. This is negligible compared with the rates of Jeans escape and volcanic evolution of hydrogen. The He^{++}-to-H^+ ratio in the solar wind is about 0.03, so the corresponding helium influx is $2 \times 10^4 cm^{-2} sec^{-1}$, a quantity far too small to influence the budget of ^{4}He. The evidence of the ^{3}He budget therefore severely constrains the possible contribution of the solar wind to atmospheric H or ^{4}He.

Let us now consider the possible contribution of interstellar gas. There is no reason to suppose that the ^{3}He-to-^{4}He ratio in the interstellar gas is different from the ratio in the solar system, so we can use the solar abundance to set a limit of $2 \times 10^4 cm^{-2} sec^{-1}$ on the source of ^{4}He that might be provided by the interstellar gas without violating constraints set by the ^{3}He budget. We can estimate the hydrogen influx from the interstellar gas directly without having to invoke ^{3}He data.

The density of interstellar hydrogen in the vicinity of the solar system is $0.1 cm^{-3}$ or less (Fahr, 1974). Its velocity relative to the sun is about 10 km sec^{-1}. The flux of interstellar hydrogen into the solar system is therefore about $10^5 cm^{-2} sec^{-1}$. The flux incident on the earth is modified by the gravitational field of the sun (Fahr, 1969) and significantly reduced as a result of photoionization of the hydrogen as it travels through the solar system (Blum and Fahr, 1970). We may take $10^5 cm^{-2} sec^{-1}$ as an upper limit on the flux of interstellar hydrogen into the atmosphere.

We find, therefore, that extraterrestrial sources of hydrogen and helium are not now important and probably never have been. Since these gases are relatively much more abundant in the universe than on earth, extraterrestrial sources of gases heavier than helium are probably even less important.

SUMMARY

The finding that extraterrestrial sources are negligible is consistent with our earlier conclusion that the atmosphere is secondary, having been released from the solid earth. We have explored the history of the release of gas, concluding that it has occurred throughout earth's history, probably at a diminishing rate, but that a significant imbalance between the rate of release and the rate of return of gas to the solid earth existed only in the first billion years or less. Accumulation of the atmosphere and ocean was therefore an early phenomenon.

We have found that water was and still is the most abundant constituent of the gases released, with carbon dioxide next in abundance. Reduced gases such as hydrogen and carbon monoxide make up only a small fraction of the gases released. One of the consequences of the inhomogeneous accretion theory for the origin of the earth is that this would have been equally true when the atmosphere was first accumulating.

We have now arrived at a description of the origin of the atmosphere that is consistent with many different pieces of evidence. In Chap. 6 we shall use the findings of this chapter in an attempt to deduce the properties of the atmosphere early in earth's history prior to the origin of life.

Chapter Six

The Atmosphere Before the Origin of Life

In this chapter and the next we shall draw on the material presented in earlier chapters to develop an account of the evolution of the atmosphere. This account will be extremely speculative, partly because there is very little evidence to constrain speculation about atmospheric evolution, and partly because many of the important ideas in this field are so new that they have not been rigorously examined. In spite of the general lack of evidence, there are aspects of atmospheric evolution that would be susceptible to quantitative theoretical investigation. The climatology and photochemistry of the primitive atmosphere are examples. While our account of atmospheric evolution may not be correct, it may furnish a framework for further study of the properties of primitive atmospheres.

In Chap. 2 and 3 we described the extent to which the composition of the modern atmosphere is controlled by biological processes acting both to produce and consume atmospheric gases. The situation must have been different before the origin of life on earth. In particular, volcanic emanations were probably the only important source of atmospheric gas. We shall therefore begin by discussing modern volcanic gases as a qualitative guide to the chemical species that may have been released to the primitive atmosphere.

The composition of the atmosphere did not directly reflect the composition of the volcanic gases, however, because of the influence of photochemical reactions, condensation and precipitation, and reactions with surface materials. We shall consider, first, the fate of the most abundant gases: water, carbon dioxide, and nitrogen. We shall find that surface temperatures, influenced by the greenhouse effect, may have evolved in such a way as to permit water to condense on Earth, but not on Venus. Mars may have been cold enough to cause both water and carbon dioxide to freeze upon the surface. The differences in surface temperatures, occasioned by the planets being at different distances from the sun, have led to very different histories of the atmospheres of Venus, Earth, and Mars. Most of this chapter, like most of this book, is devoted to Earth, but data from the other planets can provide constraints on our ideas concerning atmospheric evolution. We shall therefore discuss the evolution of the atmospheres of Venus and Mars at the end of the chapter.

Liquid water on the surface of the earth provided a medium in which carbon dioxide could dissolve and react with the rocks. Most of the earth's carbon dioxide therefore left the atmosphere to be incorporated in carbonate minerals. Nitrogen remained behind to become the most abundant atmospheric constituent.

Among the minor constituents, the most important were hydrogen and oxygen because of the influence of oxidation state on the synthesis of the organic molecules that were precursors of life. Jeans escape of hydrogen influenced the oxidation state of the atmosphere, and the rate of Jeans escape depends on the temperature structure of the atmosphere. We shall therefore discuss the temperature structure of the primitive atmosphere before estimating hydrogen and oxygen densities and their possible variations with time. We shall suggest that hydrogen constituted about 1% of the primitive atmosphere, its abundance being determined by balance between the rate of release of the gas by volcanoes and the rate of Jeans escape. Photochemical processes would have held the oxygen density very low indeed. With the passage of time, the hydrogen density would have declined and the oxygen density increased, either because of a decrease in the volcanic source of hydrogen or, after the origin of life, because of production of oxygen by green-plant photosynthesis.

The results we shall describe provide a crude indication of the composition of the prebiological atmosphere and of its development. These results summarized at the end of the section entitled "Tropospheric Photochemistry," should furnish a useful guide to the conditions under which life originated. In the final chapter we shall consider how the development of life may have affected the atmosphere.

COMPOSITION OF VOLCANIC GASES

We adopt the inhomogeneous accretion model of earth formation (see Chap. 5) and assume that conditions in the upper mantle have not changed greatly since the atmosphere began to develop. We can therefore use data on the composition of modern volcanic gases as a qualitative guide to the composition of primitive volcanic gases. The modern volcanic gases are probably derived largely from recycled sediments. This would also have been true of the volcanic gases released to the prebiological atmosphere, except during the initial period of accumulation of the atmosphere. We assume that composition is determined by chemical equilibrium with the magma from which the gases originate, and not on the previous history of the volatile constituents of the magma. For our purposes it does not matter whether the gases are recycled or are being released to the atmosphere for the first time.

TABLE 6-1. Composition of fumarolic gases other than water from Showa-shinzan, Japan, in parts per million by weight

| | Sample (temperature °C) | | |
	A1 (760°)	A3 (525°)	C3 (220°)
CO_2	29,200	25,800	13,000
CO	50	34	—
SO_2	1490	716	716
SO_3	21	11	2.7
H_2S	8.0	42	1080
S	3.7	1.8	—
Cl	728	420	433
F	238	159	35
Br	1.1	0.9	1.2
B	39	21	5.6
PO_4	2.8	3.0	0.8
NO_2	0.01	0.001	0.008
O_2	51	47	23
H_2	685	381	20
NH_3	1.3	0.8	17
N_2	567	676	1250
^{40}Ar	0.6	—	1.5
CH_4	1.5	18	—

After White and Waring (1963).

In Table 6-1 we show results of analyses of gases released from volcanic fumaroles in Japan (White and Waring, 1963). In these gases, as in almost all volcanic emanations, water vapor is by far the most abundant constituent. The constituents other than water that we wish to consider further are CO_2, CO, SO_2, SO_3, H_2S, NO_2, O_2, H_2, NH_3, N_2, ^{40}A, and CH_4.

We are confronted with the problem of estimating how much of the measured concentrations of these gases represent contamination of the samples by air. The oxygen almost certainly results from contamination, since free oxygen is not released from a magma (Holland, 1962, 1964; Fanale, 1971b). This provides a lower limit on contamination, since atmospheric oxygen may have been consumed by reaction with reduced constituents. We use the nitrogen concentration as an indication of the upper limit on contamination, since nitrogen is not likely to be consumed, and there are no abundant nitrogen compounds present in the samples that could represent the products of reaction of atmospheric nitrogen.

If we now compare the concentrations of the other gases in the sample relative to nitrogen with their concentrations in air (Table 1-2), we find that all of them are relatively more abundant in the fumarolic gas than in air, with the exception of oxygen and argon. We conclude that H_2O, CO_2,

CO, SO_2, H_2S, H_2, NH_3, and CH_4 are released from volcanoes. Nitrogen is almost certainly also released.

The significance of the very minor oxidized gases, SO_3 and NO_2, is uncertain, since these gases may have been produced by the reaction of contaminating oxygen. Although frequently present in volcanic gases today (White and Waring, 1963), we cannot be sure that they would have been present when the atmosphere lacked free oxygen.

Water and carbon dioxide are most abundant in volcanic emanations, as they are in the excess volatiles (see Chap. 5). We are interested in the nitrogen also, because of its dominance of the present atmosphere. The fact that volcanoes release much more nitrogen than ammonia suggests that ammonia has never been an abundant atmospheric constituent. The observed preponderance of nitrogen in volcanic gases is in accord with theoretical predictions of chemical equilibrium (Eugster and Munoz, 1966; Eugster, 1972). There is evidence, however, which we shall discuss below, that the primitive atmosphere contained traces of ammonia. Volcanoes may have been the source.

DEVELOPMENT OF THE ATMOSPHERIC GREENHOUSE

We assume that the atmosphere accumulated gradually on an initially cold and airless planet. When there was no atmosphere, the average surface temperature was determined by balance between the rate of absorption of solar radiation by the surface and the rate of emission of thermal radiation to space (cf. Goody and Walker, 1972). With adequate accuracy for our purposes, we may equate the rate of emission from the entire earth's surface to $4\pi R^2 \sigma T^4$, where R is the radius of the earth, σ is the Stefan–Boltzmann constant, and T is the average surface temperature. The total rate of absorption of solar energy is $\pi R^2 S(1-A)$, where $S = 1.4 \times 10^6 \, \text{erg cm}^{-2} \, \text{sec}^{-1}$ is the flux of solar energy incident on the earth, and A, the *albedo*, is the fraction of this energy that is reflected by the surface rather than being absorbed. We assume, for the time being, that S has not varied significantly over the age of the earth.

Equating absorption and emission rates we find, for the average surface temperature,

(6-1) $$T = [S(1-A)/4\sigma]^{\frac{1}{4}}$$

The present value of the Earth's albedo is 0.33, but this value is affected by the existence of oceans, vegetation, ice and snow, and clouds. Let us assume that the albedo of the Earth when it had no atmosphere was equal to the present albedo of Mars, $A = 0.17$. Then the original surface temperature was 260°K.

HEATING OF THE SURFACE BY ACCRETION

Let us consider whether solar heating or accretional heating determined surface temperatures on the primitive earth. The flux of solar radiant energy absorbed by the earth is about $10^6 \, \text{erg cm}^{-2} \, \text{sec}^{-1}$. This we equate to the rate of release of gravitational energy by infalling material, $Rg \, dm/dt$, where R cm is the radius of the earth, g cm sec^{-2} is the gravitational acceleration, and dm/dt gm cm^{-2} sec^{-1} is the rate of accretion per unit area. We find that $dm/dt = 1.6 \times 10^{-6}$ gm cm^{-2} sec^{-1} would be required to make accretional heating as large as solar heating. At this rate it would take only 10^6 years to accumulate a layer of rock 200 km thick. Evidently, accretional heating did not affect surface temperatures for a significant period of time.

Now suppose that an atmosphere composed largely of water vapor and carbon dioxide began to accumulate as a result of degassing. These molecules, principally the water vapor, would absorb some of the infrared radiation emitted by the ground. The atmosphere would reradiate this energy. Some of the atmospheric radiation would escape into space, but some of it would travel downwards to be absorbed by the ground. The additional energy would raise the average temperature of the ground.

This phenomenon, whereby the infrared opacity of the atmosphere leads to an increase in ground temperature, is known as the *greenhouse effect*. The theory of the greenhouse effect and of radiative transfer in a planetary atmosphere has been described by Goody (1964) and, in a simple approximation, by Goody and Walker (1972). The calculations for an evolving atmosphere that we are about to describe were performed by Rasool and DeBergh (1970).

The theory of the greenhouse effect relates the average ground temperature to the column density of water in the atmosphere, and thus to the partial pressure of water vapor. The influence of carbon dioxide may be neglected. Thus, for an evolving atmosphere we may plot the surface temperature against the partial pressure of water. Such a plot is shown in Fig. 6-1. As time passes and the atmosphere accumulates, we move across this figure from left to right. The surface temperature begins to increase when enough water vapor has accumulated to absorb a substantial fraction of the infrared radiation emitted by the ground. This condition is met when the vapor pressure of water is about 10^3 dyn cm^{-2}.

As Fig. 6-1 shows, the temperature–pressure curve for the earth intersects the saturated vapor pressure curve for water when about 10^4 dyn cm^{-2} of water vapor have accumulated. Further degassing does

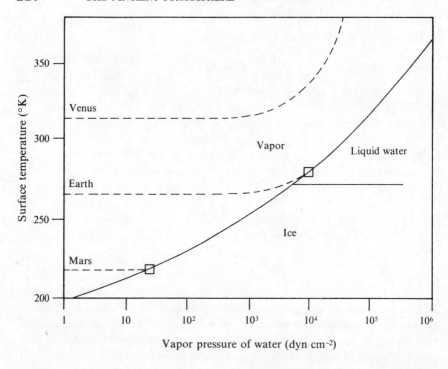

Fig. 6-1. The runaway greenhouse effect. The dashed curves show how surface temperatures increase, due to the greenhouse effect, as water vapor accumulates in the atmospheres of the inner planets. On Mars and Earth the increase is halted when the water vapor pressure is equal to the saturated vapor pressure (shown as the solid line), and freezing or condensation occurs. Temperatures are higher on Venus because Venus is closer to the sun, and saturation is never achieved: The temperature runs away. (From Goody and Walker, 1972. Copyright 1972 by Prentice-Hall, Inc., Englewood Cliffs, N.J. Used by permission of the publisher.)

not lead to an increase in the water vapor content of the atmosphere or to an increase in the surface temperature. The additional water condenses and oceans begin to accumulate. Carbon dioxide begins to dissolve in the oceans, and in due course the carbon dioxide content of the atmosphere is stabilized by carbon dioxide equilibria similar to those we discussed in Chap. 3.

We therefore assume that the water vapor and carbon dioxide contents of the primitive atmosphere were about equal to their present values after an initial accumulation period. Once water had started to condense and carbonate rocks to precipitate from the oceans, continued degassing led to an increase in the nitrogen content of the atmosphere, but caused no change in the carbon dioxide or water vapor contents.

According to this hypothesis, a similar process would have occurred on Mars. Mars is further from the sun than is the Earth, so would have started with a colder surface temperature. Frost would have begun to form after only about $20 \, \mathrm{dyn \, cm^{-2}}$ of water had accumulated in the atmosphere, as shown in Fig. 6-1, and no further increase in the water vapor content of the atmosphere or in the average surface temperature would have resulted from degassing. When about 5 mb of carbon dioxide had accumulated in the Martian atmosphere, this gas would have begun to freeze in the polar regions, forming the carbon dioxide polar caps that exist on Mars today.

The situation on Venus was quite different. Venus is closer to the sun than is the Earth, and would have started with a higher initial temperature. As shown in Fig. 6-1, its evolutionary track would never have intersected the saturated vapor pressure curve for water. The greenhouse effect was too strong. So oceans would not have condensed on Venus, and degassing may have led to a continually increasing atmosphere of water vapor and carbon dioxide and to an increasing greenhouse effect producing ever higher surface temperatures. This phenomenon is called the *runaway* greenhouse effect (Ingersoll, 1969).

We see that very different initial atmospheres could have resulted on Venus, Earth, and Mars from the release of similar mixtures of water, carbon dioxide, and nitrogen, simply because of the different distances of these planets from the sun. We shall defer discussion of the subsequent evolution of the atmospheres of Venus and Mars to the end of the chapter. Now let us consider how this description of the evolution of surface temperature must be modified to allow for possible changes in the luminosity of the sun.

EFFECT OF VARIABLE SOLAR LUMINOSITY ON THE EVOLUTION OF SURFACE TEMPERATURE

In the preceding section we assumed, for simplicity, that the luminosity of the sun, and thus the flux of solar energy incident on the earth, has not varied over the lifetime of the earth. This assumption is probably not correct.

We shall consider first the more conventional model of stellar evolution, in which the constant of gravitation is invariant and the solar luminosity increases with time. The magnitude of the increase depends on the particular model and is subject to considerable uncertainty. We shall follow Sagan and Mullen (1972) in taking 30% as a reasonable estimate of the increase in solar luminosity since the earth was formed.

If we reduce the value of S in Eq. (6-1) from $1.4 \times 10^6 \, \mathrm{erg \, cm^{-2} \, sec^{-1}}$ to $1.0 \times 10^6 \, \mathrm{erg \, cm^{-2} \, sec^{-1}}$, we obtain a revised estimate of the surface temperature of the initial airless earth of 238°K. The value is uncertain because we do not know the albedo of the airless earth. We have used a value of 0.17, which is the albedo of Mars, but Rasool and DeBergh (1970) used a value of 0.07, which is close to the albedos of Mercury and the Moon. This reduction in the albedo increases the initial surface temperature to 245°K.

As Sagan and Mullen (1972; cf. Donn, Donn, and Valentine, 1965) have pointed out, a problem now arises as the atmosphere starts to accumulate. When the atmosphere becomes saturated with water vapor, clouds will form, and ice or snow will collect on the ground. Both clouds and snow have high albedos, and their formation will markedly increase the albedo of the earth. Once the earth's atmosphere became saturated with water vapor, it is not likely that the albedo could have been lower than its present value of 0.33.

Sagan and Mullen (1972) have shown that the reduced solar luminosity combined with the increased terrestrial albedo would have led to average temperatures well below the freezing point of water. The greenhouse effect resulting from an atmosphere of water vapor and carbon dioxide would not have been large enough to support the presence of liquid water on the surface of the primitive earth. Results of their calculation of surface temperature are shown in Fig. 6-2. In this model all parameters are held constant except the solar luminosity, which increases with time at a constant rate.

The results show that surface water would have been frozen prior to about 2 billion years ago. Yet there is considerable geological evidence for the presence of liquid water on the surface at earlier times (cf. Sagan and Mullen, 1972). Most compelling, perhaps, is the fact that active life is not possible at these low temperatures (Stanier et al., 1970, p. 315), but fossil microorganisms of age greater than 2 billion years have been discovered.

We must bear in mind that the temperature we calculate from an expression like Eq. (6-1) represents a poorly defined average over both time and space. Some parts of the globe will always be warmer than this average, some parts will always be colder, and some parts will be warmer some of the time and colder the rest of the time. In principle, therefore, there could have been warm regions where water was liquid and life was active, even if the average temperature were below the freezing point. The model of local warm regions in a largely frozen planet is not very attractive, however, and we shall assume that the average surface temperature was above freezing throughout geological history.

Sagan and Mullen (1972) have investigated the possibility that the greenhouse effect was augmented by the presence of additional gases in the primitive atmosphere. The greenhouse effect produced by a carbon dioxide and water vapor atmosphere is limited because neither gas

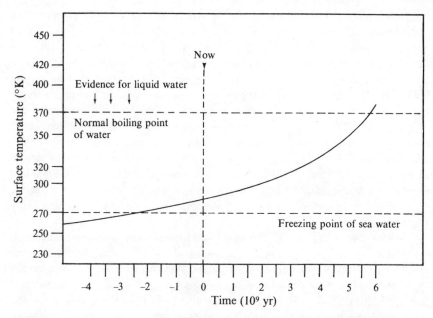

Fig. 6-2. Evolution of the surface temperature of the earth assuming that the solar luminosity has increased, at a constant rate, by 30% since the origin of the solar system. The albedo has been taken to be 0.35, the emissivity of the surface to be 0.9, and atmospheric composition has been assumed invariant and equal to that of the present day. (After Sagan and Mullen, 1972.)

absorbs significantly at wavelengths between 8 and 13 μ ($1\mu = 10^{-4}$ cm), so infrared radiation emitted by the ground at these wavelengths can escape directly to space. What is needed to close this window in the absorption spectrum is a gas that absorbs strongly between 8 and 13 μ.

Sagan and Mullen (1972) have found that CO, SO_2, O_3, NO_2, NO, CH_4, and H_2S are all ineffective in this regard. On the other hand, NH_3 absorbs strongly in this atmospheric window, and would serve to keep surface temperatures above the freezing point of water if it were present in the primitive atmosphere with a mixing ratio of only a few parts per million.

A theoretical investigation of the abundance of ammonia in the primitive atmosphere would be most valuable. At present we can only discuss the problem in general terms. Ammonia in the primitive atmosphere would have been destroyed rapidly by photolysis (Ferris and Nicodem, 1972) and by reaction with OH (McConnell, 1973). We shall present, below, an estimate of the source of ammonia required to maintain a mixing ratio of a few parts per million.

Once life had arisen there should have been little difficulty in maintaining atmospheric ammonia at a sufficient concentration. Fermenting microorganisms in an anaerobic world would have produced ammonia in abundance (Stanier et al., 1970, p. 192). One possibility, therefore, is that the earth was initially frozen and that it warmed up only after life had developed in the warmer equatorial regions and had begun to evolve ammonia. Bada and Miller (1968), however, have argued that ammonia concentrations in the range suggested by Sagan and Mullen would have been necessary for the inorganic synthesis of amino acids, prior to the origin of life. They have further argued that sufficient ammonia could be expected on the basis of chemical equilibrium in the atmosphere–ocean–sediment system. There may therefore have been a photochemical source of ammonia. A third possibility is that volcanoes released ammonia fast enough to maintain the required concentration. We shall examine some of these suggestions below, in the section on Tropospheric Photochemistry.

Alternatively, however, the solar luminosity may not have increased with time. According to some cosmological theories, the universal constant of gravitation is decreasing. If these theories are correct, it appears that the solar luminosity will have decreased also (Pochoda and Schwarzschild, 1964; Roeder and Demarque, 1966). In this case the problem is to avoid a runaway greenhouse effect on earth in order to preserve liquid water at the surface. The runaway greenhouse theory is subject to a number of uncertainties, but it appears that a factor-of-two increase in the solar flux would cause the oceans to evaporate (Ingersoll, 1969). The predicted solar evolution depends on poorly determined cosmological parameters, and is subject to other uncertainties as well. Most of the models evaluated by Roeder and Demarque (1966) predict solar fluxes incident on earth 4.5 billion years ago that were less than twice as large as the present value.

In order to progress we must decide, at least tentatively, whether the primitive atmosphere was hotter, colder, or the same temperature as the present atmosphere. There is some basis for excluding the higher temperature option. Miller and Orgel (1974) argue that a cold ocean would have been more favorable for the origin of life than a hot one. Tentatively, therefore, we conclude that solar luminosity is increasing and that the primitive earth would have been quite cold had there not been ammonia in the atmosphere. Whether there was enough ammonia to yield temperatures like those of the present day, or whether it was colder, we do not know. We shall arbitrarily make the convenient assumption that the average surface temperature was the same before the origin of life as it is today. In the next section we will present a qualitative description of the variation of temperature with height in the primitive atmosphere. The temperature profile requires consideration because of its influence on the rate of escape of hydrogen.

TEMPERATURE STRUCTURE OF THE PRIMITIVE ATMOSPHERE

After the initial period of accumulation, the primitive atmosphere would have contained about as much nitrogen, water, and carbon dioxide as the modern atmosphere. Most of earth's nitrogen is in the atmosphere today (see Chap. 3) and presumably would also have been before the origin of life. The water vapor content of the atmosphere depends on the saturated vapor pressure at the average surface temperature. Balance between silicate weathering and silicate reconstitution should have maintained the carbon dioxide partial pressure at a level not very different from its present value. The most striking feature of the prebiological atmosphere, therefore, was the absence of oxygen, which is almost entirely a product of green-plant photosynthesis (see Chap. 3).

Let us now consider temperatures in an atmosphere much like our present one, but lacking oxygen. (Reference may be made to the description of thermal structure in Chap. 1.) Oxygen has no direct effect on temperatures in the troposphere and lower stratosphere because it neither absorbs nor emits radiation in the far infrared. Temperatures in the troposphere depend on absorption and emission of infrared radiation by water vapor and on upward transport of heat by atmospheric motions. We therefore expect that temperatures would have declined at approximately the adiabatic rate (cf. Goody and Walker, 1972) from the ground to the tropopause. In the lower stratosphere they would have been approximately constant.

In the present-day atmosphere the temperature maximum at the stratopause results from the absorption by ozone of solar near-ultraviolet radiation. Heat is removed from the upper stratosphere and mesosphere by emission of infrared radiation, principally by carbon dioxide and ozone. Water vapor plays a minor role at these heights because it is confined to the troposphere by the tropopause cold trap. In an atmosphere containing little oxygen, there would have been little ozone and therefore no temperature maximum resulting from absorption by ozone.

On the other hand, McGovern (1969) has suggested that ammonia in quantities as small as 1 ppm could absorb significant solar energy in the stratosphere, serving to warm the stratopause as ozone does today. This possibility has not been quantitatively examined, so we have little choice but to ignore it. We shall therefore assume that the temperature was more or less constant between the stratopause at a height of about 15 km and the mesopause at a height of about 90 km. Mars and Venus, which lack significant amounts of ozone, appear to have such isothermal middle atmospheres today (McElroy, 1969b; cf. Ingersoll and Leovy, 1971; Hunten and Goody, 1969).

For our purposes, the temperature profile in the middle atmosphere is important mainly for its influence on eddy mixing coefficients and the upward transport of hydrogen from the tropopause to the exosphere. An isothermal middle atmosphere would still have been stable against convection, but less stable than the present-day upper stratosphere with its positive temperature gradient. We estimate that average eddy mixing coefficients in the middle atmosphere would have been somewhat higher than they are today, but not a lot higher.

The present-day thermosphere is heated by the absorption of extreme-ultraviolet solar radiation by molecular oxygen and nitrogen and atomic oxygen. It is cooled by conduction of heat downward into the mesosphere. In the absence of oxygen, an effective heat source would still have been provided by absorption by nitrogen. The temperature of the exosphere would have been different, but probably not sufficiently different to alter the rate of Jeans escape of hydrogen. The escape flux would still have been approximately equal to the limiting flux for diffusion through the bottom of the diffusive region. This limiting flux is not very sensitive to temperature, as we showed in Chap. 4.

In summary, we find that the primitive atmosphere may have had a hot thermosphere, a more or less isothermal mesosphere and stratosphere, and a troposphere much like the present one. The temperature and height of the tropopause are important because they control the mixing ratio of water vapor in the upper atmosphere and thus the rate of destruction of water by photolysis followed by escape of hydrogen (see Chap. 4). Unfortunately, we can say little definite about the tropopause in the absence of detailed theoretical study. In the simplest theory (Goody and Walker, 1972), the tropopause temperature depends on the incident flux of solar energy and on the albedo. It may therefore have been lower when the sun was less luminous; but it may equally well have been higher as a result of direct absorption of solar ultraviolet energy by ammonia. The height of the tropopause depends on the surface temperature and the tropopause temperature because the two are connected by an adiabat (see Chap. 1). In the discussion of the oxidation state of the atmosphere that follows, we shall assume that the water vapor mixing ratio in the upper atmosphere had approximately its present-day value, but we must remember that it might have been either higher or lower.

HYDROGEN AND OXYGEN IN THE PRIMITIVE ATMOSPHERE

Conflicting views have been presented on the oxygen content of the primitive atmosphere. Berkner and Marshall (1964, 1965, 1966, 1967) considered the oxygen source provided by photodissociation of water vapor followed by Jeans escape of hydrogen. They argued that the oxygen partial pressure would have been about 10^{-3} of its present level, just large enough to shield tropospheric water vapor from photodissociation. Brinkmann (1969) added to this model a simple, first-order process describing the consumption of oxygen by surface rocks. He concluded that photodissociation of water vapor followed by Jeans escape of hydrogen could have maintained oxygen levels as great as 0.25 of the present level even in the absence of photosynthesis. Both studies erred in assuming that the rate of escape of hydrogen is equal to the rate of photolysis of water. We showed in Chap. 4 that such is not the case. Here we shall suggest that the oxygen level in the primitive atmosphere was very much lower than either of these estimates.

Let us consider the balance in the primitive atmosphere between rates of supply and rates of loss for both oxygen and hydrogen (for convenience we call all reduced gases hydrogen). We assume that escape to space was a sink for hydrogen but not for oxygen (Fig. 6-3). We assume that there

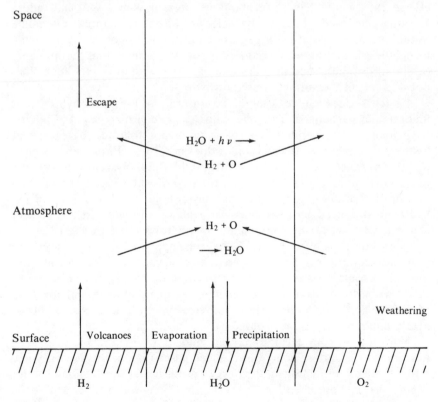

Fig. 6-3. Oxidation-reduction balance in the prebiological atmosphere. The reactions are schematic.

was a surface source of reduced gases (mostly hydrogen) provided by volcanoes, but no surface sink. We also assume that weathering reactions furnished a surface sink for oxygen, but that there was no surface source. Finally, we assume that photochemical reactions between reduced gases and oxygen were more rapid than any of these sources and sinks.

Our approach will be to consider how the atmosphere might have responded to changes in the surface source of hydrogen. For a start, let us consider the response of the rate of escape of hydrogen to space (Fig. 6-4a).

If the surface source of reduced gases was very small, there would have been little hydrogen in the atmosphere, the dominant hydrogen compound in the upper atmosphere would have been water vapor, and most escaping hydrogen would have been furnished by photolysis of water vapor. As demonstrated in Chap. 4, the rate of escape of hydrogen is controlled by upward transport and is proportional to the mixing ratio of hydrogen and its compounds above the tropopause. If the stratosphere was approximately as moist then as it is now, we expect that the hydrogen escape rate would have had about its present value. We shall adopt 10^8 atoms cm^{-2} sec^{-1} (about 10^{10} moles yr^{-1}) as a very approximate guide to the rate of escape of hydrogen from a primitive atmosphere containing negligible concentrations of reduced gases. We assume that the humidity of the stratosphere, and thus the rate of destruction of water vapor, does not depend on the surface source of hydrogen.

As the surface source is increased, the rate of escape of hydrogen at first remains unchanged. The reduced gases are consumed by atmospheric oxygen produced by photolysis of water vapor followed by escape of hydrogen. (Photolysis of water vapor does not, by itself, provide a source of oxygen. Most water molecules disrupted by photolysis are restored by recombination reactions. Overall destruction of water vapor and production of oxygen occurs only when photolysis is followed by escape of the hydrogen produced by photolysis. Destruction of water, in the present discussion, refers to the combination of photolysis and escape.) Further increase, however, overwhelms the oxygen supply, and the rate of escape of hydrogen must increase to maintain a balance between production and loss. The escape rate begins to increase when the mixing ratio of reduced gases in the lower stratosphere becomes comparable to the mixing ratio of water. For very large surface sources, the hydrogen escape rate is nearly equal to the surface source.

Now let us consider the oxygen budget of the atmosphere as a function of the surface source of reduced gases (Fig. 6-4b). For very low rates of release of hydrogen, the oxygen produced by destruction of water vapor must be consumed by weathering reactions with surface materials. As the source of hydrogen is increased it consumes more and more of the oxygen, and the surface sink declines. The decline is rapid when the

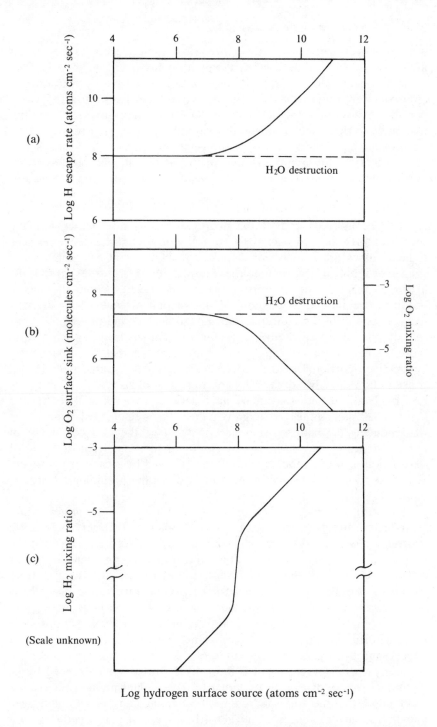

Fig. 6-4. Schematic representation of the variation of the hydrogen escape rate, the oxygen surface sink, and the hydrogen mixing ratio with the surface source of hydrogen.

source of hydrogen exceeds the source of oxygen because oxygen reacts more readily with reduced gases than with rocks.

If we assume that the rate of consumption of oxygen in weathering is approximately proportional, in the primitive atmosphere, to the oxygen partial pressure, then the behavior of the surface sink provides a qualitative guide to the behavior of the oxygen mixing ratio. A very speculative scale of mixing ratio is shown on the right of Fig. 6-4b. It assumes that the weathering rate is proportional to the oxygen mixing ratio and that the constant of proportionality may be deduced from the modern weathering rate (see Chap. 3). There are many reasons why neither of these assumptions is likely to be correct.

Now let us consider how the hydrogen mixing ratio varies with the surface source of hydrogen (Fig. 6-4c). For low surface sources, the reduced gases are rapidly consumed in reactions with atmospheric oxygen. For simplicity, we assume that the reaction is first order, so that the hydrogen mixing ratio is proportional to the source. When the source becomes comparable to the rate of destruction of water, however, the oxygen content of the atmosphere begins to decline and the photochemical lifetime of hydrogen becomes longer. The hydrogen mixing ratio therefore increases more rapidly than the surface source. The mixing ratio becomes proportional to the source again when the source is large and consumption of reduced gases by reaction with oxygen is negligible.

The details of the behavior we have described are speculative, but the main conclusion is fairly strong. Whether the prebiological atmosphere was reducing or oxidizing depends on whether the rate of release of reduced gases from volcanoes was greater than or less than the rate of destruction of water vapor by photolysis followed by escape of hydrogen. We must try to decide which was the case for the prebiological atmosphere.

The strongest evidence is provided by conditions for the origin of life. A reducing atmosphere is required (cf. Miller and Orgel, 1974). We conclude, therefore, that the volcanic source of reduced gas exceeded the rate of destruction of water. This conclusion is consistent with our estimate in Chap. 5, that the present rate of release of hydrogen from volcanoes is approximately equal to the present rate of escape of hydrogen. It seems likely that the volcanic source of gases was larger in the past.

It is convenient to consider the transition from a reducing atmosphere to an oxidizing one in terms of the balance between sources and sinks of oxygen and hydrogen. This transition could have resulted from a decline in the hydrogen source, but it is more likely to have been caused by an additional source of oxygen provided by green-plant photosynthesis.

Suppose that the atmosphere initially contained an excess of hydrogen and that a surface source of oxygen (the excess of photosynthesis over respiration and decay) was added to the system; this surface source of

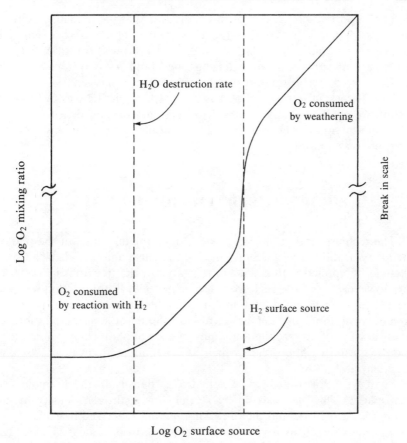

Fig. 6-5. Schematic representation of the dependence of oxygen mixing ratio on the surface source of oxygen (photosynthesis minus respiration and decay).

oxygen would gradually have increased in magnitude while the surface source of hydrogen remained constant (Fig. 6-5). At first the extra oxygen would have been readily consumed by photochemical reactions with the volcanic hydrogen. Assuming a first-order loss process, the oxygen density would have been proportional to the total oxygen source (destruction of water plus surface source). When the surface source of oxygen was comparable to the surface source of hydrogen, however, the hydrogen density would have decreased and the photochemical lifetime of oxygen would have increased. The oxygen density would therefore have increased more rapidly that the surface source, and weathering would have become increasingly important as an oxygen sink. Eventually, when the oxygen source was much greater than the hydrogen source, oxygen would have been consumed largely by weathering, as it is today.

The important conclusion of this analysis is that the oxidation state of the atmosphere is a sensitive function of the difference between the oxygen source and the hydrogen source. Thus, a rapid transition from a reducing atmosphere to an oxidizing one could have resulted from a gradual increase with time in the surface source of oxygen.

Our discussion in this section has been largely qualitative. In the next section we shall set limits on the hydrogen and oxygen densities in the prebiological atmosphere and then discuss briefly the sources and sinks of atmospheric ammonia.

TROPOSPHERIC PHOTOCHEMISTRY

The techniques that have been developed for the study of the photochemistry of the modern atmospheres of the earth and planets (see Chap. 2) could be applied with profit to the study of primitive atmospheres. As yet, however, very little such work has been done. In this section we shall use photochemical considerations to estimate the oxygen density in the prebiological atmosphere. Our calculation should be considered more as an illustration of the problems and possibilities of photochemical investigation of primitive atmospheres than as a definitive study of the oxygen abundance.

First, we need to estimate the hydrogen density in the prebiological atmosphere. This we shall do by relating the hydrogen density to the hydrogen escape rate and then invoking the limit, derived in Chap. 5, on the total amount of hydrogen lost by the earth over geological time. We assume that the volcanic source of hydrogen was considerably larger than the oxygen source provided by photolysis of water and escape of hydrogen. The volcanic source was therefore balanced by escape, and the hydrogen mixing ratio achieved the value that would just provide for upward transport and escape of hydrogen at the required rate. In Chap. 4 we showed that the hydrogen escape flux, under a wide range of conditions, is given approximately by $2 \times 10^{13} f$ atoms cm^{-2} sec^{-1}, where f is the ratio of hydrogen atoms (including combined hydrogen) to all other molecules in the lower stratosphere. We can use this expression to calculate the hydrogen mixing ratio corresponding to a given source of hydrogen.

We limit the hydrogen source according to considerations presented in Chap. 5 of the level of oxidation of the surface layers of the earth. We found that escape over the age of the earth has removed less than one-third of the hydrogen content of the ocean (less than 5×10^{22} gm H$_2$). This value sets an upper limit on the duration of any rate of release of hydrogen that we care to assume. Representative values are shown in

TABLE 6-2.   Hydrogen concentrations and lifetimes of prebiological atmospheres corresponding to different rates of release of hydrogen from volcanoes

HYDROGEN ABUNDANCE (percent H_2)	VOLCANIC SOURCE (molecules cm^{-2} sec^{-1}) (gm yr^{-1})		LIFETIME (10^6 yrs)
10	1.8×10^{12}	10^{15}	53
1	2.0×10^{11}	10^{14}	480
0.1	2.0×10^{10}	10^{13}	4800

From Walker (1976a). Copyright 1976 by John Wiley and Sons, Inc., New York. Used by permission of the publisher.

Table 6-2. We see that a prebiological atmosphere containing 10% hydrogen could have been maintained by a volcanic source of 10^{15} gm yr^{-1}, but that such an atmosphere could not have survived for more than 53 million years without violating the constraint on the total hydrogen lost from the earth. An atmosphere containing 1% hydrogen, on the other hand, could have been maintained for a geologically significant period of time, and an atmosphere with 0.1% hydrogen could have survived indefinitely. Let us select the atmosphere with a hydrogen mixing ratio of 10^{-3} for further discussion. Even this low value requires a volcanic source of hydrogen that is approximately 100 times as large as the present-day source.

We shall now argue that the oxygen concentration in this atmosphere would have been vanishingly small. Oxygen was produced by photolysis of water vapor followed by escape of hydrogen. We assume that the oxygen source was 5×10^7 molecules cm^{-2} sec^{-1}. Most of the oxygen would have been produced in the photochemically active upper levels of the atmosphere and most of it was probably consumed in the upper levels also. We, however, are interested in the oxygen abundance in the troposphere, so let us assume that all of the oxygen was transported to the troposphere, where it was consumed by photochemical reactions. This should give us an upper limit on tropospheric oxygen. The reactions we consider are listed in Table 6-3.

TABLE 6-3.   Photochemical reactions in the prebiological troposphere

REACTION	RATE	REFERENCE
$H_2O + h\nu \rightarrow OH + H$	(see text)	
$OH + H_2 \rightarrow H_2O + H$	$6.8 \times 10^{-12} \exp(-2020/T)$	Greiner (1969)
$H + H + M \rightarrow H_2 + M$	$8.0 \times 10^{-33}(300/T)^{0.6}$	Ham et al. (1970)
$H + O_2 + M \rightarrow HO_2 + M$	$1.8 \times 10^{-32} \exp(340/T)$	Kurylo (1972)
$HO_2 + H \rightarrow OH + OH$	$10^{-10} \exp(-330/T)$	Schofield (1967)
$NH_3 + OH \rightarrow NH_2 + H_2O$	$10^{-11} \exp(-1270/T)$	McConnell (1973)

We assume that photochemical activity in the troposphere was initiated by photolysis of water vapor. A very rough estimate of the rate of this process in the primitive troposphere is $2.5 \times 10^{12} \, cm^{-2} \, sec^{-1}$. This is the flux, averaged over the globe, of solar photons with wavelengths between 1840 A and 2000 A. The cross section for absorption by water is negligibly small at wavelengths longer than 2000 A (Hudson, 1974). Photons with wavelengths shorter than 1840 A would be absorbed at heights above the troposphere in an atmosphere like the modern one, but lacking oxygen and ozone. If either ammonia or sulfur dioxide achieved mixing ratios of 10^{-7} or greater in the primitive stratosphere, however, they would have shielded the troposphere from ultraviolet radiation in the neighbourhood of 2000 A. We may, therefore, be overestimating the rate of photochemical processes in the primitive troposphere.

In the atmosphere we have described, the photodissociation of water—

$$(6\text{-}2) \qquad\qquad H_2O + h\nu \rightarrow OH + H$$

—will be followed most frequently by

$$(6\text{-}3) \qquad\qquad OH + H_2 \rightarrow H_2O + H$$

and

$$(6\text{-}4) \qquad\qquad H + H + M \rightarrow H_2 + M$$

where M is any atmospheric molecule or atom. The closed cycle of reactions is illustrated in Fig. 6-6. For the rough photochemical calculations that follow, we shall ignore the temperature dependence of the rate coefficients, using a temperature of 288°K throughout, and we shall assume that the mixing ratios of all constituents are independent of altitude. Neither of these approximations is accurate, but they will suffice for the rough estimation of surface densities.

We take $[M] = [N_2] = 2 \times 10^{19} \, cm^{-3}$ and $[H_2] = 2 \times 10^{16} \, cm^{-3}$. Equilibrium between these three reactions, with a photolysis rate of $2.5 \times 10^{12} \, cm^{-2} \, sec^{-1}$, gives $[OH] = 4.8 \times 10^4 \, cm^{-3}$ and $[H] = 7.4 \times 10^9 \, cm^{-3}$ at the ground.

Oxygen would have been consumed in this atmosphere by

$$(6\text{-}5) \qquad\qquad H + O_2 + M \rightarrow HO_2 + M$$

followed by

$$(6\text{-}6) \qquad\qquad HO_2 + H \rightarrow OH + OH$$

and (twice), by reaction (6-3),

$$OH + H_2 \rightarrow H_2O + H$$

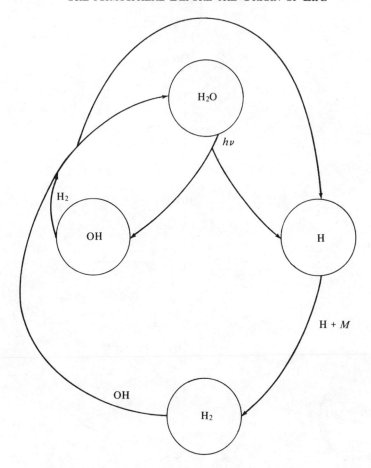

Fig. 6-6. Reactions in the primitive atmosphere initiated by photolysis of H_2O.

The reaction chain is illustrated in Fig. 6-7. Its overall effect is

$$(6\text{-}7) \qquad\qquad 2H_2 + O_2 \rightarrow 2H_2O$$

With $[H] = 7.4 \times 10^9 \, \text{cm}^{-3}$ and oxygen consumed at a rate of $5 \times 10^7 \, \text{cm}^{-2} \, \text{sec}^{-1}$, we calculate that in equilibrium $[O_2] = 2.1 \times 10^4 \, \text{cm}^{-3}$ and $[HO_2] = 5.1 \times 10^2 \, \text{cm}^{-3}$.

This calculation is subject to severe uncertainties, principally in the rate of photolysis of water vapor. It does, however, indicate that photochemical reactions could have maintained vanishingly low oxygen densities in the prebiological atmosphere, even if the atmosphere were only weakly reducing (a hydrogen mixing ratio of only 10^{-3}).

We can use the results of our photochemical calculation to investigate the ammonia budget of the atmosphere. Ammonia would have been

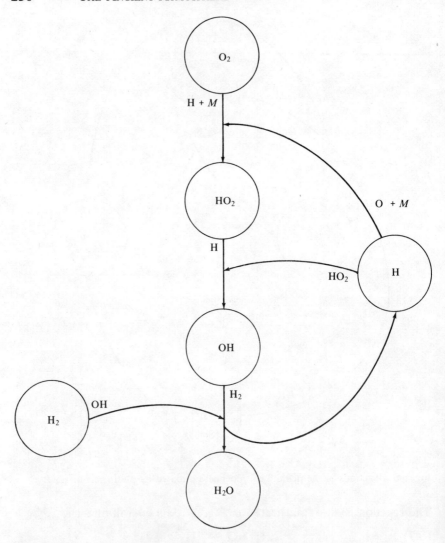

Fig. 6-7. Chain of reactions that destroys O_2 in the primitive atmosphere.

destroyed by reaction with OH—

(6-8) $$NH_3 + OH \rightarrow NH_2 + H_2O$$

—as well as by photolysis (Abelson, 1965, 1966; Ferris and Nicodem, 1972). Neglect of photolysis gives us an upper limit on the photochemical lifetime of ammonia. With $[OH] = 4.8 \times 10^4 \, \text{cm}^{-3}$, this upper limit is 1.7×10^8 sec, or about 5.5 yr. If the ammonia mixing ratio was 10^{-6}, as required to maintain surface temperatures above freezing, the height-integrated rate of destruction of ammonia would have been about

$10^{11}\, cm^{-2}\, sec^{-1}$. This is bigger than the volcanic source of hydrogen required to maintain the hydrogen mixing ratio at 10^{-3}. It does not seem likely, therefore, that volcanoes could have provided the ammonia in the prebiological atmosphere.

The problem in synthesizing ammonia photochemically in the primitive atmosphere would have been to break the nitrogen molecule into its constituent atoms. Atomic nitrogen, once produced, would quite probably have reacted with hydrogen in a weakly reducing atmosphere to produce ammonia. But photochemical disruption of the nitrogen molecule requires very energetic solar photons and would probably have occurred only in the mesosphere and thermosphere of the primitive atmosphere. It is likely that any ammonia synthesized at these high altitudes would have been destroyed by further photochemical reactions before it could be transported into the troposphere (cf. Strobel et al., 1970; Strobel, 1971; McConnell and McElroy, 1973).

The problem of synthesizing ammonia is very similar to the problem of synthesizing hydrocarbons as precursors of organic molecules. For hydrocarbon synthesis the difficult step is to break the carbon–oxygen bond in the carbon monoxide molecule. Once a carbon atom had been released, it is quite likely that it would have reacted with hydrogen, but it is possible that hydrocarbons synthesized high in the atmosphere would have been oxidized by reaction with hydroxyl radicals before they could be transported to the troposphere.

Until some quantitative photochemical theory has been applied to these problems, we can do little more than speculate. One possibility that has been suggested in connection with the origin of life (cf. Miller and Orgel, 1974) is that lightning flashes provided the energy necessary to disrupt nitrogen and carbon monoxide molecules in the troposphere. Whether lightning could have generated ammonia fast enough to maintain a mixing ratio of 10^{-6}, we do not know.

So far in this chapter, we have been concerned with deducing the properties of the atmosphere in which life evolved. Our very tentative findings are that tropospheric temperatures were more or less as they are today, possibly cooler, and that the nitrogen, water vapor, and carbon dioxide abundances were also approximately equal to their present values. The hydrogen mixing ratio probably did not exceed about 1%; the abundance of oxygen was negligible. There is indirect evidence for ammonia mixing ratios of about 10^{-6}, but the source of this ammonia has not been identified; volcanic emanations appear to be inadequate.

We shall not attempt to describe the processes by which life may have originated on earth. Much has been written on this subject (Oparin, 1938, 1961, 1972; Rutten, 1971; Gabel and Ponnamperuma, 1972; Miller and Orgel, 1974; Calvin, 1969, 1975). In Chap. 7 we shall consider how life and its evolution may have affected the atmosphere. Before taking up this subject, let us describe how atmospheric evolution on Venus and Mars—presumably unaffected by life—has differed from that on Earth.

VOLATILE ELEMENTS ON THE INNER PLANETS

According to the inhomogeneous accretion theory described in Chap. 5, volatile elements condensed from the solar nebula only when low temperatures had been achieved and more refractory elements had already accreted to form protoplanets. Accretional energy had already caused the protoplanets to melt and differentiate. The volatile elements were added to the planets as surface veneers, accreted at least in part after the dissipation of the nebula.

The theory therefore provides us with an approximate guide to the composition of the volatile elements on the different planets. All other things being equal, Venus and Mars would have been veneered by the same material that veneered the Earth. It provides no guide, however, to the amounts of volatile material accreted by each planet. These would have depended on the statistics of accretion. We shall have to turn to the atmospheres themselves for information on amounts. Let us first consider composition.

All other things were not equal, of course. Venus was closer to the sun than was Earth, and Mars was further away. It is likely, therefore, that the nebula was warmer at the orbit of Venus and colder at the orbit of Mars at the time that the nebula dispersed (Lewis, 1974). It is therefore possible that more volatiles had condensed near Mars and fewer had condensed near Venus. In particular, water may not yet have entered into hydrated silicate minerals near Venus, but may already have done so near Earth and Mars (Lewis, 1972). The probable existence of a temperature gradient therefore introduces an uncertainty into the composition of the volatiles. Keeping this uncertainty in mind, let us see what can be learned from the volatiles on Earth.

In Table 6-4 we present estimates by Turekian and Clark (1975) of the abundances of a number of volatiles in the surface layers of the Earth. These are not the same as the excess volatiles discussed in Chap. 5. Excess volatiles are the volatiles that have been released from the Earth to form atmosphere, hydrosphere, and sedimentary rocks. The values in Table 6-4 represent the total amounts of these elements accreted by the Earth.

Table 6-4 also shows how the terrestrial abundances, when normalized to carbon, compare with the abundances in carbonaceous chondrites. Carbonaceous chondrites are the best samples we have of the

TABLE 6-4. Abundances of low-temperature condensate elements in the earth and in carbonaceous chondrites

	Moles of Element					Ratio Atoms Element:Atoms C (including ^{20}Ne, ^{84}Kr, and ^{132}Xe only)			Earth Compared with Carbonaceous Chondrites $(a)/(b)$
	Hydro-sphere	Atmos-phere	Crust	Upper mantle	Total	Earth total (a)	Atmosphere, hydrosphere, and crust	Carbonaceous chondrites (b)	
H	183×10^{21}	→	→	→	183×10^{21}	8.1	24.4	7.8	1.0
C	→	→	7.5×10^{21}	15×10^{21}	22.5×10^{21}	1	1	1	1
N	→	2.84×10^{20}	3.4×10^{20}	4.2×10^{20}	7.6×10^{20}	3.3×10^{-2}	3.8×10^{-2}	8.1×10^{-2}	0.4
Ne	→	3.3×10^{15}	→	→	3.3×10^{15}	13×10^{-8}	40×10^{-8}	0.35×10^{-8}	37
^{40}Ar	→	1.7×10^{18}	0.1×10^{18}	→	1.8×10^{18}	—	—	—	—
^{36}Ar	→	5.7×10^{15}	→	→	5.7×10^{15}	25×10^{-8}	76×10^{-8}	1.27×10^{-8}	20
Kr	→	2.1×10^{14}	→	→	2.1×10^{14}	0.53×10^{-8}	1.6×10^{-8}	0.016×10^{-8}	33
Xe	→	0.16×10^{14}	1.5×10^{14}	→	1.7×10^{14}	0.20×10^{-8}	0.6×10^{-8}	0.013×10^{-8}	15
Pb	→	→	1.64×10^{18}	0.53×10^{18}	2.2×10^{18}	9.6×10^{-3}	17×10^{-3}	7.2×10^{-3}	1.3

After Turekian and Clark (1975).
Arrows indicate that small amounts of the element have been pooled into the larger adjacent reservoir.

material that contributed to the volatile-rich veneer (see Chap. 5). In order to facilitate the comparison, the last column of the table presents ratios of the terrestrial abundances, relative to carbon to the abundances in carbonaceous chondrites. Hydrogen, nitrogen, and lead are as abundant on Earth, relative to carbon, as in the carbonaceous chondrites, but the inert gases are enriched on Earth. If the carbonaceous chondrites are the remnants of the material that carried volatiles into the Earth, they have apparently lost inert gases during the life of the solar system. The Earth is probably a more reliable guide than the carbonaceous chondrites to the relative abundances of the volatiles to be expected on Venus and Mars. Let us suppose that Venus and Mars accreted volatile-rich material of about the same composition as Earth (possibly with less hydrogen as water of hydration on Venus) and consider the subsequent evolution of their atmospheres.

EVOLUTION OF THE ATMOSPHERE OF VENUS

We can expect the thermal histories and therefore the degassing histories to have been very similar on Venus and Earth because of the similarities in the masses and mean densities of the two planets (Hanks and Anderson, 1969). We therefore assume that the atmosphere accumulated rapidly, possibly during accretion of the surface layers. Even if Venus did not accrete water of hydration, the gases released to the primitive atmosphere would have contained as much water as carbon dioxide because both gases would have been produced by oxidation of the hydrocarbons contained in the volatile-rich material accreted by the planet (see Chap. 5). Today, however, there is very little water in the atmosphere of Venus (see Chap. 1).

We shall take up the fate of water on Venus below. But first, let us consider the present atmosphere (see Chap. 1). Compared to Earth, Venus has a massive atmosphere, with a surface pressure of 100 bars, and a high surface temperature of 750°K (cf. Hunten, 1971). These two facts are related. Because of its opacity in the far infrared, the massive atmosphere prevents the hot surface from cooling by emission of radiation to space. The high temperature, in turn, maintains the massive atmosphere by preventing volatile constituents from combining with material in the solid phase.

Two mechanisms may contribute to the high surface temperatures. One is the greenhouse effect, in which visible sunlight heats the surface of the planet (Pollack, 1969). The other is the general circulation of the atmosphere, resulting from horizontal variations in the rate of absorption of solar energy at higher levels of the atmosphere (Goody and Robinson, 1966). According to the circulation model, the high temperatures result from compression of air as it descends through the atmosphere. The relative importance of direct solar heating and temperature increase by compression is not known at present. It is, however, clear that circulation plays an important role in governing temperatures. Both the cloud tops (Goody, 1965) and the surface (Sinclair et al., 1970) are very nearly isothermal, providing evidence of the importance of horizontal motions. The temperature profile below the clouds is nearly adiabatic, providing evidence of the importance of vertical motions.

Reactions between gases and rocks are expected to be rapid at the high surface temperature of Venus (Palm, 1969; Mueller, 1970; Mueller and Kridelbaugh, 1973). It has therefore been suggested that the atmosphere is in chemical equilibrium with the surface (Mueller, 1964, 1969; Lewis, 1968, 1970, 1971). This may be true for some constituents, although there are difficulties with the equilibrium hypothesis (Walker, 1975b). It appears, however, that the temperature is too high to permit the major constituent, carbon dioxide, to react with any mineral that is likely to be present on the surface of Venus (Orville, 1974). It is therefore possible that all of the carbon dioxide that has ever been released from the interior of Venus is still in the atmosphere. The mass of the atmosphere may therefore be indicative of the total amount of degassing on the planet.

There are about 1.2×10^{22} moles of carbon dioxide in the atmosphere of Venus. This is somewhat greater than the amount of carbon in the crust of the Earth (Table 6-4), but less than the amount in the crust and upper mantle combined. Degassing may have released more volatiles on Venus than on Earth, but not by a great amount. If we assume that the upper mantle of Venus has retained about as much carbon dioxide as has been released to the atmosphere, we find that Venus and Earth accreted approximately equal amounts of volatiles (cf. Sagan, 1962).

If carbon dioxide has not reacted with the surface of Venus, it is possible to estimate the abundances of a number of undetected gases from the data of Table 6-4 on the abundances, relative to carbon, of terrestrial volatiles. The estimates of Turekian and Clark (1975) are given in Table 6-5. In making these estimates, Turekian and Clark have assumed that the upper mantle of Venus contains about as much carbon

TABLE 6-5. Predicted composition of the atmosphere of Venus

	CO_2	N_2	^{40}Ar	^{36}Ar	Ne	Kr	Xe
Percent by volume	98.12	1.86	2×10^{-2}	7.5×10^{-5}	4.3×10^{-5}	2.7×10^{-6}	2.0×10^{-6}

After Turekian and Clark (1975).

and nitrogen as the atmosphere, that the potassium content of the crust of Venus is comparable to that of the crust of the Earth, that the isotopic ratios of the inert gases are the same on Venus as on Earth, and that the high surface temperature prevents the adsorption of xenon. Measurement of the concentrations of some of these gases in the atmosphere of Venus will provide useful constraints on our understanding of atmospheric evolution.

There remains the problem of the loss of water from Venus, and this we shall now discuss. The idea that escape of hydrogen has been much more rapid from Venus than from the Earth is an old one (see references in Ingersoll and Leovy, 1971). We shall argue that more rapid escape follows quite plausibly from the fact that Venus is closer to the sun.

As we described earlier in this chapter, release to the atmosphere of water and carbon dioxide on Venus would probably have led to a runaway greenhouse effect, producing surface temperatures too high to permit the condensation of water (Sagan, 1960; Gold, 1964; Ingersoll, 1969; Rasool and DeBergh, 1970). It is possible that the runaway would have been deferred for a time while solar luminosity increased (Pollack, 1971), but it is equally possible that ammonia in the primitive atmosphere of Venus compensated for the presumed lower solar luminosity, as it may have done on Earth.

What can we say about the properties of the atmosphere after accumulation was complete and after runaway had occurred? We have already suggested that the partial pressure of carbon dioxide might have been the same then as it is now, 100 bars, corresponding to 1.2×10^{22} moles of carbon dioxide in the atmosphere. We expect that there would have been at least an equal number of moles of water vapor, contributing a partial pressure of 41 bars. Indeed there may have been as much water as there is in the terrestrial ocean, 10^{23} moles, contributing a partial pressure of 260 bars. We anticipate that the rate of loss of hydrogen from such an atmosphere would have been large unless the water vapor was confined to the lower atmosphere by condensation. We shall now argue that confinement does not occur in a sufficiently humid atmosphere (Ingersoll, 1969).

In a convecting atmosphere with abundant water vapor (a mixing ratio greater than 0.1), the temperature and the water vapor mixing ratio decrease very slowly with altitude. In such an atmosphere, there can be an unsaturated region near the ground where temperature decreases at the dry adiabatic lapse rate, and the water vapor mixing ratio is constant. Above the saturation level, the release of latent heat by condensing water holds the lapse rate to a very small value. The water vapor pressure decreases at approximately the rate given by the barometric law, and the water vapor mixing ratio declines slowly. Ingersoll has shown that the tropopause (where convection ceases) in such an atmosphere occurs at great heights and that the water vapor mixing ratio at the tropopause and above is almost the same as the mixing ratio near the ground. A cold trap, such as that which confines water to the terrestrial troposphere, would therefore not have existed in the primitive, wet atmosphere of Venus.

Water vapor, then, would have been a dominant component of the upper atmosphere of Venus as well as of the lower atmosphere. In the upper atmosphere water would have been subject to destruction by photolysis and the escape of hydrogen. If we assume that the rate of escape was limited by upward transport, as in the modern terrestrial atmosphere, we can use results derived in Chap. 4 to estimate the escape flux as

$$(6\text{-}9) \qquad\qquad F_e \sim 3 \times 10^{13} f/(1+f)$$

where f is twice the ratio of the water vapor number density to the carbon dioxide number density. This expression yields an escape flux of at least 2×10^{13} atoms $cm^{-2} sec^{-1}$, since the value of f was at least two.

Two limiting processes other than transport must be considered, however. First, hydrogen cannot escape faster than it is produced by photolysis of water vapor; the escape flux therefore cannot exceed the flux of solar photons able to dissociate water. This flux is about 10^{13} $cm^{-2} sec^{-1}$ (cf. Hunten, 1973). A more severe constraint is imposed by consideration of the effect on exospheric temperature of the energy carried away by escaping hydrogen atoms (McElroy, 1974). A hydrogen atom requires 0.6 eV of kinetic energy in order to escape from Venus's gravitational field. An escape flux of 10^{13} atoms $cm^{-2} sec^{-1}$ would therefore carry about 6×10^{12} eV $cm^{-2} sec^{-1}$ or 9.7 erg $cm^{-2} sec^{-1}$ out of the thermosphere. But the thermosphere is heated by extreme-ultraviolet radiation from the sun with a flux of only about 1 erg $cm^{-2} sec^{-1}$ (Hinteregger, 1970). In the absence of additional sources of heat, the energy carried away by escaping hydrogen atoms must have been less than this.

The problem deserves theoretical investigation. What probably happened was that exospheric temperatures were depressed to the point where Jeans escape rather than upward transport was the process that limited the rate of loss of hydrogen. This appears to be the case in the modern atmosphere of Venus, as we described in Chap. 4, but the exospheric temperature today is low because carbon dioxide is the principal atmospheric constituent, whereas in the primitive atmosphere it was low, we are suggesting, because of loss of kinetic energy to space.

Even with a low exospheric temperature the escape flux would have been large. It had to be large in order to carry away enough energy to depress the temperature. We shall take 10^{12} atoms $cm^{-2} sec^{-1}$ as a very rough estimate of the energy-limited escape flux. The flux would have been independent of the abundance of water vapor until the mixing ratio fell so low that transport became the limiting factor ($f \lesssim 0.03$).

At this rate, it would take only 30 million years to destroy 10^{22} moles of water and only 300 million years to destroy an amount of water equal to the mass of the terrestrial ocean. Even if the escape flux did not achieve the limit established above, it appears likely that most of Venus' water was lost in less than a billion years. The oxygen left behind by the escaping hydrogen was presumably consumed in reactions with the surface (Dayhoff et al., 1967). The rate at which the surface could have consumed oxygen would have been limited by the rate of exposure to the atmosphere of fresh, unweathered rock (Chap. 3). Substantial amounts of oxygen could have accumulated in the atmosphere during the period of dissipation of the water unless this rate was very large (Walker, 1975b).

In summary, it is possible that Venus and Earth accreted volatiles in approximately the same amounts, although Venus may have lacked water of hydration. The histories of their atmospheres diverged, however, and their present atmospheres are very different, because Venus is too close to the sun to permit water to condense on the surface. Water was rapidly destroyed, as a result, by photolysis followed by escape of hydrogen, the oxygen being consumed by reactions with surface material. In the presence of high surface temperatures and in the absence of liquid water, carbon dioxide remained in the atmosphere rather than entering the solid phase, as it has on Earth. The high surface temperature is maintained by the greenhouse effect, caused by the massive atmosphere, and also by atmospheric circulation.

EVOLUTION OF THE ATMOSPHERE OF MARS

The surface pressure on Mars is only 6 mb, less than 1% of the surface pressure on Earth (Chap. 1). The first thing we must try to understand is why there is so little atmosphere. One possibility is that Mars accreted very little volatile material during the late stages of the evolution of the solar system. This is not very plausible. Earth and Venus appear to have accreted roughly similar amounts, and condensed volatiles may have been relatively more abundant at the orbit of Mars because of lower temperatures. The smaller mass of Mars may have reduced the efficiency with which it captured volatile-rich material, but this factor seems unlikely to have made so large a difference to the volatile input.

A second possibility is that Mars accreted abundant volatiles, but has retained most of them within the solid planet. Because of its smaller mass, Mars may always have had a colder interior than Earth or Venus (Hanks and Anderson, 1969), and surface temperatures during accretion may also have been lower. There may therefore have been less degassing during accretion and less subsequent degassing as well. Fanale (1971b) has examined the problem. He favors extensive degassing during accretion on Mars as on Earth.

A third possibility is that Mars released abundant volatiles, as Earth and Venus did, but that these have largely been removed from the atmosphere, as has happened on Earth (Levine, 1976; Fanale, 1976; Owen, 1976). This possibility we want to explore further (see Fig. 6-8).

First, let us note that Mars is unique among the inner planets in that surface temperatures are sufficiently low to cause the major atmospheric

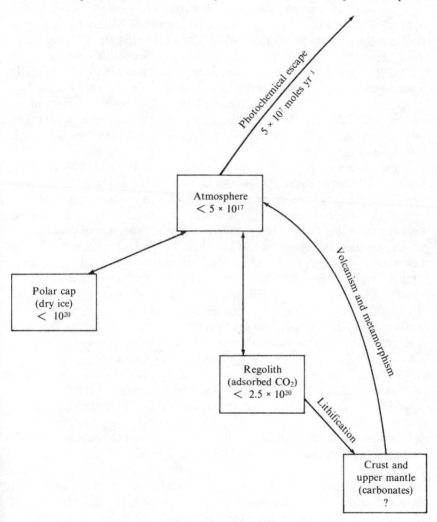

Fig. 6-8. Reservoirs of carbon dioxide on Mars (in moles). Similar reservoirs and processes affect water vapor in the Martian atmosphere.

constituent, carbon dioxide, to condense (Leighton and Murray, 1966). This fact has a major influence on Martian climate and on changes in the atmosphere with time. We shall describe some of these climatic phenomena below. For the time being, we must remember that there may be substantial amounts of carbon dioxide frozen in the polar caps. The largest estimate of this amount indicates enough carbon dioxide to provide a surface pressure of 1 bar (Sagan et al., 1973), but other estimates are much lower (Murray and Malin, 1973; Ingersoll, 1974).

The low temperatures on Mars also make possible the adsorption of large amounts of carbon dioxide and water onto the surfaces of soil particles (Fanale and Cannon, 1971b). The amount of these gases contained in the *regolith* (soil) depends on its depth. If this depth is 1 km, which seems like a reasonable upper limit, the regolith could contain about 1% as much water as the terrestrial ocean and about 5% as much carbon dioxide as terrestrial carbonate rocks (Fanale and Cannon, 1974). Evidently the polar caps and the regolith could contain much more volatile material than the atmosphere, but there must have been other sinks for atmospheric gases if Mars degassed anything like as much material as Earth.

An obvious sink for carbon dioxide and water is provided by reaction with the rocks (Huguenin, 1976a,b). Most of the carbon dioxide released by degassing on Earth has entered into the solid phase of the planet in the form of carbonate minerals. This tendency should have been stronger on Mars because the surface temperature is lower. The partial pressure of carbon dioxide is higher on Mars than on Earth, however, so the atmosphere of Mars cannot be in equilibrium with the solid phase. It is possible that reactions between gases and rocks are too slow on Mars to permit the attainment of equilibrium, or it is possible that tectonic activity and erosion have not exposed enough fresh rock to the atmosphere to consume all of the carbon dioxide. Nevertheless, Mars may have degassed as much carbon dioxide as Earth; most of the carbon dioxide may now be in the crust of Mars, in the form of carbonate minerals.

The fate of water may have been similar. Most terrestrial water is on the surface (Turekian and Clark, 1975) because temperatures below the surface are too high to permit the accumulation of large deposits of hydrated silicate minerals. The crust of Mars may have absorbed much more water than the crust of the Earth because of lower Martian temperatures (Pollack et al., 1970a,b; Houck et al., 1973).

The loss of atmospheric gases to space by Jeans escape and by photochemical escape is another sink that must be considered. We discussed these processes in Chap. 4 and derived rates of destruction of water and of carbon dioxide. We also noted that these rates were not likely to have varied with time, provided that carbon dioxide has always been the dominant atmospheric constituent and the flux of solar ionizing radiation has been constant. Assuming constant rates of destruction, we deduced that Mars has lost about 10^{25} molecules cm^{-2} of water and about 10^{23} molecules cm^{-2} of carbon dioxide over the lifetime of the planet. There are at present only 5×10^{19} molecules cm^{-2} of water in the atmosphere of Mars and about 2×10^{23} molecules cm^{-2} of carbon dioxide. For comparison, there are about 10^{28} molecules cm^{-2} of water in the terrestrial hydrosphere. Evidently, loss of gas to space has been significant if there has been very little degassing on Mars, but not if there has been about as much on Mars as on Earth.

Although there is currently very little nitrogen in the atmosphere of Mars (Dalgarno and McElroy, 1970; Nier et al., 1976; Owen and Biemann, 1976), we cannot use this fact to set a limit on the total amount of degassing. As we showed in Chap. 4, photochemical escape of nitrogen is rapid. The escape flux is proportional to the nitrogen mixing ratio as long as carbon dioxide is the dominant atmospheric constituent; the residence time is less than 600 million years. Mars may therefore have lost as much as 1 bar of nitrogen in 4.5 billion years.

This brings us to inert gases as possible indicators of the amount of degassing on Mars (Owen, 1974; Levine and Riegler, 1974). Inert gas concentrations have recently been measured on Mars by the Viking spacecraft (Owen and Biemann, 1976; Biemann et al., 1976), but questions concerning possible adsorption of inert gases by the low temperature regolith have not yet been resolved. Taken at face value, the Viking data seem to indicate that Mars has degassed less, relative to its mass, than have Earth and Venus, but that the amount of degassing greatly exceeds the mass of the modern atmosphere. Most of the water and carbon dioxide released from the interior of Mars may have been consumed by reaction with the surface, possibly during an early period of rapid accretion and degassing. Most of the nitrogen may have been lost to space as a result of the low value of gravitational acceleration on Mars. There may, however, have been a time in the past when the atmosphere was predominantly nitrogen. The present abundance of carbon dioxide and water in the atmosphere, polar cap, and regolith combined may be determined by a kinetic balance between the rate of consumption of volatiles in reactions with fresh rock and the rate of return of these volatiles to the atmosphere by volcanoes.

Let us now consider the partitioning of the volatiles between the atmosphere, polar cap, and regolith. We consider the regolith first. The total amount of gas it contains depends on its depth and on its average temperature. The depth is unknown, but the temperature can be estimated (Fanale and Cannon, 1974). It is also possible to estimate how much gas can be exchanged between atmosphere and regolith during periodic temperature changes (diurnal, seasonal, or longer term). This depends on the depth to which the periodic temperature fluctuations penetrate and the ease with which gaseous volatiles diffuse through the porous material of the regolith. Obviously more exchange is possible the longer the period of the temperature variation. Fanale and Cannon find that exchange of gas between atmosphere and regolith affects the mass of the atmosphere for long-term variations in temperature (to be discussed below). Exchange may not be important on time scales as short as a day or a year.

Now let us consider the polar cap. Carbon dioxide at a partial pressure of 6 mb freezes at a temperature of 145°K. No point on the surface of Mars can have a temperature lower than this, or the atmosphere would freeze. During the winter, the polar region radiates to space more energy than it receives from the sun. Its temperature falls to 145°K, and carbon dioxide begins to condense. Latent heat of condensation prevents further fall in temperature. The polar cap grows as fast as the latent heat is radiated into space. During the summer, the situation is reversed. The polar cap absorbs more solar energy than it radiates; the additional heat causes carbon dioxide to evaporate and the polar cap to shrink. Growth and decay of the polar caps may cause a seasonal change of as much as 20% in the mass of the Martian atmosphere (Leighton and Murray, 1966). Water vapor also tends to freeze in the polar winter region. Seasonal variations of the water content of the Martian atmosphere are therefore expected, and water is indeed observed to be most abundant in the summer hemisphere (Schorn et al., 1969).

The northern polar cap does not disappear, even at the height of the northern summer (cf. Goody and Walker, 1972). Let us assume, for purposes of discussion, that the permanent cap is carbon dioxide (Leighton and Murray, 1966) although this has been contested (Ingersoll, 1974). If there is always solid carbon dioxide at the north pole, the temperature there must always be close to 145°K. Knowledge of the temperature makes it possible to calculate the total amount of energy radiated to space over the course of a year. Subject to some uncertainty in emissivity and albedo, this turns out to be very close to the total amount of solar energy absorbed during the course of a year (Cross, 1971), so the permanent polar cap is neither growing nor shrinking.

Suppose, now, that annual absorption of solar energy were to increase for some reason. To maintain energy balance, the average temperature of the cap would have to be higher. But a higher temperature cap can only be in equilibrium with a larger partial pressure of carbon dioxide. The cap would therefore evaporate, adding carbon dioxide to the atmosphere, until a new equilibrium was achieved or until the cap had completely disappeared. If the permanent polar cap is carbon dioxide, therefore, it controls the amount of carbon dioxide in the atmosphere of Mars. The partial pressure of carbon dioxide is the saturated vapor pressure at the temperature of the permanent cap; the temperature is determined by the average flux of solar energy (*insolation*) into the polar region (Leighton and Murray, 1966).

Sagan et al. (1973) have suggested several possible causes for changes in insolation (cf. Murray et al., 1973). One that appears to be well established is changes in the *obliquity* (the angle between the spin axis and the normal to the orbital plane). Perturbations of Mars' orbit by the sun and other planets cause the obliquity to change from a minimum of 15° to a maximum of 35° with a periodicity of about 10^6 yr (Ward, 1973, 1974). The temperature of the north pole varies, as a result, from about 130°K to about 160°K, and the equilibrium partial pressure of carbon dioxide varies from a few tenths of a millibar to about 30 mb (Ward et al., 1974). There may not be enough carbon dioxide in the permanent polar cap to supply the 30 mb atmosphere, but release of gas from the regolith would probably serve the same function (Fanale and Cannon, 1974).

This is what is known as climatic change on Mars. It is very likely that perturbations of the orbit of Mars cause the surface pressure to oscillate by a factor of about ten with a period of about a million years. We are at present at an intermediate stage of this oscillation; the obliquity is 25.2°.

There is also, however, the possibility of climatic instability on Mars (Sagan et al., 1973; Gierasch and Toon, 1973). In discussing the energy balance of the polar regions, we neglected the transport of energy from low latitudes to high latitudes by atmospheric motions. Such transport appears to be negligible at present because the density is low and therefore the heat capacity of the atmosphere is low (Gierasch and Goody, 1967; Leovy and Mintz, 1969; cf. Goody and Walker, 1972). But if the density were to increase as a result of warming of the polar regions, transport of energy from equator to pole might become more effective. The effect of such transport would be to decrease the temperature difference between the equator and the pole, causing further increase in polar temperatures and further increase in the carbon dioxide partial pressure (at least until all of the carbon dioxide was in the atmosphere).

The proposed instability has been investigated by Sagan et al. (1973). They find that for a given level of polar insolation there are two stable climatic regimes on Mars. One has low polar temperatures, a low surface pressure, and little horizontal transport of heat. The other has a small variation of temperature with latitude and a high surface pressure. Polar temperatures in the hypothetical high pressure regime are about 190°K, and surface pressures are close to 1 bar. The high pressure regime is possible, of course, only if the supply of carbon dioxide in the polar cap or regolith is adequate. If it is not, the unstable runaway stops only when all of the available carbon dioxide is in the atmosphere. Sagan et al. have suggested a number of mechanisms that could cause the climate of Mars to move from one stable regime to another. They have also described a number of lines of evidence suggesting that the Martian atmosphere has been denser in the past than it is at present.

So the Martian atmosphere and climate may oscillate substantially on a time scale of a million years or so. They may also have undergone very large excursions, the surface pressure achieving a value comparable to that of the terrestrial atmosphere. Climatic change and climatic instability are both a consequence of the fact that the major atmospheric constituent condenses at prevailing Martian surface temperatures. No comparable phenomena occur in the atmospheres of Venus or Earth.

COMPARISON OF ATMOSPHERIC EVOLUTION ON VENUS, EARTH, AND MARS

The very tentative conclusion of our comparison of the evolution of the atmospheres of the inner planets is that these atmospheres started out more or less the same but ended up being very different because of the different distances of the planets from the sun. High temperatures on Venus kept the major volatiles, water and carbon dioxide, in the atmosphere. Water was quickly destroyed, as a result, leaving a massive atmosphere of carbon dioxide. Low temperatures on Mars caused the major volatiles to enter the solid phase. Nitrogen was lost to space because of the low gravity on Mars, and a very tenuous atmosphere was left behind. Temperatures on Earth permitted carbon dioxide to enter the solid phase, avoiding a massive, Venus-like atmosphere, while preserving liquid water on the surface. Nitrogen remained behind in the atmosphere to become the dominant constituent. The weakly reducing atmosphere, liquid water, and clement temperatures permitted life to originate on Earth. The subsequent evolution of the terrestrial atmosphere has been dominated by biological processes. These are the subject of the final chapter of this book.

Chapter Seven
Evolution of the Atmosphere

It is convenient to think of atmospheric evolution in terms of four stages. The first stage was the origin of the atmosphere, which we described in Chap. 5. This was the stage in which the atmosphere grew to approximately its present mass. We have suggested that this stage lasted for less than 100 million years, but the evidence is not conclusive.

The second stage preceded the origin of life on earth. This stage, the prebiological or *chemical era*, was discussed in Chap. 6. Physical and chemical processes governed the composition of the atmosphere and determined the environmental conditions under which life originated.

After life had originated, biological processes became the dominant factor in determining atmospheric composition. The third stage represents the transition from control by physical and chemical processes to control by biological processes. We shall call it the *microbial era*. During this stage, organisms were gradually developing their metabolic capabilities and becoming increasingly powerful agents of geochemical change. Atmospheric composition was evolving in response to the introduction of one new metabolic process after another. This stage culminated in the development of green-plant photosynthesis and the rise of atmospheric oxygen. The biological developments important to global geochemical mass balance had all occurred by the end of the third stage, late in the Precambrian.

During the fourth stage, atmospheric composition was controlled largely by biological processes; it did not change radically because major new metabolic capabilities did not arise. Climate and the distribution of continents changed, however, and these changes affected atmospheric composition by changing the partial pressures at which sources and sinks of atmospheric gases are in balance. The fourth stage therefore represents a period during which the atmosphere changed, not because of biological changes or changes in the processes that control atmospheric composition, but because of geological factors that affected the distribution of geochemically active organisms. We shall call this stage the *geological era*.

The microbial and geological eras are the subject of this chapter. We shall begin with a very tentative account of the order in which organisms may have acquired important new metabolic capabilities and the changes in atmospheric composition that may have resulted. Then we shall turn to geological and paleontological evidence to attempt to establish a time scale for the microbial era of atmospheric evolution. A summary appears at the end of the section entitled "Geologic History of the Early Atmosphere."

We shall begin our discussion of the geological era of atmospheric evolution by trying to decide when oxygen first achieved approximately its present abundance. Next we shall consider what limits can be set on the variability of atmospheric composition during the Phanerozoic (see Chap. 1 for a description of the geological time scale). Then we shall examine the geological history of the last 100 million years in order to illustrate how geological change may have affected the atmosphere. There turn out to be serious uncertainties concerning the atmospheric effects of geological change, even in the recent past. We are therefore forced to concentrate on some possible effects, while assuming that others are less important. This we shall do in the concluding sections of the chapter, in which we present a tentative, qualitative account of the Phanerozoic history of atmospheric oxygen.

THE MICROBIAL ERA

Our starting point is the prebiological atmosphere described in Chap. 6. This atmosphere contained nitrogen, water, and carbon dioxide in approximately their present amounts. It also contained hydrogen, at approximately the 1% level, and very little oxygen. Abiotic processes in this atmosphere and the ocean led to the synthesis of organic molecules and eventually to the origin of life. These processes have been extensively reviewed (Calvin and Calvin, 1964; Commoner, 1965; Gabel and Ponnamperuma, 1972; Haldane, 1964; Oparin, 1938, 1961, 1972; Ponnamperuma and Gabel, 1968).

We want to consider what happened after life had begun. For this purpose we now identify some of the more important developments in the evolution of microbial metabolism, drawing on work by Hall (1971), Margulis (1970), and Uzzell and Spolsky (1974). Then we shall try to decide how these developments may have affected the atmosphere. Our findings are summarized in Fig. 7-1.

The first organisms presumably had very limited metabolic capabilities. They would have built themselves mostly out of abiotically synthesized organic molecules available in the environment and derived what little energy they needed from reactions between available organic molecules. Their biosynthetic capabilities increased as one essential compound after another was depleted in the environment. Increased biosynthesis would have called for an increased energy source, but these

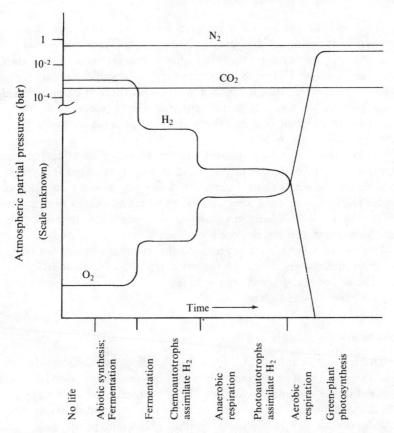

Fig. 7-1. Postulated evolution of prokaryote metabolism and its effect on atmospheric composition.

primitive organisms were still limited to the derivation of energy from fermentation reactions of abiotically synthesized organic molecules (fermentation is discussed in the next section).

There was a big step forward when a new energy source became available. We shall assume that the first organism to be free of dependence on fermentation as its source of energy was a chemoautotroph that derived energy from the reaction between hydrogen and carbon dioxide, like modern methane bacteria (see Chap. 1).

The first autotroph would have flourished as it metabolized abundant supplies of atmospheric hydrogen and carbon dioxide. All the heterotrophs would have flourished also because the autotroph would have synthesized abundant new supplies of organic molecules. The new organisms may, however, have been so successful as to depress the level of atmospheric hydrogen to the point where it was no longer useful as an energy source. Further development of life awaited a new energy source.

The new energy source may have been photosynthesis. *Bacterial photosynthesis* is simpler than green-plant photosynthesis and would have been the first to develop. Bacterial photosynthesis requires a reduced compound such as hydrogen, methane, hydrogen sulfide, or an organic molecule to serve as electron donor in the reduction of carbon dioxide. When reduced compounds were depleted, a new electron donor was required. Green-plant photosynthesis made it possible to use abundant water for this function.

Oxidation of water by photosynthesis had a side effect of great importance both to the atmosphere and to life. It released free oxygen. As a result, the atmosphere accumulated oxygen, possibly quite rapidly, *aerobic respiration* opened abundant new sources of energy to organisms, and any of the old anaerobic organisms that could not tolerate oxygen were forced into the restricted environments they inhabit today.

The rise of atmospheric oxygen and the evolution of aerobic respiration were the closing events of the microbial era of atmospheric evolution. Let us now examine in greater detail the various evolutionary stages that we have outlined above.

Fermentation

The most widespread of all energy-releasing metabolic pathways is fermentation, and it seems fairly clear that this must have been the energy source for the original organisms. Fermentation derives energy from oxidation–reduction reactions in which organic compounds serve as electron donors as well as electron acceptors. Different compounds usually perform these different functions, although both may be derived from a single fermentable compound, such as sugar. A large number of fermentations are known that consume many different classes of organic compounds. Among the gaseous products that fermentation might have released to the atmosphere are carbon dioxide, hydrogen, ammonia, and methane.

A typical modern fermentation, such as the alcoholic fermentation of sugar, involves many intermediate steps. The first fermentations probably had only one or two steps. Still, to get a clearer idea of the effects of fermentation, let us consider the alcoholic fermentation, which converts sugar to alcohol and carbon dioxide:

$$(7\text{-}1) \qquad C_6H_{12}O_6 \rightarrow 2C_2H_5OH + 2CO_2$$

The carbon dioxide is more oxidized than the sugar; the alcohol is more reduced (see oxidation and reduction in Chap. 1). Fermentation has thus caused no change in the overall level of oxidation. Fermentation can derive energy only from organic compounds that are neither highly reduced nor highly oxidized.

Fermentation is relatively inefficient as an energy source. Fermentation of a given amount of sugar provides less than 10% of the energy that is released by aerobic respiration of the same amount of sugar. Thus, much less growth is possible for a fermenting organism given a certain amount of sugar than for a respiring organism.

If the first organisms were fermenting heterotrophs, they were entirely dependent on the supply of organic compounds provided by abiotic processes, including photochemical reactions in the atmosphere. The organisms would have extracted as much energy as they could from newly synthesized organic material, producing highly oxidized and highly reduced products that were of no further biological use. Some of these waste products were gases that returned to the atmosphere to be converted abiotically back to organic products of intermediate levels of oxidation. Others must in time have found their way into sediments, to be lost to the biosphere.

As long as life was able only to rearrange organic molecules that were synthesized abiotically, it would not have been a geochemical force and would not have had any significant effect on the composition of the atmosphere. This situation changed when life developed the ability to synthesize organic molecules from inorganic compounds.

The First Autotrophs

We shall follow Margulis (1970) in assuming that the first autotroph used chemical energy rather than light. Pigments and the mechanism of photosynthesis were probably later developments (Hall, 1971). The electron acceptor used by the first chemoautotroph may have been carbon dioxide; the electron donor may have been hydrogen. Both gases were relatively abundant in the primitive atmosphere. Stephenson (1949, p. 55) mentions a bacterium that synthesizes acetate from carbon dioxide and hydrogen:

$$(7\text{-}2) \qquad 2CO_2 + 4H_2 \rightarrow CH_3COOH + 2H_2O$$

(Stanier et al., 1970, p. 646). Modern methane bacteria are anaerobic autotrophs that derive energy from the reaction

$$(7\text{-}3) \qquad CO_2 + 4H_2 \rightarrow CH_4 + 2H_2O$$

These are examples of the kinds of reactions that might have been used by the first autotrophic organisms.

Autotrophs should have been able to synthesize organic molecules from hydrogen and carbon dioxide much more rapidly than abiotic processes, so a greatly expanded level of biological activity became possible. This activity constituted a drain on atmospheric hydrogen because much of the organic matter would not have been mineralized by fermentation but would have been incorporated into sediments instead. There would have been a drain on carbon dioxide also, but we assume that abiotic processes, such as those described in Chap. 3, would have maintained the carbon dioxide partial pressure. Hydrogen, however, was provided only by volcanoes, with a little recycling as a product of fermentation, and was lost from the atmosphere by escape to space as well as by incorporation into organic matter not susceptible to fermentation.

The level of biological activity was therefore limited by the volcanic supply of hydrogen. The new sink may have lowered the hydrogen content of the atmosphere, permitting the oxygen content to rise (see Chap. 6). The hydrogen content could not have been lowered very far, however, or the autotrophs would have starved (or suffocated). It would be interesting to find out how low the hydrogen partial pressure can fall before methane bacteria cease to grow and reproduce.

Even before hydrogen became limiting, organisms may have suffered from a shortage of fixed nitrogen. Fixed nitrogen was originally provided mostly as ammonia, by abiotic processes including photochemical reactions in the atmosphere. This source should have been sufficient when similar abiotic processes were the only source of the organic molecules that sustained life. But once there was a vastly more efficient biological source of organic molecules, the abiotic source of fixed nitrogen may have been inadequate compared with the loss of nitrogen in organic compounds incorporated into sediments. Nitrogen fixation is widespread among prokaryotic organisms and may have been introduced at this stage in the evolution of life. It would not have influenced the atmosphere significantly. Nitrogen fixation is metabolically expensive, so organisms fix nitrogen only when their nutrient supply is deficient in nitrogen compounds. The primitive nitrogen fixers presumably operated just fast enough to replace the fixed nitrogen that was lost to the biosphere by burial in sediments and by photochemical conversion to N_2.

Once autotrophic organisms had fully exploited the chemical energy of volcanic hydrogen, the stage was set for a new metabolic development. The only remaining source of abundant energy was sunlight, so it is likely that the next major development was photosynthesis. Photosynthetic organisms, unlike all of the organisms we have so far introduced, must be exposed to sunlight. This may have been a problem because the surface of the earth may have been bathed in lethal fluxes of ultraviolet radiation. We shall examine the problem of ultraviolet exposure in the next section, before continuing our description of the evolution of microbial metabolism and its effect upon the atmosphere.

Exposure of Primitive Organisms to Ultraviolet Radiation

The genetic material of the cell, DNA, is extremely susceptible to damage by ultraviolet radiation at wavelengths around 2500 A (Smith,

1969). Sagan (1973) has estimated that typical contemporary microorganisms would be killed in a matter of seconds if exposed to the full intensity of solar radiation in this wavelength region. Contemporary organisms, of course, are not exposed to solar ultraviolet radiation because atmospheric ozone absorbs this radiation in the stratosphere. But atmospheric ozone is produced from atmospheric oxygen (see Chap. 2). If oxygen was as rare in the primitive atmosphere as we have estimated (Chap. 6), there would not have been enough ozone to shield the surface from solar ultraviolet radiation (Ratner and Walker, 1972).

Prior to the origin of photosynthesis, irradiation of the surface of the earth need not have hindered life. The earliest heterotrophs and chemoautotrophs could have lived in the depths of the sea or in the mud at the bottom of shallow water, where they would not have been exposed to sunlight. Indeed, abundant ultraviolet light probably benefited early life by initiating the photochemical reactions responsible for abiotic synthesis of organic molecules. Photosynthetic organisms, however, must have been exposed to sunlight at visible and infrared wavelengths. We must consider how they might have been protected from sunlight at ultraviolet wavelengths.

One possibility is that solar ultraviolet radiation was absorbed by some atmospheric gas other than ozone. Sagan (1973) has examined the absorption by methane, ammonia, water vapor, carbon dioxide, nitrogen, hydrogen, and hydrogen sulfide, concluding that none of these gases could have shielded the surface from lethal ultraviolet radiation. He has also considered absorption by a number of organic molecules that might have been present in the primitive atmosphere, concluding that none of these would have been effective at abundance levels that might reasonably have existed. One gas that does absorb strongly at wavelengths as long as 3200 A is sulfur dioxide (Thompson et al., 1963; Hudson, 1971), but it is unlikely that sulfur dioxide was a significant constituent of the primitive reducing atmosphere. We can conclude, tentatively, that the atmosphere did not shield the surface from solar ultraviolet radiation before the rise of oxygen and the development of the ozone layer.

Some support for the idea that ultraviolet radiation at one time penetrated to the surface of the earth comes from laboratory investigations of the effects of such radiation on organisms (Kelner, 1969). These investigations have shown that many microorganisms possess elaborate mechanisms that serve, as far as we know, only to repair damage to DNA caused by ultraviolet light. These repair mechanisms cannot protect the organism from a massive dose of ultraviolet, but they can undo much of the damage done by a moderate dose (Sagan and Pollack, 1974). But modern microbes are almost never exposed to ultraviolet light (outside the laboratory), so it is not clear why the repair mechanisms exist. One possibility is that they evolved at a very early stage in the development of life, when there was ultraviolet radiation at the surface of the earth.

Presumably, then, the first photosynthetic organisms must have protected themselves by living underneath a shield of some substance that
transmitted visible light but absorbed the harmful ultraviolet light. Water
is a possible shielding substance, but water does not absorb very strongly
in the ultraviolet range, and several tens of meters of pure water would
have been required (Berkner and Marshall, 1964; Sagan, 1973). A more
satisfactory shielding material has been proposed by Sagan (1973). He
suggests that the primitive microbes were protected by a layer of purine
and pyrimidine bases, the components of DNA that absorb strongly in the
ultraviolet and are therefore responsible for the susceptability of DNA to
ultraviolet damage. A protective layer of organic material containing a
few percent of purines and pyrimidines would have had to be only about
100μ thick. Sagan suggests that this may have been the size of the cells
of the first microbes to live in sunlight. The simplest eukaryotic cells are
this large, but prokaryotic cells are smaller. An alternative possibility is
that there was a thin film of organic matter on the surface of the sea,
which absorbed the incident ultraviolet radiation. A third possibility is
that the first photosynthesizers lived beneath a protective layer made up
of the remains of their brethren. Stromatolites are among the oldest
biological remains yet discovered (J. W. Schopf, 1972; Barghoorn, 1971).
They appear to be fossilized algal mats. An algal mat is structurally well
suited to providing ultraviolet screening for individual algae a short
distance below the surface.

We shall assume, therefore, that microorganisms learned to protect
themselves from ultraviolet radiation by surrounding themselves with
organic matter including those constituents of DNA that absorb most
strongly in the ultraviolet range. Once the problem of ultraviolet irradiation had been solved, life was able to utilize the energy of sunlight. We
shall now discuss the introduction of photosynthesis and its possible effect
on the atmosphere.

Bacterial Photosynthesis

The development of pigments and photosynthesis permitted organisms, for the first time, to use the energy of sunlight directly. Bacterial
photosynthesis, which does not release oxygen, is a simpler process than
green-plant photosynthesis and almost certainly preceded it (Margulis,
1970; Miller and Orgel, 1974).

The first photosynthesizers may have derived the carbon for biosynthesis from organic molecules in the environment. They may therefore
have been heterotrophs. The new source of energy would still have given
them a competitive advantage because it would have allowed them to
metabolize organic molecules that other organisms had to discard. Photosynthesis may therefore have made it possible to clean up the organic

debris in the ocean and to reduce the drain of organic matter into the sediments.

Probably all modern photosynthetic bacteria are autotrophs, however, and are able to derive their carbon from carbon dioxide. To reduce the carbon dioxide they use reduced compounds as electron donors. Hydrogen, hydrogen sulfide, thiosulfate, and organic molecules can all be oxidized by modern purple and green bacteria.

Oxidation of sulfide by photosynthetic sulfur bacteria would have provided a source of sulfate for the first time. Sulfate reduction may have been added to the metabolic repertoire soon after bacterial photosynthesis to take advantage of this sulfate. We described sulfate reducing bacteria in Chap. 1, in connection with the sulfuretum. These are heterotrophs that obtain energy by oxidizing organic molecules using sulfate as electron acceptor. In the process hydrogen sulfide is restored to the environment.

Although the reduced compounds that were destroyed in bacterial photosynthesis were largely restored by metabolic processes such as fermentation and sulfate reduction there would still have been a steady drain of reduced organic compounds to the sediments. Such a drain survives even under the oxidizing atmosphere that we have today. Ferrous iron might have been an abundant reducing constituent of the primitive ocean, arising from the weathering of igneous rocks. There do not, however, appear to be any modern photosynthetic organisms that use ferrous iron as electron donor (Stanier et al., 1970). Iron may be less useful as an electron donor than the other reduced compounds we have mentioned because the oxidized form is insoluble and may therefore be difficult to eliminate from the cell.

Volcanoes, therefore, provided the only source of reduced compounds to replace those lost to the sediments. So life was dependent on the volcanic supply of hydrogen after the introduction of bacterial photosynthesis, as it had been before. Photosynthesis permitted life to expand, nevertheless; abundant energy rendered more efficient the biological recycling of reduced compounds.

Bacterial photosynthesis should also have depressed atmospheric hydrogen to lower levels than before. Presumably a higher partial pressure of hydrogen is needed by organisms that derive energy from its oxidation than by organisms that command abundant energy and need the hydrogen only as a source of electrons. It would be interesting to learn how low a hydrogen pressure can be tolerated by photosynthetic bacteria.

A reduction in the concentration of hydrogen would have permitted the oxygen concentration to rise. There was still no source of oxygen other than photolysis of water vapor followed by escape to space of hydrogen. Most oxygen was still consumed by photochemical reactions with atmospheric hydrogen. Thus, lower hydrogen concentrations permitted higher oxygen concentrations. Further research might make it possible to estimate the hydrogen and oxygen concentrations in the atmosphere after the origin of bacterial photosynthesis.

The next major metabolic development freed life from its dependence on volcanoes as a source of reduced compounds. This was green-plant photosynthesis, in which water replaces a reduced compound as the electron donor that makes possible the reduction of carbon dioxide. We call the process green-plant photosynthesis to distinguish it from bacterial photosynthesis, but it is not practised only by green plants. Prokaryotic *blue–green algae* are green-plant photosynthesizers; they may resemble the organisms that first possessed this capability.

Green-Plant Photosynthesis

It is not a trivial step to switch from hydrogen as electron donor to water. A considerable input of energy is required to dissociate the water into hydrogen and oxygen. To provide the energy and perform the dissociation, green-plant photosynthesizers are equipped with a second type of photochemical reaction center in addition to the center that performs functions resembling those of bacterial photosynthesis. The mechanisms of green-plant photosynthesis are therefore considerably more complex than those of bacterial photosynthesis.

There is another problem that had to be overcome by green-plant photosynthesizers and indeed by all aerobic organisms. The transformation of water to oxygen in photosynthesis or oxygen to water in respiration involves compounds of intermediate valence states such as HO_2, H_2O_2, and OH (Fridovich, 1975). These compounds, particularly the hydroxyl radical, are very reactive and are likely to attack any organic compound in the cell. Before they could start to work with the oxidation and reduction of oxygen, therefore, organisms had to develop mechanisms to hold down the concentrations of these reactive intermediaries within their cells. Modern organisms contain a number of enzymes[*] that serve specifically to control these concentrations. Enzymes with similar functions must have been possessed by the first green-plant photosynthesizers.

Once organisms had learned to oxidize water and to protect themselves from the toxic intermediary products, life was freed from its dependence on volcanoes as a source of reduced compounds. A considerable expansion of biological activity should have occurred, with photo-autotrophs providing an abundant supply of organic matter to support the heterotrophs.

---

[*] An enzyme is a complex organic molecule (a protein) that catalyses a particular biochemical reaction, greatly enhancing the rate of the reaction.

Green-plant photosynthesis provided a new source of atmospheric oxygen, a source that was potentially much larger than the abiotic source of photolysis followed by escape of hydrogen. We considered the effect upon the atmosphere of an additional oxygen source in Chap. 6. The oxygen partial pressure would have risen slowly at first, as the extra oxygen reacted with atmospheric hydrogen. When the oxygen source became comparable to the hydrogen source, however, the rise of atmospheric oxygen would have been rapid. The partial pressure of hydrogen would have fallen rapidly at the same time. Thus, green-plant photosynthesis could have led to a geologically rapid transition from a weakly reducing atmosphere to an oxidizing one.

The rise of oxygen necessitated adjustment on the part of organisms that had evolved in an anaerobic world. Either they had to learn to tolerate the new gas, or else they had to retire to restricted environments which were free of oxygen. Modern anaerobes, which cannot tolerate oxygen, possess all of the metabolic capabilities we have described so far, with the exception of green-plant photosynthesis. They flourish in environments, such as the mud on the bottoms of lakes, that are kept free of oxygen by an abundant supply of organic matter.

However, the rise of oxygen presented life with new metabolic opportunities as well as with problems. Among these new opportunities was aerobic respiration, in which organisms derive energy for growth from the oxidation of organic molecules by oxygen. Aerobic respiration extracts much more energy from a given amount of organic matter than any of the chemical processes we have discussed so far. Respiring organisms can therefore prosper at much lower concentrations of organic matter than can, for example, fermenting organisms.

Aerobic respiration could serve as an energy source for autotrophs as well as heterotrophs. There are many modern microbes that derive energy from the oxidation of inorganic compounds and extract carbon from carbon dioxide (Stanier et al., 1970, p. 598). *Hydrogen bacteria* derive energy from the reaction

(7-4) $$2H_2 + O_2 \rightarrow 2H_2O$$

Nitrifying bacteria oxidize ammonia to nitrite:

(7-5) $$2NH_3 + 3O_2 \rightarrow 2NO_2^- + 2H^+ + 2H_2O$$

and nitrite to nitrate:

(7-6) $$2NO_2^- + O_2 \rightarrow 2NO_3^-$$

The *colorless sulfur bacteria* oxidize sulfide, sulfur, and thiosulfate:

(7-7) $$2H_2S + O_2 \rightarrow 2S + 2H_2O$$

(7-8) $$2S + 2H_2O + 3O_2 \rightarrow 2SO_4^{--} + 4H^+$$

(7-9) $$S_2O_3 + H_2O + 2O_2 \rightarrow 2SO_4^{--} + 2H^+$$

The *iron bacteria* oxidize ferrous iron:

(7-10) $$4Fe^{++} + 4H^+ + O_2 \rightarrow 4Fe^{+++} + 2H_2O$$

These organisms presumably caused a rapid reduction in the abundance of reduced inorganic compounds in the biosphere and they have served to hold these abundances low ever since oxygen rose to the level that made aerobic respiration possible. As Hall (1971) has pointed out, the resultant scarcity of reduced inorganic species has left chemoautotrophy with a minor role in modern metabolism.

The nitrifying bacteria performed another geochemically important function. They provided the first abundant source of nitrate in the biosphere. Many organisms have adjusted to this change by developing the ability to use nitrate as a source of cellular nitrogen. The ability to use ammonia is more widespread, however, suggesting that ammonia preceded nitrate as the dominant form of fixed nitrogen in the biosphere.

The rise of nitrate led to the development in many organisms of anaerobic respiration using nitrate as electron acceptor. The process, called *denitrification*, was invoked in Chap. 3 as a source of atmospheric N_2. An example is provided by the oxidation of acetate:

(7-11) $$5CH_3COOH + 8NO_3^- \rightarrow 10CO_2 + 4N_2 + 6H_2O + 8OH^-$$

When the nitrate concentration is high, denitrifying bacteria may release nitrous oxide (see Chap. 2). The nitrous oxide can be further reduced to molecular nitrogen if nitrate is in short supply. Some denitrifyers can use inorganic electron donors, such as hydrogen or sulfur:

(7-12) $$5S + 6NO_3^- + 2H_2O \rightarrow 5SO_4^{--} + 3N_2 + 4H^+$$

(Stanier et al., 1970, p. 214).

All of the denitrifying bacteria are capable of aerobic respiration and use nitrate as electron acceptor only if oxygen is not available. On the other hand, anaerobic respirers that use sulfate or carbon dioxide as electron acceptors are obligate anaerobes. The implication is that denitrification was a late evolutionary development that originated at about the same time as aerobic respiration, accompanying the rise of atmospheric oxygen. Olson (1970) has discussed possible relationships between nitrate and oxygen respiration and the evolution of photosynthesis.

Green-plant photosynthesis and the respiration processes that it made possible were the last of the geochemically important developments of microbial metabolism. Life was freed from its dependence on the volcanic supply of reduced compounds, and some new factor must have served to limit the overall abundance of organisms. The new limiting factor may have been the supply of the nutrient element phosphorus. Phosphorus appears to limit the biological productivity of the modern ocean, as we described in Chap. 3.

The atmosphere, after the introduction of green-plant photosynthesis, became more or less modern, in that its composition was subject to control by the very processes that operate today. Subsequent evolution of the atmosphere was caused more by changes in geology and climate than by changes in the metabolic capabilities of organisms. We shall discuss the geological era of atmospheric evolution below. First, let us attempt to date some of the events in the microbial era that we have been describing.

Geologic History of the Early Atmosphere

We shall consider, first, the fossil record left by early organisms. It turns out that fossils provide useful constraints on atmospheric oxygen in the late Precambrian, but tell us little about the timing of the events that preceded the rise of oxygen. Evidence about earlier metabolic evolution is furnished by data on the isotopic composition of carbon in Precambrian sediments; we shall discuss this evidence next. Probably the best record of the composition of the Precambrian atmosphere is provided by the sedimentary rocks themselves. We conclude this section with a review of this record. Our tentative time scale for the evolution of metabolism and the Precambrian atmosphere is illustrated in Fig. 7-2.

The Precambrian fossil record is of little use for our purposes because all but the very latest Precambrian organisms were microbes, microbes show very little relationship between form and metabolic function, and the process of fossilization at best preserves only form. Thus, although microfossils as old as 3.2 billion years have been discovered (cf. J. W. Schopf, 1972), we really do not know how the oldest microorganisms lived. The evidence of stromatolites is somewhat less equivocal (J. W. Schopf et al., 1971). These structures resemble modern algal mats. For some time they were interpreted as the products of blue–green algae (which practice green-plant photosynthesis), but recent evidence suggests that similar structures can be built by photosynthetic bacteria (Doemel and Brock, 1974). Stromatolites, which date back almost 3 billion years, therefore provide evidence for photosynthesis, but not necessarily for a biological source of oxygen.

Microfossils are more abundant in the Proterozoic (younger than 2.5 billion years) and the identification of some of them as representing blue–green algae may be more secure (J. W. Schopf, 1972, 1976). There is independent evidence, which we shall discuss below, for the existence of green-plant photosynthesis by that time.

The fossil record of the late Proterozoic (younger than 1 billion years) provides evidence of the existence of free oxygen. Microfossils in the 0.9 billion-year-old Bitter Springs formation have been interpreted as eukaryotic (Schopf, 1972). Most eukaryotes are aerobic, so their presence is indicative of free oxygen in the atmosphere, or at least of the existence

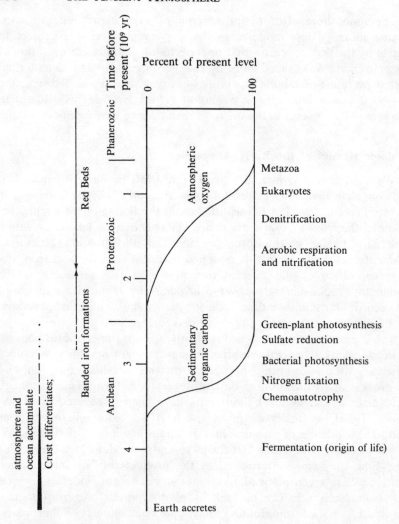

Fig. 7-2. Tentative time scale for the evolution of metabolism and of atmospheric oxygen.

of oxygen oases, local areas of above-average oxygen concentration surrounding algal colonies (Fischer, 1965).

Rhoads and Morse (1971) have compared the early Metazoan fossil record and the conditions under which Metazoa live in modern waters containing little or no oxygen. (*Metazoa* are multicelled animals.) They find that Metazoa are absent in waters and sediments where the oxygen concentration is less than 0.1 ml/l. The absence of Metazoa is characteristic of the fossil record prior to about 0.8 billion years ago. At oxygen concentrations somewhat higher, but below 1.0 ml/l, they find that small, soft-bodied species such as worms are present, but no shell-forming animals. The fauna resemble those discovered in the 650 million-year-old Ediacara formation, and burrows made by similar organisms are present in older rocks. At higher oxygen concentrations in modern waters (6 ml/l), there exists a wide range of shell-forming Metazoa, as is the case in the Phanerozoic fossil record.

The analysis of Rhoads and Morse can probably be used to set lower limits on the oxygen abundance—about 2% of the present level 800 million years ago, and about 20% of the present level at the beginning of the Cambrian, 580 million years ago. It is also possible, but not certain, that these times represent the first attainment of these oxygen levels. This possibility is discussed below. For the time being we shall use the Metazoan evidence simply to establish these lower limits on atmospheric oxygen.

Now let us consider the evidence provided by carbon isotope studies (Oehler et al., 1972). We used data on carbon isotopes in Chap. 3, in our discussion of the oxygen budget of the atmosphere. Carbon isotopes are useful because organic carbon is isotopically lighter (contains less ^{13}C) than the carbon dioxide from which it is synthesized. Organisms assimilate the light isotope ^{12}C more rapidly than the heavy isotope ^{13}C. Presumably this fractionation occurs in the metabolic pathway which organisms use to assimilate carbon dioxide and is independent of the source of energy or of reducing power. This pathway is called the *Calvin Cycle* and is common to all autotrophs (Stanier et al., 1970, p. 229). We shall therefore assume that isotopically light organic carbon is a product of autotrophic organisms, not restricted to green-plant photosynthesizers (Degens, 1969). There is no reason to suppose that abiotic synthesis of organic molecules causes significant isotopic fractionation.

Very little isotopic enrichment occurs when carbonate rocks are precipitated from sea water. The production by organisms of isotopically light organic carbon can therefore be detected by comparison of the isotopic composition of contemporaneous reduced carbon and carbonate in sedimentary rocks. Such a comparison has been made by Eichmann and Schidlowski (1975) for a large number of Precambrian rocks of different ages. They find that the isotopic difference between reduced and oxidized carbon is essentially the same in all of the rocks studied, extending back in time more than 3 billion years. This very important result indicates that by 3 billion years ago autotrophic organisms had replaced abiotic synthesis as the major source of organic compounds.

Evidence of an even earlier origin of autotrophy is provided by another aspect of the carbon isotope data. As we explained in Chap. 3, the accumulation of isotopically light organic carbon in the sedimentary reservoir should cause the remaining oxidized carbon to become isotopically heavier (Broecker, 1970a; Schidlowski et al., 1975). Carbonates deposited during a period of growth of the organic carbon reservoir should therefore show an increasing abundance of ^{13}C. But the data of Schidlowski et al. (1975) show no secular change in the isotopic composition of carbonates dating back more than 3 billion years. The implication of these data is that the mass of organic carbon of biogenic origin in sedimentary rocks had grown to more or less its present value by 3 billion years ago.

There are about 10^{21} moles of organic carbon in sedimentary rocks today (see Chap. 3). It would have taken the primitive autotrophs a significant time to produce this much carbon, so autotrophic organisms must have originated even earlier than 3 billion years ago. Let us estimate how much earlier.

Prior to the onset of green-plant photosynthesis, which we assume to be a later development, both chemoautotrophs and photoautotrophs required reduced compounds as electron donors. We have suggested above that volcanoes were the ultimate source of these reduced compounds. Thus, organic compounds could not be buried in sediments faster than reduced gases, principally hydrogen, were released from volcanoes. What is more, organisms had to compete for volcanic hydrogen with escape to space and with reduction of oxygen produced by photolysis of water. Since the rate of escape of hydrogen is proportional to the hydrogen mixing ratio (see Chap. 4), we can estimate how the rate of burial of organic carbon would have varied with the hydrogen mixing ratio if the volcanic source and the rate of destruction of water remained constant.

We show such an estimate in Fig. 7-3, where we have assumed that volcanoes produced 5×10^{12} moles of hydrogen per year. This is the hydrogen production rate that we estimated in Chap. 6 to be required in order to maintain a hydrogen mixing ratio of 10^{-3} in the prebiological atmosphere. The figure shows that most of the reducing power of volcanic gases would have gone into sedimentary organic carbon once organisms were able to reduce the hydrogen mixing ratio significantly below its prebiological level. It seems clear that organisms must have been able to do this, or there would have been no competitive advantage in developing a metabolic use for hydrogen.

We may therefore take the rate of burial of organic carbon to have been approximately equal to the rate of release of volcanic hydrogen. At the rate we have assumed for the volcanic source, which is no more than an assumption, it would have taken 400 million years to accumulate the sedimentary carbon reservoir. Very tentatively, therefore, we date the onset of autotrophic metabolism and the first reduction in the partial pressure of atmospheric hydrogen at 3.5 billion years ago.

A question now arises, why the organic carbon reservoir did not continue to grow after it reached approximately its present size approximately 3 billion years ago? Nowadays it does not grow because organic

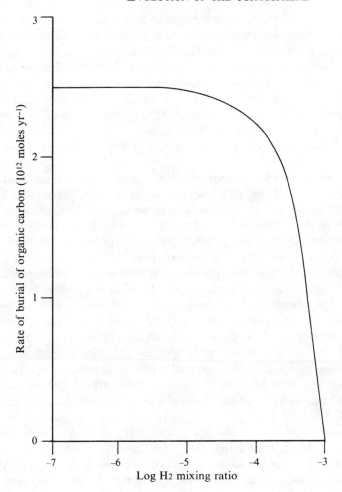

Fig. 7-3. Rate of burial of organic carbon as a function of the mixing ratio of hydrogen in the atmosphere for a volcanic source of hydrogen equal to 5×10^{12} moles yr^{-1}. It is assumed that reduced compounds are removed from the atmosphere only by escape of hydrogen to space and by burial of organic carbon.

carbon is oxidized when sediments are exposed to the atmosphere by uplift and erosion. But there is evidence, to be described below, that the atmosphere was still without oxygen 3 billion years ago, and we have identified no biological processes that would have diminished the total store of reduced carbon in the absence of respiration. Perhaps the volcanic source of hydrogen had decreased to negligibly small values by 3 billion years ago, or perhaps organic carbon was oxidized by metamorphic processes within the earth as fast as it was added to sediments. The question has to remain unanswered for the time being.

The carbon isotope data therefore suggest that autotrophic metabolism, but not necessarily photosynthesis, began much earlier than 3 billion years ago. Stromatolites suggest that photosynthesis, but not necessarily green-plant photosynthesis was in existence almost 3 billion years ago. Sulfur isotope measurements provide further evidence of bacterial photosynthesis at about this time (Perry et al., 1971; Garrels et al., 1973). We turn now to geological evidence concerning the existence of green-plant photosynthesis and atmospheric oxygen.

Red beds become abundant in the geological record younger than about 2 billion years ago (Cloud, 1968, 1972), but they are virtually absent among older sediments. Red beds appear to be formed when iron is weathered and deposited under oxidizing conditions (Van Houten, 1973). They provide the best indication we have of the first appearance of oxygen in the atmosphere. But there seems to be no evidence of what oxygen partial pressure is required for the formation of red beds.

Evidence of an anoxic atmosphere prior to 2 billion years is provided by detrital uraninite and pyrite in Witwatersrand conglomerates and detrital siderite in iron formations in Labrador (Garrels et al., 1973). These minerals are unstable in the presence of oxygen and should not, in principle, have survived erosion and transport in an oxygen rich atmosphere. According to Holland (1975), an upper limit of about 1% on the oxygen mixing ratio is consistent with the existence of detrital uraninite. According to Muir (1975), detrital pyrite is common even today. The evidence provided by these minerals is therefore not strong (Davidson, 1965). The evidence of the *banded iron formations* is more convincing (Cloud, 1973).

Banded iron formations are sedimentary rocks of almost pure iron oxide and silica containing very little detrital material. The banding consists of thin layers alternately silica rich and iron rich which are, in some formations, continuous over horizontal distances of hundreds of kilometers. Banded iron formations are common among sedimentary rocks with ages between 2 and 3 billion years. They are particularly abundant towards the end of this period and are very rare in younger rocks. The discussion that follows refers to the Proterozoic formations, which differ significantly from the Archean ones.

The absence of banded iron formations dating from after free oxygen appeared in the atmosphere is easy to understand. The horizontal extent of banded iron formations and their freedom from detrital material indicate that the iron and silica they contain was transported in solution. Soluble ferrous iron reacts readily with oxygen to form insoluble ferric iron. Therefore the banded iron formations were deposited at a time when the atmosphere lacked oxygen (MacGregor, 1927). The disappearance of banded iron formations and the appearance of redbeds about 2 billion years ago therefore marks the beginning of the aerobic atmosphere.

Banded iron formations also provide evidence for the existence of green-plant photosynthesis. We described above how iron could be transported in solution under an anoxic atmosphere, but we did not describe what would have caused the iron to precipitate. The mechanism that caused precipitation was presumably oxidation. Much of the iron in banded iron formations is ferric.

It is possible that there were photosynthetic organisms in the Precambrian that used ferrous iron as electron donor, but the absence of modern organisms with this ability makes it unlikely. It is more likely that the iron was oxidized inorganically by oxygen produced either by green-plant photosynthesis or by photolysis of water followed by escape of hydrogen to space.

Estimates of the rate of deposition in banded iron formations suggest that the abiotic source would have been inadequate. Trendall and Blockley (1970) estimate that a single formation, the Hamersley Group of Western Australia, accumulated iron at the rate of 2×10^{13} gm yr^{-1} (Holland, 1973b). This would have called for an oxygen supply, for this one formation, of 10^{11} moles yr^{-1}, which is ten times our estimate of the abiotic source of oxygen (see Chap. 6). The only alternative to destruction of water vapor as a source of oxygen is green-plant photosynthesis. Thus, the banded iron formations indicate that green-plant photosynthesis originated well before 2 billion years ago.

If this conclusion is correct, there is a question of why free oxygen did not appear in the atmosphere until about 2 billion years ago. The banded iron formations themselves may well be the answer (Garrels et al., 1973; Schidlowski et al., 1975). All of the oxygen produced during the first few hundred million years of green-plant photosynthesis may have been used in oxidizing abundant reduced compounds, principally iron, in sea water. It is possible that oxygen began to accumulate in the atmosphere only after the oceans had been swept free of ferrous iron.

A very tentative chronology of the microbial era of atmospheric evolution follows. During the first billion years of earth's history, abiotic processes were the only source of organic molecules and life had no significant effect on the composition of the atmosphere. Autotrophy originated about 3.5 billion years ago, making possible much more rapid synthesis of organic compounds and causing a significant reduction in the concentration of hydrogen in the atmosphere. Bacterial photosynthesis arose about 3 billion years ago, causing a further reduction in atmospheric hydrogen. The shortage of hydrogen led to the development of a process that could use abundant water as electron donor, namely, green-plant photosynthesis. This may have been about 2.5 billion years ago. About 2 billion years ago oxygen first appeared in the atmosphere in geochemically significant amounts. The subsequent history of the oxygen partial pressure will be discussed below as part of our description of the geological era of atmospheric evolution.

THE GEOLOGICAL ERA

We should, in principle, be able to reconstruct the geological history of the atmosphere from the geological record combined with an understanding of the processes that control atmospheric composition. We are still a long way from achieving this goal, however. Our understanding of the controlling processes is largely qualitative. The models described in Chap. 3 are barely able to make quantitative predictions for the present, let alone the past. In addition, our knowledge of geologic history is still far from certain. Interpretations of the geologic record are frequently vague and often conflicting. The account of the history of the atmosphere that we are about to present will therefore be qualitative and very speculative. We shall concentrate on oxygen as the most interesting of the atmospheric gases, at least as far as life is concerned. Factors that control the concentrations of carbon dioxide and nitrogen were described in Chap. 3. These gases are likely to have varied less with time than has oxygen.

The situation of the minor atmospheric gases is different from that of oxygen, nitrogen, and carbon dioxide. The photochemistry of the minor gases is reasonably well understood. We described it in Chap. 2. It would be easy to calculate the densities of these gases as functions of time if the major gas concentrations were known and if the biological sources of several of the gases could be specified. Very little work has been done in this area, but there are opportunities for interesting research. Possible variations in trace gas densities can be explored even in the absence of reliable geological information. We shall describe one such study in the next section, where we discuss the rise of atmospheric oxygen.

After this we shall turn to the history of the atmosphere during the Phanerozoic. First we shall attempt to set limits on possible excursions in atmospheric composition in order to constrain the speculation that follows. Then we shall examine the geological record of the last 100 million years and try to deduce how it has affected the atmosphere. Because of the presumed relationship between atmospheric oxygen and oceanic phosphorus, we shall need to discuss the processes that may control the phosphorus content of the ocean.

The geological record becomes increasingly obscure as we go back in time. We shall therefore attempt only a very general description of atmospheric history during the Paleozoic and Mesozoic. The chapter ends with a brief review of atmospheric evolution.

The Rise of Oxygen

We must now try to decide when atmospheric oxygen first achieved an abundance close to that of the present day. The difficulty is that the only datum point between the appearance of red beds 2 billion years ago (corresponding to the first appearance of oxygen in the atmosphere) and the present is the appearance of eukaryotes, Metazoa, and shell-forming Metazoa in the late Precambrian. The biological evidence has already been used to set lower limits on the oxygen partial pressure. The question to be discussed in this section is whether the biological record also provides an upper limit on atmospheric oxygen.

It is possible to interpret the Metazoan fossil record as indicating low oxygen levels at the beginning of the Phanerozoic (Nursall, 1959). The evidence is the sudden appearance, at the base of the Cambrian, of abundant organisms with preservable hard parts. Shell-forming animals have not been found in the Precambrian, but there must have been a diverse fauna of soft-bodied organisms just waiting for the opportunity to make shells.

Towe (1970) has argued that the formation of hard tissue is expensive in terms of oxygen consumption and would have been possible only when oxygen levels had risen to sufficiently high values. He suggests, therefore, that oxygen was still increasing at the end of the Precambrian. The argument has been further developed by Rhoads and Morse (1971), who derived the lower limits on oxygen abundance that we cited above and suggested that these limits were the actual oxygen concentrations.

Indirect support for increasing oxygen in the Phanerozoic has been furnished by Crimes (1974). He finds that the diversity of Metazoa living at the bottom of the sea in deep water increased markedly from the Cambrian to the Devonian, and then increased again at the beginning of the Cretaceous (see Chap. 1 for the geologic time scale). He argues that these increases are not an artifact of preservation nor the result of competition among shallow water fauna. It is possible that deep-ocean water contained too little oxygen at the beginning of the Paleozoic to support a diversity of life at the bottom of the sea, and that this situation has gradually ameliorated with the passage of time.

An alternative interpretation of the relationship between atmospheric oxygen and the diversity of life has been presented by Berkner and Marshall (1964, 1965, 1966, 1967). They suggest that oxygen levels in the Precambrian were too low to provide enough ozone in the atmosphere to shield the surface from lethal ultraviolet radiation, and that life expanded enormously at the beginning of the Cambrian when an adequate ozone screen first developed. The calculation of ozone levels in the atmosphere as a function of oxygen partial pressure is a photochemical study that merits attention. A very simple model of ozone photochemistry has been investigated by Ratner and Walker (1972). It was found that an adequate ozone screen is produced when the oxygen partial pressure is less than 10^{-3} of its present level. If this result is correct, the ozone screen was established prior to 1 billion years ago, when the Metazoan record indicates that the oxygen partial pressure was at least ten times as large. It seems unlikely, therefore, that ultraviolet radiation restricted life in the late Precambrian.

In terms of our understanding, described in Chap. 3, of the processes that control the oxygen level in the present atmosphere, it is hard to explain why oxygen pressures should have remained low for almost 2 billion years after the introduction of green-plant photosynthesis. It is also hard to understand why there should have been a secular increase of oxygen during the Phanerozoic. We shall therefore abandon the interpretation of the Cambrian expansion of life in terms of rising oxygen partial pressure. This we do reluctantly, because there are no other data points between the initiation of red bed formation and the present. Alternative explanations of the burgeoning of life in the Cambrian have been offered (cf. Valentine and Moores, 1972).

We shall assume, therefore, that atmospheric oxygen rose fairly rapidly after its first appearance in the atmosphere. It might have taken a few hundred million years to oxidize reduced minerals at the surface of the earth. We assume that the atmosphere has been essentially modern for at least the last billion years, with its composition controlled by the processes that operate today. We shall now consider geological evidence that limits the amplitude of excursions in composition during the Phanerozoic, before we attempt to reconstruct the Phanerozoic history of the atmosphere.

Limits on Excursions in Atmospheric Composition

The fossil record provides the most reliable limits on possible excursions in atmospheric oxygen. Shell-forming Metazoa have been abundant since the beginning of the Cambrian; therefore, the oxygen partial pressure during this period has not fallen below 20% of its present level. The permissible lower limit on oxygen partial pressure has increased gradually through time as organisms less tolerant to low oxygen levels have evolved and survived. Organisms are intolerant of high oxygen partial pressures also (Fridovich, 1975), so the fossil record sets an upper limit on the oxygen partial pressure. What this upper limit may be, is not clear, but Lovelock and Lodge (1972) suggest that fires would destroy all standing vegetation if the atmosphere were 25% oxygen. Evidently, oxygen can have risen little above its present level since land plants evolved in the Silurian. Possible excursions in atmospheric oxygen during the Phanerozoic are therefore quite limited in amplitude.

The fossil record can also be used to set limits on the densities of atmospheric trace gases. An illustration of the possibilities is provided by a study that we described in Chap. 2. From the fossil record of *Pinus Ponderosa*, Chameides and Walker (1975) deduced an upper limit on the ozone density at the ground during the last 30 million years. Then they used a theoretical model of ozone photochemistry to deduce corresponding upper limits on the densities of methane, hydrogen, and carbon monoxide, as well as the flux of methane from the ground to the atmosphere. More such studies of the constraints on atmospheric composition imposed by the requirements of particular organisms should be possible in the future.

Limits on possible excursions of the composition of the ocean and of atmospheric carbon dioxide can be derived from an examination of the record of marine evaporite deposits. Holland (1972) finds that the partial pressure of carbon dioxide could, under special circumstances, have varied by as much as factor of 100, either up or down, without having an obvious effect on marine evaporites. A more probable upper limit on carbon dioxide variations during the Phanerozoic is a factor of four. There are few evaporites of Precambrian age, so Holland's analysis cannot be extended back beyond about 700 million years.

The Last 100 Million Years

With these modest constraints on possible atmospheric variations let us turn, now, to the geological history of the atmosphere. We shall begin with the recent past, for which the geological record is best understood. We shall describe geological changes during the last 100 million years, drawing largely on a paper by Hays and Pitman (1973), and will identify several changes that may have affected the oxygen content of the atmosphere. These include changes in sea level relative to the continents, in global average temperature, and in the total length of shoreline. The changes in atmospheric oxygen produced by these geological changes are not all of the same sign, unfortunately, and it is not clear which effects have predominated. The situation is further complicated by possible effects of geological change on the phosphate concentration in sea water, which also influences the equilibrium oxygen partial pressure (see Chap. 3). We therefore discuss possible changes in oceanic phosphate, concluding that they may have tended to counteract the effects of geological change on atmospheric oxygen.

There was a period of unusually rapid sea-floor spreading (see Chap. 1) during the upper Cretaceous, extending from about 110 to 85 million years ago. The volume of a midocean ridge is a function of spreading rate because the material of the ridge contracts as it cools. Temperatures decrease with distance from the ridge crest more rapidly when spreading is slow than when it is fast. The ridges therefore had an unusually large volume during the period of rapid spreading, and sea water was displaced onto the continents. At the height of the *transgression*, which is the name given to an advance of the oceans over the continents, as much as 40% of the present continental area was covered by shallow inland seas. Since the close of the episode of rapid spreading the ridges have decreased in volume and the seas have withdrawn from the continents (*regression*). The sea level at present is unusually low.

One hundred million years ago the continents were grouped into one or two supercontinents (see Fig. 7-4). These were broken up and dispersed during the episode of rapid spreading, causing climatic changes that have been discussed by Hays and Pitman and by others (cf. Brooks, 1951). Most important of the climatic changes for our purposes was a decrease in temperature that culminated in the Pleistocene glacial ages.

Let us consider, first, the effect of this temperature decrease on atmospheric oxygen. The solubility of oxygen in sea water increases as temperature decreases. Accordingly, cooling would have caused an increase in the oxygen content of deep-sea water which would have led to a contraction of the area of the sea floor underlain by anaerobic sediments (see Chap. 3). This in turn should have caused a decrease in the rate of fossilization of organic carbon and thus in the net source of atmospheric oxygen. All other things being equal, therefore, cooling should have decreased the partial pressure of oxygen (Walker, 1974).

Next let us consider the effect of the breakup of the supercontinents and the resultant increase in the total amount of shoreline in the world. We showed in Chap. 3 that organic carbon is buried in sediments mainly in near-shore areas where the rate of accumulation of sediments and the rate of photosynthesis are high. The breakup of the supercontinents should therefore have increased the rate of burial of carbon, leading to an increase in atmospheric oxygen. We have now identified two changes in the recent geological past that should have had opposite effects on atmospheric oxygen. These two effects illustrate the difficulty of deducing the history of atmospheric composition.

There is another factor that we have not yet evaluated. The advance of the oceans across the continents may have a much larger effect on the oxygen budget than changes in either the temperature or the length of shoreline. Many of the shallow seas that resulted should have exhibited an *estuarine* type of circulation, in which the supply of fresh water by precipitation and runoff from the land exceeds the rate of evaporation. In this type of circulation, low density, low salinity water runs out of the shallow sea at the surface and is replaced by an inward flow at greater depths. The surface water is relatively depleted in nutrients, as a result of biological processes, while the deeper water has a relatively high nutrient content. The circulation therefore leads to a concentration of nutrients in the shallow sea (Redfield et al., 1963). The Black Sea is a striking modern example of an estuarine circulation leading to high nutrient densities, very high organic productivity, and a low oxygen content in the deeper waters.

An *antiestuarine* circulation results when evaporation exceeds the rate

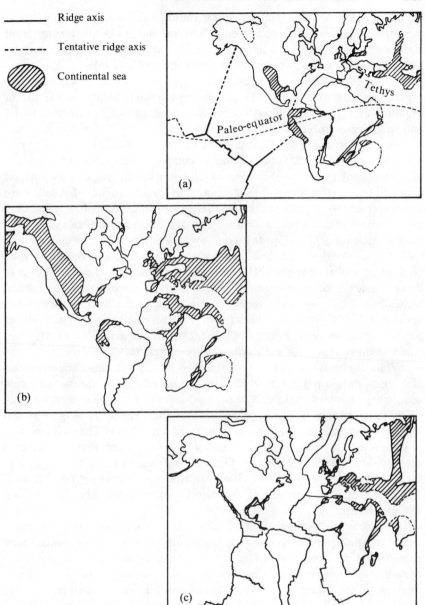

Ridge axis

Tentative ridge axis

Continental sea

Fig. 7-4. Arrangement of the continents and extent of epicontinental seas for three geological epochs: (a) Cenomanian, –100 to –94 million years; (b) Santonian-Campanian, –85 to –70 million years; (c) Eocene, –50 to –40 million years. The arrangement of the ridge axes at present is shown in (c), where they are plotted with respect to North America. (From Hays and Pitman, 1973. Copyright 1973 by Macmillan (Journals) Ltd., London. Used by permission of the publisher.)

of supply of fresh water. Nutrient-poor water flows into the shallow sea at the surface. Evaporation increases its salinity and hence its density, until it sinks and flows out again at greater depths, where the nutrient concentration is higher. The Mediterranean is a modern example of an antiestuarine circulation. It is characterized by low nutrient levels and low organic productivity. Anaerobic sediments and significant fossilization of organic carbon are not likely to occur in shallow seas characterized by antiestuarine circulation.

We do not know what proportion of ancient shallow seas exhibited estuarine rather than antiestuarine circulation, though a study of the distribution of evaporite deposits, which form in arid regions, might cast some light on this question. It is not likely, however, that all were antiestuarine, so we may conclude that there were more, highly productive, shallow seas during the transgression than there are today.

By providing more widespread areas of anaerobic sediment and carbon burial, these shallow seas would have increased the oxygen content of the atmosphere. The effect of increasing oxygen would be to drive the areas of carbon burial back off the more exposed continental slopes into the poorly oxygenated shallow seas. In equilibrium, the rate of carbon burial must equal the rate of carbon weathering, but a higher oxygen concentration is required to achieve this equilibrium at times when shallow seas with high nutrient concentrations are abundant.

When the seas withdrew from the land during the subsequent regression, they should have left behind large areas of poorly consolidated new sediments, many of them rich in organic carbon. Erosion and weathering of these sediments would have accelerated the decline in oxygen caused by the disappearance of the shallow, productive seas. There is, in fact, evidence that the rate of erosion today is two or three times as large as the average Phanerozoic rate (Garrels and Mackenzie, 1971, p. 260; Gregor, 1968, 1970; Judson, 1968; Judson and Ritter, 1964; Menard, 1961); this may be a result of the current regression of the oceans from the continents.

We therefore predict a correlation between sea level and the abundance of atmospheric oxygen. The mechanisms of control of atmospheric oxygen that we described in Chap. 3 may have yielded high oxygen levels during transgressions and low oxygen levels during regressions. If this analysis of the effect of transgression is correct, and if shallow seas are more important for oxygen than coastlines, then oxygen has been declining during the recent past from a maximum about 80 million years ago at the height of the upper Cretaceous transgression.

The situation is complicated, however, by possible changes in the phosphate concentration in the sea. All other things being equal, an increase in phosphate will lead to an increase in atmospheric oxygen (see Chap. 3). Phosphate is supplied to the sea mainly in solution in river water. Enhanced erosion associated with the present regression may have enhanced the rate of supply. We must therefore discuss the processes that remove phosphate from the sea. The residence time of oceanic phosphorus is about 10^5 yr (Broecker, 1971), so the phosphate concentration can respond readily to changes in the rates of supply or removal.

Berner (1973) has argued that phosphate is removed from sea water at a geochemically significant rate by reaction with poorly crystallized, hydrous, ferric oxides in iron-rich volcanogenic sediments. The rate of this removal process is governed by the rate of supply of fresh volcanogenic sediments; it would have been enhanced during the period of rapid sea-floor spreading in the upper Cretaceous.

There is, in addition, considerable evidence that phosphate can precipitate from interstitial sediment water to form sedimentary carbonate–apatite deposits known as *phosphorites* (Tooms et al., 1969). These deposits form in waters of less than 1000 m depth, where sediment accumulation rates are low and upwelling or estuarine conditions produce unusually high organic productivity (Veeh et al., 1973). Decay of organic matter in the sediments leads to high concentrations of phosphate in the interstitial waters. Removal of phosphorus by this mechanism may have been particularly rapid in the highly productive shallow seas that we assume to have accompanied the Upper Cretaceous transgression (Piper and Codispoti, 1975).

Changes in oceanic phosphate may therefore have worked to cancel the processes that were changing atmospheric oxygen. Phosphorus may have declined during the transgression as a result of rapid sea-floor spreading and the presence of widespread areas of anaerobic sediments. During regression the phosphorus supply may have been enhanced by more rapid erosion. Perhaps it is not surprising that atmospheric oxygen has not varied much since the present-day controlling processes were established.

Let us now summarize our tentative conclusions about the recent geological history of atmospheric oxygen. During the last 100 million years, the breakup and separation of the continents has increased the total length of shoreline, causing the oxygen level to increase. The redistribution of the continents, however, has caused the global temperature to fall, decreasing the oxygen level. The decline in the rate of sea-floor spreading has caused sea level to fall relative to the continents, leading to the disappearance of inland seas with estuarine circulations and to a further reduction in atmospheric oxygen. On the other hand, the decreasing rate of spreading and the falling sea level have both caused oceanic phosphate and therefore atmospheric oxygen to increase. In short, our consideration of the recent past has enabled us to identify a number of possible causes of change in the oxygen partial pressure, but not to decide whether oxygen has increased, decreased, or remained constant.

In order to draft a tentative history of atmospheric oxygen during the Phanerozoic, we shall have to arbitrarily assume that one effect is predominant. Although phosphorus variations may serve to decrease the amplitude of oxygen variations we shall assume that the direct effect of widespread anaerobic sediments is dominant, with oxygen increasing during transgression and decreasing during regression. This is no more than an assumption. There is no direct evidence of Phanerozoic fluctuations in atmospheric oxygen.

Phanerozoic History of Atmospheric Oxygen

We assume that changes in the atmosphere during the Phanerozoic resulted from the same geological processes as did changes during the last 100 million years. Episodes of rapid sea-floor spreading led to transgressions of shallow seas over the continents. Continental drift led to changes in climate and in oceanic circulation. The climatic record is too uncertain to be of much use to us. The direct record of sea-floor spreading does not extend far enough back in time. Even the record of transgression and regression is confused by problems of preservation of older sediments. We shall therefore consider, first, evidence provided by sulfur isotope studies.

The sulfur isotopes have properties that resemble those of the carbon isotopes. Evaporite deposits of different ages preserve a record of the isotopic composition of oceanic sulfate at the time they were deposited (Holser and Kaplan, 1966). The isotopic composition of oceanic sulfate changes either as a result of river supply of freshly weathered sulfate or as a result of sulfate reduction by organisms in anaerobic environments (Rees, 1970). Sulfate-reducing bacteria produce sulfide that is isotopically lighter (richer in ^{32}S compared to ^{34}S) than ambient sulfate. Much of the sulfide reacts with iron in the sediments to form insoluble iron sulfide (Berner, 1970, 1971a).

Thus, when anaerobic environments are rare we can expect little sulfate reduction. Weathering of isotopically light sulfide should cause oceanic sulfate to become isotopically lighter. Conversely, abundant anaerobic sediments should cause sulfide deposition to exceed sulfide weathering. Oceanic sulfate should get heavier (Holland, 1973c).

Changes in the ratio of total sulfide to sulfate need not affect atmospheric oxygen directly. Garrels and Perry (1974) have shown how sulfur can be transferred from reduced to oxidized reservoirs without changing the composition of the atmosphere. Walker (1974) has argued that since organic carbon is oxidized when sulfate is reduced, the sulfide simply substitutes for organic carbon in the sediments, leading to no net consumption or release of oxygen.

We therefore cannot interpret imbalances in the sulfur cycle as directly causing change in atmospheric oxygen. Rather, we wish to use the sulfur isotope record as evidence of the presence or absence of the anaerobic sediments that promote increase in oxygen. The suggestion is that abundant anaerobic sediments cause an increase in atmospheric oxygen and an increase in sulfate reduction. The increase in sulfate reduction, in turn, causes oceanic sulfate to become isotopically heavier.

Some support for this interpretation is provided by studies by Berner (1971b, 1972) of the modern sulfur budget of the ocean. He finds that sulfide weathering is currently more rapid than sulfide deposition, an imbalance attributed to the relative rarity of anaerobic sediments today. This conclusion coincides with the deduction we made above from the modern emergence of the continents from the sea. Presumably oceanic sulfate is presently becoming isotopically lighter and atmospheric oxygen is decreasing in abundance, but the changes are too slow to be directly detectable.

The Phanerozoic history of sulfate isotope composition is shown in Fig. 7-5a. The significance of the fine structure is uncertain, but the overall impression is of abundant anaerobic sediments from the Cambrian to the Devonian, a minimum in the rate of sulfate reduction in the Permian, and a gradual recovery to levels intermediate between those of the two extremes. Very roughly, this corresponds to the record of marine transgressions shown in Fig. 7-5b. Although the transgression curve does not show the steady decline from the Cambrian to the Permian that appears in the sulfur isotope data, it is possible that the extent of the early Paleozoic seas has been underestimated because of subsequent erosion of the sediments they deposited.

Some support for the association of sulfur isotopes, transgression, and carbon burial is provided by data on the average carbon content of sedimentary rocks on the Russian platform. These data, based on analyses of nearly 26,000 samples of different ages, are shown in Fig. 7-5c.

In the absence of any better evidence we assume that the oxygen level was high during transgressions, when abundant anaerobic sediments also caused oceanic sulfate to become isotopically heavier. The oxygen level was low during regressions, when there were few areas of carbon burial and sulfate reduction. The Phanerozoic history of atmospheric oxygen therefore looks like some combination of the curves of Fig. 7-5. The oxygen level was high during the early Paleozoic, dropping to a minimum early in the Mesozoic. It then recovered to a maximum in the late Cretaceous, and has since declined. The amplitude of these supposed fluctuations is not known. It may be possible, in the future, to derive some evidence from the fossil record.

Atmospheric Change and the Evolution of Life

It is well established that there have been marked fluctuations in the diversity of life during the Phanerozoic (cf. Newell, 1963). Figure 7-6

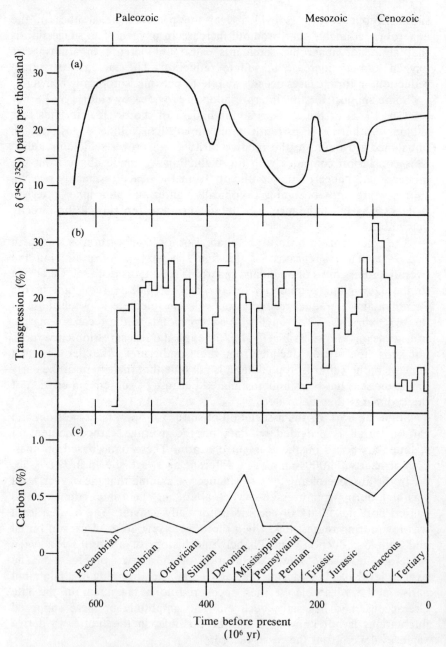

Fig. 7-5. Data that may relate to the oxygen content of the atmosphere: (a) Sulfur isotope composition of evaporites; (b) Extent of epicontinental seas; (c) Average organic carbon content of Russian sedimentary rocks. (Data from Holser and Kaplan, 1966; Damon, 1971; Ronov, 1958.)

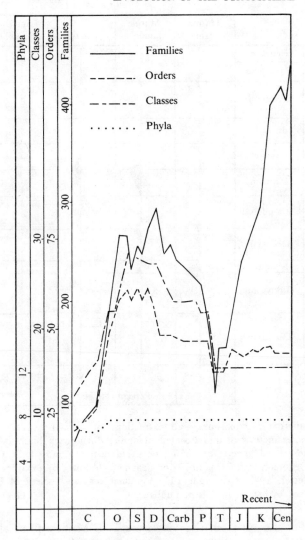

Fig. 7-6. Diversities of higher taxa of well-skeletonized benthic marine invertebrates. (From Valentine, J.W., *Science*, Vol. 180, pp. 1078-1079, 8 June 1973. Copyright 1973 by the American Association for the Advancement of Science. Used by permission of the publisher.)

shows the diversity of marine *benthic* invertebrates (bottom-dwellers) on various taxonomic levels as a function of time. Figure 7-7 shows how this kind of record can be interpreted in terms of periods of *extinction* followed by periods of *expansion*.

McAlester (1970) has suggested that some extinctions may have resulted from oxygen stress at times when the oxygen partial pressure was

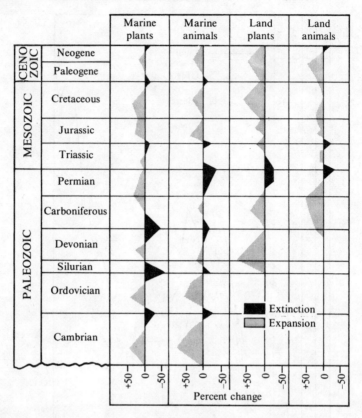

Fig. 7-7. Phanerozoic expansions and extinctions of life. The curves show net expansions and extinctions of animal and plant families during each Phanerozoic epoch. Paleozoic intervals of general extinction occurred near the end of the Cambrian, Ordovician, Devonian, and Permian Periods; each was followed by a major expansion. (From McAlester, 1973. Copyright 1973 by the Canadian Society of Petroleum Geologists. Used by permission of the publisher.)

changing. (For dissent from this view, see T. J. M. Schopf et al., 1971.) Extinctions were particularly severe at the end of the Permian (the close of the Paleozoic). According to the indicators shown in Fig. 7-5 this was a time of particularly low levels of oxygen. Our version of the history of atmospheric oxygen is therefore consistent with McAlester's suggestion.

Many other environmental stresses would have been present during regression. Valentine and Moores (1972) have discussed, among other factors, the decrease in provinciality and increase in seasonal fluctuations of climate, causing instability of the food supply. It is probable that mass extinctions were caused by many such stresses operating together. Changes in atmospheric composition may have been a factor, but probably not the only one.

Another way in which biological extinctions may have resulted from atmospheric change has recently been suggested. Atmospheric ozone would largely disappear if some cataclysm greatly increased the concentration of nitrogen oxides in the stratosphere (see Chap. 2). Depressed ozone would result in enhanced fluxes of lethal solar ultraviolet radiation at the surface. Ruderman (1974) has suggested enhanced cosmic ray fluxes caused by supernova explosions of nearby stars as one way of increasing stratospheric nitrogen oxides. Reid et al. (1976) have suggested *solar proton events* (enhanced fluxes of energetic particles from the sun) coinciding with reversals of the earth's magnetic field as another way. Hays (1971) has found fossil evidence of extinctions at the times of field reversals.

The ideas of McAlester, Ruderman, and Reid and others all emphasize what has been a major thesis of this book: the close interaction of the atmosphere and life. Our very speculative and incomplete account of the evolution and geological history of the atmosphere is probably incorrect in many respects. Much of the book, however, has been devoted to the processes that control atmospheric composition. The goal has been to call attention to the possibility of change in the atmosphere and to illustrate the interrelationship of atmospheric evolution with the evolution of the crust, the oceans, and life.

REVIEW

It has been our goal, in this book, to demonstrate that the processes that control atmospheric composition are understood well enough to permit a tentative description of the early history of the atmosphere, as well as to establish the likelihood of atmospheric change during Phanerozoic time. This chapter has shown, however, that it is one thing to say that the atmosphere may have changed and quite another to say how and when it has done so. The atmospheric history that is summarized below, although vague and qualitative, must still be regarded as extremely speculative.

The atmosphere originated in a geologically short period of time as a result of outgassing of the solid earth. The mass of the atmosphere has not changed much since the end of this early outgassing period. During the chemical era which preceded the origin of life, the atmosphere was much as it is today, except that it lacked oxygen and contained as much as 1% hydrogen. The effect of life, during the succeeding microbial era, was to gradually decrease the hydrogen content of the atmosphere and increase the oxygen content. This process culminated in the transition to an essentially modern, oxidizing atmosphere after the development of green-plant photosynthesis.

Abundances of the major constituents (nitrogen, oxygen, and carbon dioxide) of this essentially modern atmosphere have been controlled by geological and biological processes that have not changed in nature for perhaps 2 billion years. While oxygen and possibly carbon dioxide may have fluctuated in response to changes in geological conditions, there has been no secular change in atmospheric composition since the beginning of what we have called the geological era. Fluctuations in atmospheric oxygen may have yielded a minimum near the end of the Paleozoic, about 250 million years ago, and a maximum in the Cretaceous, about 100 million years ago.

Densities of many minor constituents of the atmosphere are controlled by photochemical processes and depend on biological sources of a few key gases (methane and oxides of nitrogen). Changes in conditions at the surface of the earth could have caused large changes in the concentrations of methane, carbon monoxide, hydrogen, and ozone. Nothing is known about such changes, however, apart from an upper limit that applies only to the last 30 million years.

I hope that this book has shown that the atmosphere has a geological history, just as do the crust of the earth, the ocean, and life. We know very little about this history because we do not yet know how to read the record, wherever it is preserved. Earth's history must ultimately embrace atmospheric history, if only because of the obvious influence of atmospheric evolution on the evolution of life.

References

ABELSON, P. H. (1965): Abiotic synthesis in the Martian environment. *Proc. U.S. Nat. Acad. Sci.*, 54, 1490–1497.

ABELSON, P. H. (1966): Chemical events on the primitive earth. *Proc. U.S. Nat. Acad. Sci.*, 55, 1365–1372.

ACKERMAN, M. (1971): Ultraviolet solar radiation related to mesospheric processes. In G. Fiocco, ed., *Mesospheric Models and Related ·Experiments* (D. Reidel, Dordrecht, Holland), pp. 149–159.

ALDAZ, L. (1969): Flux measurements of atmospheric ozone over land and water. *J. Geophys. Res.*, 74, 6943–6946.

ALEXANDER, M. (1961): *Introduction to Soil Microbiology* (John Wiley and Sons, New York).

ANDERSON, D. E. and C. W. HORD (1971): Mariner 6 and 7 ultraviolet spectrometer experiment: Analysis of hydrogen Lyman-alpha data. *J. Geophys. Res.*, 76, 6666–6673.

ANDERSON, L. G. (1976): Atmospheric chemical kinetics data survey. *Rev. Geophys. Space Phys.*, 14, 151–171.

ANHAEUSSER, C. R. (1972): The evolution of the early Precambrian crust of Southern Africa. *Information Circular No. 70* (Economic Geology Research Unit, University of the Witwatersrand, Johannesburg, South Africa).

ARMSTRONG, R. L. (1968): A model for the evolution of strontium and lead isotopes in a dynamic earth. *Rev. Geophys.*, 6, 175–199.

ARMSTRONG, R. L. (1971): Isotopic and chemical constraints on models of magma genesis in volcanic arcs. *Earth Planet. Sci. Lett.*, 12, 137–142.

ARMSTRONG, R. L. and J. A. COOPER (1971): Lead isotopes in island arcs. *Bull. Volcanol.*, 35, 27–63.

ARMSTRONG, R. L. and S. M. HEIN (1973): Computer simulation of Pb and Sr isotope evolution of the earth's crust and upper mantle. *Geochim. Cosmochim. Acta*, 37, 1–18.

ARNOLD, P. W. (1954): Losses of nitrous oxide from soil. *J. Soil Sci.*, 5, 116–128.

ARRHENIUS, G. (1950): Carbon and nitrogen in subaquatic sediments. *Geochim. Cosmochim. Acta*, 1, 15–21.

ARRHENIUS, G., B. R. DE, and H. ALFVEN (1974): Origin of the ocean. In E. Goldberg, ed., *The Sea* (Wiley–Interscience, New York), Vol. 5, pp. 839–861.

ASTON, B. (1924): The rarity of the inert gases on the earth. *Nature*, 114, 786.

AXFORD, W. I. (1968): The polar wind and the terrestrial helium budget. *J. Geophys. Res.*, 73, 6855–6859.

BADA, J. L. and S. L. MILLER (1968): Ammonium ion concentration in the primitive ocean. *Science*, 159, 423–425.

BAINBRIDGE, A. E. and L. E. HEIDT (1966): Measurements of methane in the troposphere and lower stratosphere. *Tellus*, 18, 221–225.

279

BAME, S. J., A. J. HUNDHAUSEN, J. R. ASBRIDGE, and I. B. STRONG (1968): Solar wind ion composition. *Phys. Rev. Lett.*, 20, 393–395.

BANKS, P. M. and T. E. HOLZER (1968): The polar wind. *J. Geophys. Res.*, 73, 6846–6854.

BANKS, P. M. and T. E. HOLZER (1969a): Features of plasma transport in the upper atmosphere. *J. Geophys. Res.*, 74, 6304–6316.

BANKS, P. M. and T. E. HOLZER (1969b): High-latitude plasma transport: The polar wind. *J. Geophys. Res.*, 74, 6317–6332.

BANKS, P. M. and G. KOCKARTS (1973): *Aeronomy* (Academic Press, New York).

BARGHOORN, E. S. (1971): The oldest fossils. *Scientific American*, 224, no. 5, 30–42.

BARTHOLOMEW, W. V. and F. V. CLARK (1965): *Soil Nitrogen* (American Society of Agronomy, Madison, Wisconsin).

BATES, D. R. (1959): Some problems concerning the terrestrial atmosphere above about the 100 km level. *Proc. Roy. Soc. London*, A253, 451–462.

BATES, D. R. and P. B. HAYS (1967): Atmospheric nitrous oxide. *Planet. Space Sci.*, 15, 189–197.

BATES, D. R. and M. NICOLET (1950): The photochemistry of atmospheric water vapor. *J. Geophys. Res.*, 55, 301–327.

BECKER, R. H. (1973): Oceanic growth models. *Science*, 182, 601–602.

BELTON, M. J. S., D. M. HUNTEN, and M. B. McELROY (1967): A search for an atmosphere on Mercury. *Astrophys. J.*, 150, 1111–1124.

BERKNER, L. V. and L. C. MARSHALL (1964): The history of growth of oxygen in the earth's atmosphere. In P. J. Brancazio and A. G. W. Cameron, eds., *The Origin and Evolution of Oceans and Atmospheres* (John Wiley and Sons, New York), pp. 102–126.

BERKNER, L. V. and L. C. MARSHALL (1965): On the origin and rise of oxygen concentration in the earth's atmosphere. *J. Atmos. Sci.*, 22, 225–261.

BERKNER, L. V. and L. C. MARSHALL (1966): Limitation on oxygen concentration in a primitive planetary atmosphere. *J. Atmos. Sci.* 23, 133–143.

BERKNER, L. V. and L. C.. MARSHALL (1967): The rise of oxygen in the earth's atmosphere with notes on the Martian atmosphere. *Advances Geophys.*, 12, 309–331.

BERNER, R. A. (1970): Sedimentary pyrite formation. *Amer. J. Sci.*, 268, 1–23.

BERNER, R. A. (1971a): *Principles of Chemical Sedimentology* (McGraw-Hill, New York).

BERNER, R. A. (1971b): Worldwide sulfur pollution of rivers. *J. Geophys. Res.*, 76, 6597–6600.

BERNER, R. A. (1972): Sulfate reduction, pyrite formation, and the oceanic sulfur budget. In D. Dyrssen and D. Jagner, eds., *The Changing Chemistry of the Oceans* (John Wiley and Sons, New York), pp. 347–361.

BERNER, R. A. (1973): Phosphate removal from sea water by adsorption on volcanogenic ferric oxides. *Earth Planet. Sci. Lett.*, 18, 77–86.

BIEMANN, K., T. OWEN, D. R. RUSHNECK, A. L. LAFLEUR, and D. W. HOWARTH

(1976): The atmosphere of Mars near the surface: Isotope ratios and upper limits on noble gases. *Science*, 194, 76–78.

BLUM, P. W. and H. J. FAHR (1970): Interaction between interstellar hydrogen and the solar wind. *Astron. Astrophys.*, 4, 280–290.

BOLIN, B. (1970): The carbon cycle, *Scientific American*, 223, No. 3, 125–132.

BOWEN, H. J. M. (1966): *Trace Elements in Biochemistry* (Academic Press, New York).

BREEDING, R. J., J. P. LODGE, J. B. PATE, D. C. SHEESLEY, H. B. KLONIS, B. FOGLE, J. A. ANDERSON, T. R. ENGLERT, P. L. HAAGENSON, R. B. McBETH, A. L. MORRIS, R. POGUE, and A. F. WARTHURG (1973): Background trace gas concentrations in the central United States. *J. Geophys. Res.*, 78, 7057–7064.

BRINKMANN, R. T. (1969): Dissociation of water vapor and evolution of oxygen in the terrestrial atmosphere. *J. Geophys. Res.*, 74, 5355–5368.

BRINKMANN, R. T. (1970): Departures from Jeans escape rate for H and He in the earth's atmosphere. *Planet. Space Sci.*, 18, 449–478.

BRINKMANN, R. T. (1971a): More comments on the validity of Jeans escape rate. *Planet. Space Sci.*, 19, 791–794.

BRINKMANN, R. T. (1971b): Mars: Has nitrogen escaped? *Science*, 174, 944–945.

BRINTON, H. C. and H. G. MAYR (1971): Temporal variations of thermospheric hydrogen derived from *in situ* measurements. *J. Geophys. Res.*, 76, 6198–6201.

BROECKER, W. S. (1970a): A boundary condition on the evolution of atmospheric oxygen. *J. Geophys. Res.*, 75, 3553–3557.

BROECKER, W. S. (1970b): Man's oxygen reserves. *Science*, 168, 1537–1538.

BROECKER, W. S. (1971): A kinetic model for the chemical composition of sea water. *Quaternary Res.*, 1, 188–207.

BROECKER, W. S., R. D. GERARD, M. EWING, and B. C. HEEZEN (1961): Geochemistry and physics of ocean circulation. In M. Sears, ed., *Oceanography* (American Association for the Advancement of Science, Washington, D.C.) pp. 201–322.

BROECKER, W. S., Y.-H. LI, and T.-H. PENG (1971): Carbon dioxide—man's unseen artifact. In Donald W. Hood, ed., *Impingement of Man on the Oceans* (Wiley–Interscience, New York), pp. 287–324.

BROOKS, C. E. P. (1951): Geological and historical aspects of climatic change. In T. F. Malone, ed., *Compendium of Meteorology* (American Meteorological Society, Boston), pp. 1004–1018.

BROWN, H. (1952): Rare gases and the formation of the earth's atmosphere. In G. P. Kuiper, ed., *The Atmospheres of the Earth and Planets* (Univ. of Chicago Press, Chicago), 2nd ed., pp. 258–266.

BURNS, R. C. and R. W. F. HARDY (1975): *Nitrogen Fixation in Bacteria and Higher Plants* (Springer-Verlag, New York).

CAHILL, L. J. (1965): The magnetosphere. *Scientific American*, 212, No. 3, 58–68.

CALVERT, J. G., J. A. KERR, K. L. DEMERJIAN, and R. D. McQUIGG (1972): Photolysis of formaldehyde as a hydrogen atom source in the lower atmosphere. *Science*, 175, 751–752.

CALVIN, M. (1969): *Chemical Evolution* (Oxford University Press, New York).

CALVIN, M. (1975): Chemical evolution. *Amer. Scientist*, 63, 169–177.

CALVIN, M. and G. J. CALVIN (1964): Atom to Adam. *Amer. Scientist*, 52, 163–186.

CAMERON, A. G. W. (1968): A new table of abundances of the elements in the solar system. In L. H. Ahrens, ed., *Origin and Distribution of the Elements* (Pergamon Press, New York), pp. 125–143.

CAMERON, A. G. W. (1973): Accumulation processes in the primitive solar nebula. *Icarus*, 18, 407–450.

CAMERON, A. G. W. and M. R. PINE (1973): Numerical models of the primitive solar nebula. *Icarus*, 18, 377–406.

CANALAS, R. A., E. C. ALEXANDER, and O. K. MANUEL (1968): Terrestrial abundance of noble gases. *J. Geophys. Res.* 73, 3331–3334.

CHAMBERLAIN, J. W. (1963): Planetary coronae and atmospheric evaporation. *Planet. Space Sci.*, 11, 901–960.

CHAMBERLAIN, J. W. (1969): Escape rate of hydrogen from a carbon dioxide atmosphere. *Astrophys. J.*, 155, 711–714.

CHAMBERLAIN, J. W. and F. J. CAMPBELL (1967): Rate of evaporation of a non-Maxwellian atmosphere. *Astrophys. J.*, 149, 687–705.

CHAMBERLAIN, J. W. and G. R. SMITH (1971): Comments on the rate of evaporation of a non-Maxwellian atmosphere. *Planet. Space Sci.*, 19, 675–684.

CHAMEIDES, W. (1974): *A Photochemical Theory of Tropospheric Ozone*, (Ph.D. dissertation, Yale University).

CHAMEIDES, W. (1975): Tropospheric odd nitrogen and the atmospheric water vapor cycle. *J. Geophys. Res.*, 80, 4989–4996.

CHAMEIDES, W. L. and D. H. STEDMAN (1977): Tropospheric ozone: Coupling transport and photochemistry. *J. Geophys. Res.*, 82, 1787–1794.

CHAMEIDES, W. L., D. H. STEDMAN, R. R. DICKERSON, D. W. RUSCH, and R. J. CICERONE (1977): NO_x production in lightning. *J. Atmos. Sci.* 34, 143–149.

CHAMEIDES, W. and J. C. G. WALKER (1973): A photochemical theory of tropospheric ozone. *J. Geophys. Res.*, 78, 8751–8760.

CHAMEIDES, W. and J. C. G. WALKER (1975): Possible variation of ozone in the troposphere during the course of geologic time. *Amer. J. Sci.*, 275, 737–752.

CHAMEIDES, W. L. and J. C. G. WALKER (1976): A time-dependent photochemical model for ozone near the ground. *J. Geophys Res.*, 81, 413–420.

CHAPMAN, D. J. and R. D. TOCHER (1966): Occurrence and production of carbon monoxide in brown algae. *Can. J. Bot.*, 44, 1438–1442.

CHAPMAN, S. (1926): Ionization in the upper atmosphere. *Quart. J. Roy. Meteorol. Soc.*, 52, 225–236.

CHAPMAN, S. (1930): A theory of upper-atmospheric ozone. *Mem. Roy. Meteorol. Soc.*, 3, 103–125.

CHAPMAN, S. (1931): The absorption and dissociative or ionizing effects of monochromatic radiation in an atmosphere on a rotating earth, 2, Grazing incidence. *Proc. Phys. Soc. London*, 43, 483–501.

CHAPMAN, S. and T. G. COWLING (1970): *The Mathematical Theory of Nonuniform Gases* (Cambridge University Press, London), 3rd ed.

CHASE, C. G. and E. C. PERRY (1972): The oceans: Growth and oxygen isotope evolution. *Science*, 177, 992–994.

CHASE, C. G. and E. C. PERRY (1973): Oceanic growth models: Reply. *Science*, 182, 602–603.

CICERONE, R. J., R. S. STOLARSKI, and S. WALTERS (1974): Stratospheric ozone destruction by man-made chlorofluoromethanes. *Science*, 185, 1165–1167.

CLARK, S. P. (1971): *Structure of the Earth* (Prentice-Hall, Englewood Cliffs, New Jersey).

CLARK, S. P., K. K. TUREKIAN, and L. GROSSMAN (1972): Model for the early history of the earth. In E. C. Robertson, ed., *The Nature of the Solid Earth* (McGraw-Hill, New York), pp. 3–18.

CLARKE, F. W. (1924): *The Data of Geochemistry* (5th ed.). *U.S. Geol. Survey Bull.*, 770, 518.

CLOUD, P. E. (1968): Atmospheric and hydrospheric evolution on the primitive Earth. *Science*, 160, 729–736.

CLOUD, P. (1972): A working model of the primitive earth. *Amer. J. Sci.*, 272, 537–548.

CLOUD, P. (1973): Paleoecological significance of the banded-iron formation. *Econ. Geol.*, 68, 1135–1143.

COLE, K. D. (1966): Theory of some quiet magnetospheric phenomena related to the geomagnetic tail. *Nature*, 211, 1385–1387.

COLEGROVE, F. D., F. S. JOHNSON, and W. B. HANSON (1966): Atmospheric composition in the lower thermosphere. *J. Geophys. Res.*, 71, 2227–2236.

COMMONER, B. (1965): Biochemical, biological, and atmospheric evolution. *Proc. U.S. Nat. Acad. Sci.*, 53, 1183–1194.

COMMONER, B. (1970): Threats to the integrity of the nitrogen cycle: Nitrogen compounds in soil, water, atmosphere and precipitation. In S. F. Singer, ed., *Global Effects of Environmental Pollution* (Springer-Verlag, New York), pp. 70–95.

CRAIG, H. and W. B. CLARKE (1970): Oceanic ^{3}He: Contribution from cosmogenic tritium. *Earth Planet. Sci. Lett.*, 9, 45–48.

CRAIG, H. and R. F. WEISS (1971): Dissolved gas saturation anomalies and excess helium in the ocean. *Earth Planet. Sci. Lett.*, 10, 289–296.

CRIMES, T. P. (1974); Colonization of the early ocean floor. *Nature*, 248, 328–330.

CROSS, C. A. (1971): The heat balance of the Martian polar caps. *Icarus*, 15, 110–114.

CRUIKSHANK, D. P., D. MORRISON, and K. LENNON (1973): Volcanic gases: Hydrogen burning at Kilauea Volcano, Hawaii. *Science*, 182, 277–279.

CRUTZEN, P. J. (1970): The influence of nitrogen oxides on the atmospheric ozone content. *Quart. J. Roy. Meteorol. Soc.*, 96, 320–325.

CRUTZEN, P. J. (1971): Ozone production rates in an oxygen–hydrogen–nitrogen oxide atmosphere. *J. Geophys. Res.*, 76, 7311–7327.

CRUTZEN, P. J. (1973): Gas-phase nitrogen and methane chemistry in the atmosphere. In B. M. McCormac, ed., *Physics and Chemistry of Upper Atmospheres* (D. Reidel, Dordrecht, Holland), pp. 110–124.

CRUTZEN, P. J. (1974a): Photochemical reactions initiated by and influencing ozone in the troposphere. *Tellus*, 26, 47–57.

CRUTZEN, P. (1974b): A review of upper atmospheric photochemistry. *Can. J. Chem.*, 52, 1569–1581.

DALGARNO, A. (1969): Infrared day and night airglow of the earth's upper atmosphere. *Phil. Trans. Roy. Soc. London Ser. A Math. Phys. Sci.*, 264, 153–160.

DALGARNO, A. and M. B. McELROY (1970): Mars: Is nitrogen present? *Science*, 170, 167–168.

DAMON, P. E. (1971): The relationship between late Cenozoic volcanism and tectonism and orogenic–epeirogenic periodicity. In K. K. Turekian, ed., *The Late Cenozoic Glacial Ages* (Yale Univ. Press, New Haven, Conn.), pp. 15–35.

DAVIDSON C. F. (1965): Geochemical aspects of atmospheric evolution. *Proc. U.S. Nat. Acad. Sci.*, 53, 1194–1205.

DAWSON, J. B. (1971): Advances in kimberlite geology. *Earth Sci. Rev.*, 7, 187–214.

DAYHOFF, M. O., R. V. ECK, E. R. LIPPINCOTT, and C. SAGAN (1967): Venus: Atmospheric evolution. *Science*, 155, 556–558.

DEEVEY, E. S. and M. STUIVER (1964): Distribution of natural isotopes of carbon in Linsley Pond and other New England lakes. *Limnol. Oceanogr.*, 9, 1–11.

DEGENS, E. T. (1969): Biogeochemistry of stable carbon isotopes. In G. Eglinton and M. T. J. Murphy, eds., *Organic Geochemistry; Methods and Results* (Springer-Verlag, Berlin), pp. 304–329.

DETWILER, G. R., D. L. GARRETT, J. D. PURCELL, and R. TOUSEY (1961): The intensity distribution in the ultraviolet solar spectrum. *Ann Geophys.*, 17, 9–18.

DEUSER, W. G. (1971): Organic-carbon budget of the Black Sea. *Deep-Sea Res.*, 18, 995–1004.

DEUSER, W. G., E. T. DEGENS, G. R. HARVEY, and M. RUBIN (1973): Methane in Lake Kivu: New data bearing on its origin. *Science*, 181, 51–54.

DOEMEL, W. N. and T. D. BROCK (1974): Bacterial stromatolites: Origin of laminations. *Science*, 184, 1083–1085.

DOLE, S. H. (1970): *Habitable Planets for Man* (American Elsevier, New York), 2nd ed.

DONAHUE, T. M. (1966): The problem of atomic hydrogen. *Ann. Geophys.*, 22, 175–188.

DONN, W. L., B. D. DONN, and W. G. VALENTINE (1965): On the early history of the earth. *Geol. Soc. Amer. Bull.*, 76, 287–306.

DUEDALL, I. W. and A. R. COOTE (1972): Oxygen distribution in the Pacific Ocean. *J. Geophys. Res.*, 77, 2201–2203.

DUNKIN, D. B., F. G. FEHSENFELD, A. L. SCHMELTEKOPF, and E. E. FERGUSON (1968): Ion–molecule reaction studies from 300°K to 600°K in a temperature-controlled, flowing afterglow system. *J. Chem. Phys.*, 49, 1365–1371.

DYMOND, J. and L. HOGAN (1973): Noble gas abundance patterns in deep sea basalts—primordial gases from the mantle. *Earth Planet Sci. Lett.*, 20, 131–139.

EHHALT, D. H. (1967): Methane in the atmosphere. *J. Air Pollut. Control*, 17, 518–519.

EHHALT, D. H. and L. E. HEIDT (1973): Vertical profiles of CH_4 in the troposphere and stratosphere. *J. Geophys. Res.*, 78, 5265–5271.

EHHALT, D. H., L. E. HEIDT, and E. A. MARTELL (1972): The concentration of atmospheric methane between 44 and 62 kilometers altitude. *J. Geophys. Res.*, 77, 2193–2196.

EICHER, D. L. (1968): *Geologic Time* (Prentice-Hall, Englewood Cliffs, New Jersey).

EICHMANN, R. and M. SCHIDLOWSKI (1975): Isotopic fractionation between coexisting organic carbon–carbonate pairs in Precambrian sediments. *Geochim. Cosmochim. Acta*, 39, 585–595.

EMERY, K. O. (1960): *The Sea off Southern California* (John Wiley and Sons, New York).

ERIKSSON, E. (1952): Composition of atmospheric precipitation. 1. Nitrogen compounds. *Tellus*, 4, 215–232.

ERIKSSON, E. (1963): Possible fluctuations in atmospheric carbon dioxide due to changes in the properties of the sea. *J. Geophys. Res.*, 68, 3871–3876.

ERNST, W. G. (1969): *Earth Materials* (Prentice-Hall, Englewood Cliffs, New Jersey).

EUGSTER, H. P. (1972): Ammonia in minerals and early atmosphere. In R. W. Fairbridge, ed., *The Encyclopedia of Geochemistry and Environmental Sciences* (Van Nostrand, New York), pp. 29–33.

EUGSTER, H. P. and J. MUNOZ (1966): Ammonium micas: Possible sources of atmospheric ammonia and nitrogen. *Science*, 151, 683–686.

EZER, D. and A. G. W. CAMERON (1963): The early evolution of the sun. *Icarus*, 1, 422–441.

EZER, D. and A. G. W. CAMERON (1971): Pre-main sequence stellar evolution with mass loss. *Astrophys. Space Sci.*, 10, 52–70.

FABIAN, P. (1973): A theoretical investigation of tropospheric ozone and stratospheric–tropospheric exchange processes. *Pure Appl. Geophys.*, 106–108, 1044–1057.

FABIAN, P. (1974): Comments on "A photochemical theory of tropospheric ozone" by W. Chameides and J. C. G. Walker. *J. Geophys. Res.*, 79, 4124–4125.

FABIAN, P. and C. E. JUNGE (1970): Global rate of ozone destruction at the earth's surface. *Arch. Met. Geophys. Biokl. A*, 19, 161–172.

FABIAN, P. and P. G. PRUCHNIEWICZ (1973): Meridional distribution of tropospheric ozone from ground-based registrations between Norway and South Africa. *Pure Appl. Geophys.*, 106–108, 1027–1035.

FABIAN, P., P. G. PRUCHNIEWICZ, and A. ZAND (1971): Transport und Austauschvorgänge in der Atmosphäre. *Naturwissenschaft.*, 58, 541–549.

FAHR, H. J. (1969): Influence of interstellar matter on the density of atmospheric hydrogen. *Ann. Geophys.*, 25, 475–478.

FAHR, H. J. (1974): The extraterrestrial UV background and the nearby interstellar medium. *Space Sci. Rev.*, 15, 483–540.

FANALE, F. P. (1971a): A case for catastrophic early degassing of the earth. *Chem. Geol.*, 8, 79–105.

FANALE, F. P. (1971b): History of Martian volatiles: Implications for organic synthesis. *Icarus*, 15, 279–303.

FANALE, F. P. (1976): Martian volatiles: Their degassing history and geochemical fate. *Icarus*, 28, 179–202.

FANALE, F. P. and W. A. CANNON (1971a): Physical adsorption of rare gas on terrigenous sediments. *Earth Planet. Sci. Lett.*, 11, 362–368.

FANALE, F. P. and W. A. CANNON (1971b): Adsorption on the Martian regolith. *Nature*, 230, 502–504.

FANALE, F. P. and W. A. CANNON (1974): Exchange of adsorbed H_2O and CO_2 between the regolith and atmosphere of Mars caused by changes in surface insolation. *J. Geophys. Res.*, 79, 3397–3402.

FEHSENFELD, F. C. and E. E. FERGUSON (1972): Thermal energy reaction rate constants for H^+ and CO^+ with O and NO. *J. Chem. Phys.*, 56, 3066–3070.

FERGUSON, E. E. (1973): Rate constants of thermal energy binary ion–molecule reactions of aeronomic interest. *Atomic Data and Nuclear Data Tables*, 12, 159–178.

FERRIS, J. P. and D. E. NICODEM (1972): Ammonia photolysis and the role of ammonia in chemical evolution, *Nature*, 238, 268–269.

FINK, U., H. P. LARSON, G. P. KUIPER, and R. F. POPPER (1972): Water vapor in the atmosphere of Venus. *Icarus*, 17, 617–631.

FISCHER, A. G. (1965): Fossils, early life, and atmospheric history. *Proc. U.S. Nat. Acad. Sci.*, 53, 1205–1213.

FISHER, D. E. (1974): The planetary primordial component of rare gases in the deep earth. *Geophys. Res. Lett.*, 1, 161–164.

FISHER, D. E. (1976a): Rare gas clues to the origin of the terrestrial atmosphere. In B. F. Windley, ed., *The Early History of the Earth* (John Wiley and Sons, New York), pp. 547–556.

FISHER, D. E. (1976b): Trapped helium and argon and the formation of the atmosphere by degassing. *Nature*, 256, 113–114.

FITE, W. L. (1969): Positive ion reactions. *Can. J. Chem.*, 47, 1797–1808.

FRIDOVICH, I. (1975): Oxygen: Boon and bane. *Amer. Scientist*, 63, 54–59.

FRIEDMAN, H. (1960): The sun's ionizing radiations. In J. A. Ratcliffe, ed., *Physics of the Upper Atmosphere* (Academic Press, New York), pp. 133–218.

GABEL, N. W. and C. PONNAMPERUMA (1972): Primordial organic chemistry. In C. Ponnamperuma, ed., *Exobiology, Frontiers in Biology* (North-Holland, Amsterdam), Vol. 23, pp. 95–135.

GARRELS, R. M. (1965): Silica: Role in the buffering of natural waters. *Science*, 148, 69.

GARRELS, R. M. and F. T. MACKENZIE (1971): *Evolution of Sedimentary Rocks* (W. W. Norton and Co., New York).

GARRELS, R. M. and F. T. MACKENZIE (1972): A quantative model for the sedimentary rock cycle. *Marine Chemistry*, 1, 27–41.

GARRELS, R. M. and E. A. PERRY (1974): Cycling of carbon, sulfur, and oxygen through geologic time. In E. Goldberg, ed., *The Sea* (Wiley–Interscience, New York), Vol. 5, pp. 303–336.

GARRELS, R. M., E. A. PERRY, and F. T. MACKENZIE (1973): Genesis of Precambrian iron formations and the development of atmospheric oxygen. *Econ. Geol.*, 68, 1173–1179.

GEORGII, H.-W. (1963): Oxides of nitrogen and ammonia in the atmosphere. *J. Geophys. Res.*, 68, 3963–3970.

GIBBS, M. (1970): The inhibition of photosynthesis by oxygen. *Amer. Scientist*, 58, 634–640.

GIERASCH, P. and R. GOODY (1967): An approximate calculation of radiative heating and radiative equilibrium in the Martian atmosphere. *Planet. Space Sci.*, 15, 1465–1477.

GIERASCH, P. J. and O. B. TOON (1973): Atmospheric pressure variations and the climate of Mars. *J. Atmos. Sci.*, 30, 1502–1508.

GILBERT, D. L. (1972): Oxygen and life. *Anesthesiology* 37, 100–111.

GILLULY, J., J. G. REED, and W. M. CADY (1970): Sedimentary volumes and their significance. *Geol. Soc. Amer. Bull.*, 81, 353–376.

GLIKSON, A. Y. (1970): Geosynclinal evolution and geochemical affinities of early Precambrian systems. *Tectonophys.*, 9, 397–433.

GOERING, J. J., R. C. DUGDALE, and D. W. MENZEL (1966): Estimates of *in situ* rates of nitrogen uptake by *Trichodesmium sp.* in the tropical Atlantic Ocean. *Limnol. Oceanogr.*, 11, 614–620.

GOERING, J. J., F. A. RICHARDS, I. A. CODISPOTI, and R. C. DUGDALE (1973): Nitrogen fixation and denitrification in the ocean: Biogeochemical budgets. In E. Ingerson, ed., *Proceedings of Symposium on Hydrogeochemistry and Biogeochemistry*, Vol. II, *Biogeochemistry* (The Clarke Co., Washington), pp. 12–27.

GOLD, T. (1964): Outgassing processes on the Moon and Venus. In P. J. Brancazio and A. G. W. Cameron, eds., *The Origin and Evolution of Atmospheres and Oceans*, (John Wiley and Sons, New York), pp. 249–256.

GOLDMAN, A., D. G. MURCRAY, F. H. MURCRAY, W. J. WILLIAMS, and J. N. BROOKS (1973): Distribution of water vapor in the stratosphere as determined from balloon measurements of atmospheric emission spectra in the 24–29 μm region. *Appl. Optics*, 12, 1045–1053.

GOODY, R. M. (1964): *Atmospheric Radiation* (Oxford University Press, London).

GOODY, R. M. (1965): The structure of the Venus cloud veil. *J. Geophys. Res.*, 70, 5471–5481.

GOODY, R. M. and A. R. ROBINSON (1966): A discussion of the deep circulation of the atmosphere of Venus. *Astrophys. J.*, 146, 339–355.

GOODY, R. M. and J. C. G. WALKER (1972): *Atmospheres* (Prentice-Hall, Englewood Cliffs, New Jersey).

GREEN, A. E. S., T. SAWADA, B. C. EDGAR, and M. A. UMAN (1973): Production of carbon monoxide by charged particle deposition. *J. Geophys. Res.*, 78, 5284–5291.

GREEN, H. W. (1972): A CO_2 charged asthenosphere. *Nature, Phys. Sci.*, 238, 2–5.

GREGOR, C. B. (1968): The rate of denudation in post-Algonkian time. *Koninkl. Ned. Akad. Wetenschap. Proc.*, 71, 22–30.

GREGOR, B. (1970): Denudation of the continents. *Nature*, 228, 273–275.

GREGOR, B. (1971): Carbon and atmospheric oxygen. *Science*, 174, 316–317.

GREINER, N. R. (1969): Hydroxyl radical kinetics by kinetic spectroscopy. V. Reactions with H_2 and CO in the range 300–500°K. *J. Chem. Phys.*, 51, 5049–5051.

GROSSMAN, L. (1972): Equilibrium condensation in the primitive solar nebula. *Geochim. Cosmochim. Acta*, 36, 597–619.

GROSSMAN, L. and J. W. LARIMER (1974): Early chemical history of the solar system. *Rev. Geophys. Space Phys.*, 12, 71–101.

GUDIKSEN, P. H., A. W. FAIRHALL, and A. J. REED (1968): Roles of mean meridional circulation and eddy diffusion in the transport of trace substances in the lower atmosphere. *J. Geophys. Res.*, 73, 4461–4473.

HAHN, J. (1974): The North Atlantic Ocean as a source of atmospheric N_2O. *Tellus*, 26, 160–168.

HALDANE, J. B. S. (1964): Genesis of life. In D. R. Bates, ed., *The Planet Earth* (Pergamon Press, New York), pp. 325–341.

HALL, J. B. (1971): Evolution of the prokaryotes. *J. Theor. Biol.*, 30, 429–454.

HAM, D. O., D. W. TRAINOR, and F. KAUFMAN (1970): Gas phase kinetics of $H + H + H_2 \rightarrow 2H_2$. *J. Chem. Phys.*, 53, 4395–4396.

HAMMOND, A. L. and T. H. MAUGH (1974): Stratospheric pollution: Multiple threats to earth's ozone. *Science*, 186, 335–338.

HANKS, T. C. and D. L. ANDERSON (1969): The early thermal history of the earth. *Phys. Earth Planet. Inter.*, 2, 19–29.

HAYASHI, C. (1961): Stellar evolution in early phases of gravitational contraction. *Pub. Astronom. Soc. Japan*, 13, 450–452.

HAYES, J. M. (1967): Organic constituents of meteorites—a review. *Geochim. Cosmochim. Acta*, 31, 1395–1440.

HAYS, J. D. (1971): Faunal extinctions and reversals of the earth's magnetic field. *Bull. Geol. Soc. Amer.*, 82, 2433–2447.

HAYS, J. D. and W. C. PITMAN, III (1973): Lithospheric plate motion, sea level changes and climatic and ecological consequences. *Nature*, 246, 18–22.

HAYS, P. B. and V. C. LIU (1965): On the loss of gases from a planetary atmosphere. *Planet. Space Sci.*, 13, 1185–1212.

HEALD, E. F., J. NAUGHTON, and I. L. BARNES (1963): The chemistry of volcanic gases: Use of equilibrium calculations in the interpretation of volcanic gas samples. *J. Geophys. Res.*, 68, 545–557.

HERING, W. S. and T. R. BORDEN (1964): *Ozonesonde Observations over North America* (Air Force Cambridge Research Laboratories, Bedford, Massachusetts), Vols. 1, 2.

HESS, S. L. (1959): *Introduction to Theoretical Meteorology* (Holt, Rinehart, and Winston, New York).

HILLS, J. G. (1973): On the process of accretion in the formation of the planets and comets. *Icarus*, 18, 505–522.

HINTEREGGER, H. E. (1970): The extreme ultraviolet solar spectrum and its variation during a solar cycle. *Ann. Geophys.*, 26, 547–554.

HOLEMAN, J. N. (1968): The sediment yield of major rivers of the world. *Water Resources Res.*, 4, 737–747.

HOLLAND, H. D. (1962): Model for the evolution of earth's atmosphere. In A. E. J. Engle, H. L. James, and B. F. Leonard, eds., *Petrologic Studies: A Volume in Honor of A. F. Buddington* (Geological Society of America, New York), pp. 447–477.

HOLLAND, H. D. (1964): On the chemical evolution of the terrestrial and cytherean atmospheres. In P. J. Brancazio and A. G. W. Cameron, eds., *The Origin and Evolution of Atmospheres and Oceans* (John Wiley and Sons, New York), pp. 86–101.

HOLLAND, H. D. (1972): The geologic history of sea water—an attempt to solve the problem. *Geochim. Cosmochim. Acta*, 36, 637–651.

HOLLAND, H. D. (1973a): Ocean water, nutrients, and atmospheric oxygen. In *Proceedings of Symposium on Hydrogeochemistry and Biogeochemistry*, Vol. 1 (The Clarke Co., Washington), pp. 68–81.

HOLLAND, H. D. (1973b): The oceans: A possible source of iron in iron formations. *Econ. Geol.*, 68, 1169–1172.

HOLLAND, H. D. (1973c): Systematics of the isotopic composition of sulfur in the oceans during the Phanerozoic and its implications for atmospheric oxygen. *Geochim. Cosmochim. Acta*, 37, 2605–2616.

HOLLAND, H. D. (1975): Comment at a conference on The Early History of the Earth (University of Leicester, England).

HOLSER, W. J. and I. R. KAPLAN (1966): Isotope geochemistry of sedimentary sulfates. *Chem. Geol.*, 1, 93–135.

HOLZER, T. E. and W. I. AXFORD (1971): Interaction between interstellar helium and the solar wind. *J. Geophys. Res.*, 76, 6965–6970.

HOLZER, T. E., J. A. FEDDER, and P. M. BANKS (1971): A comparison of kinetic and hydrodynamic models of an expanding ion-exosphere. *J. Geophys. Res.*, 76, 2453–2468.

HOUCK, J. R., J. B. POLLACK, C. SAGAN, D. SCHAACK, and J. A. DECKER (1973): High-altitude spectroscopic evidence for bound water on Mars. *Icarus*, 18, 470–480.

HOUCK, J. R., J. B. POLLACK, D. SCHAACK, R. A. REED, and A. SUMMERS (1975): Jupiter: Its infrared spectrum from 16 to 40 micrometers. *Science*, 189, 720–722.

HUBBERT, M. K. (1969): Energy resources. In *Resources and Man* (Committee on Resources and Man of the National Academy of Sciences, National Research Council, W. H. Freeman and Co., San Francisco), pp. 157–242.

HUDSON, R. D. (1971): Critical review of ultraviolet photoabsorption cross sections for molecules of astrophysical and aeronomic interest. *Rev. Geophys. Space Phys.*, 9, 305–406.

HUDSON, R. D. (1974): Absorption cross sections of stratospheric molecules. *Can. J. Chem.*, 52, 1465–1478.

HUGUENIN, R. L. (1976a): Mars: Chemical weathering as a massive volatile sink. *Icarus*, 28, 203–212.

HUGUENIN, R. L. (1976b): Surface oxidation: A major sink for water on Mars. *Science*, 192, 138–139.

HUNTEN, D. M. (1971): Composition and structure of planetary atmospheres. *Space Sci. Rev.*, 12, 539–599.

HUNTEN, D. M. (1973): The escape of light gases from planetary atmospheres, *J. Atmos. Sci.*, 30, 1481–1494.

HUNTEN, D. M. and R. M. GOODY (1969): Venus: The next phase of planetary exploration. *Science*, 165, 1317–1323.

HUNTEN, D. M. and M. B. McELROY (1970): Production and escape of hydrogen on Mars. *J. Geophys. Res.*, 75, 5989–6001.

HUNTEN, D. M. and G. MÜNCH (1973): The helium abundance on Jupiter. *Space Sci. Rev.*, 14, 433–443.

HUNTEN, D. M. and D. F. STROBEL (1973): Production and escape of terrestrial hydrogen. *J. Atmos. Sci.*, 31, 305–317.

HUTCHINSON, G. E. (1944): Nitrogen in the biogeochemistry of the atmosphere. *Amer. Scientist*, 32, 178–195.

HUTCHINSON, G. E. (1954): The biochemistry of the terrestrial atmosphere. In G. P. Kuiper, ed., *The Earth as a Planet* (Univ. of Chicago Press, Chicago), pp. 371–433.

HUTCHINSON, G. E. (1973): Eutrophication. *Amer. Scientist*, 61, 269–279.

INGERSOLL, A. P. (1969): The runaway greenhouse: A history of water on Venus. *J. Atmos. Sci.*, 26, 1191–1198.

INGERSOLL, A. P. (1974): Mars: The case against permanent CO_2 frost caps. *J. Geophys. Res.*, 79, 3403–3410.

INGERSOLL, A. P. and C. B. LEOVY (1971): The atmospheres of Mars and Venus. *Ann. Rev. Astrophys.*, 9, 147–182.

JACOBI, W. and K. ANDRÉ (1963): The vertical distribution of radon 222, radon 220, and their decay products in the atmosphere. *J. Geophys. Res.*, 68, 3799–3814.

JAFFE, L. S. (1973): Carbon monoxide in the biosphere: Sources, distribution, and concentrations. *J. Geophys. Res.*, 78, 5293–5305.

JANSSEN, M. A., R. E. HILLS, D. D. THORNTON, and W. J. WELCH (1973): Venus: New microwave measurements show no atmospheric water vapor. *Science*, 179, 994–997.

JEANS, J. H. (1925): *The Dynamical Theory of Gases* (Cambridge Univ. Press, London).

JOHNSON, H. E. and W. I. AXFORD (1969): Production and loss of He^3 in the earth's atmosphere. *J. Geophys. Res.*, 74, 2433–2438.

JOSEPH, J. H. (1967): Diurnal and solar variations of neutral hydrogen in the thermosphere. *Ann. Geophys.*, 23, 365–374.

JUDSON, S. (1968): Erosion of the land. *Amer. Scientist*, 56, 356–374.

JUDSON, S. and D. F. RITTER (1964): Rates of regional denudation in the United States. *J. Geophys. Res.*, 69, 3395–3401.

JUNGE, C. E. (1962): Global ozone budget and exchange between stratosphere and troposphere. *Tellus*, 14, 363–377.

JUNGE, C. E. (1963): *Air Chemistry and Radioactivity* (Academic Press, New York).

JUNGE, C. E. (1972): The cycle of atmospheric gases—natural and man made. *Quart. J. Roy. Meteorol. Soc.*, 98, 711–729.

JUNGE, C. E. (1974): Residence time and variability of tropospheric trace gases. *Tellus*, 26, 477–488.

KAPLAN, L. D. (1973): Background concentrations of photochemically active trace constituents in the stratosphere and upper troposphere. *Pure Appl. Geophys.*, 106–108, 1341–1345.

KAULA, W. M. (1968): *An Introduction to Planetary Physics* (John Wiley and Sons, New York).

KEITH, M. L. and J. N. WEBER (1964): Carbon and oxygen isotopic composition of selected limestones and fossils. *Geochim. Cosmochim. Acta*, 28, 1787–1816.

KELLOG, W. W. and S. H. SCHNEIDER (1974): Climate stabilization: For better or for worse? *Science*, 186, 1163–1172.

KELNER, A. (1969): Biological aspects of ultraviolet damage, photoreactivation and other repair systems in microorganisms. In F. Urbach, ed., *The Biological Effects of Ultraviolet Radiation* (Pergamon Press, New York), pp. 77–82.

KENNEDY, G. C. and B. E. NORDLIE (1968): The genesis of diamond deposits. *Econ. Geol.*, 63, 495–503.

KOBLENTZ-MISHKE, O. J., V. V. VOLKOVINSKY, and J. G. KABANOVA (1970): Plankton primary production of the world ocean. In W. S. Wooster, ed., *Scientific Exploration of the South Pacific* (National Academy of Sciences, Washington), pp. 183–193.

KOCKARTS, G. (1971): Penetration of solar radiation in the Schumann–Runge bands of molecular oxygen. In G. Fiocco, ed., *Mesospheric Models and Related Experiments* (D. Reidel, Dordrecht, Holland), pp. 160–176.

KOCKARTS, G. (1972): Distribution of hydrogen and helium in the upper atmosphere. *J. Atmos. Terr. Phys.*, 34, 1729–1743.

KOCKARTS, G. (1973): Helium in the terrestrial atmosphere. *Space Sci. Rev.*, 14, 723–757.

KOYAMA, T. (1963): Gaseous metabolism in lake sediments and paddy soils and the production of atmospheric methane and hydrogen. *J. Geophys. Res.*, 68, 3971–3973.

KU, T-L., W. S. BROECKER, and N. OPDYKE (1968): Comparison of sedimentation rates measured by paleomagnetic and the ionium methods of age determination. *Earth Planet. Sci. Lett.*, 4, 1–16.

KUENEN, P. H. (1950): *Marine Geology* (John Wiley and Sons, New York).

KUIPER, G. P. (1949): Planetary atmospheres and their origin. In G. P. Kuiper, ed., *The Atmospheres of the Earth and Planets* (Univ. of Chicago Press, Chicago), pp. 306–405.

KUMAR, S. and D. M. HUNTEN (1974): Venus: An ionospheric model with an exospheric temperature of 350°K. *J. Geophys. Res.*, 79, 2529–2532.

KUMMLER, R. H. and T. BAURER (1973): A temporal model of tropospheric carbon–hydrogen chemistry. *J. Geophys. Res.*, 78, 5306–5316.

KURYLO, M. J. (1972): Absolute rate constant for the reaction $H + O_2 + M \rightarrow HO_2 + M$ over the temperature range 203–404°K. *J. Phys. Chem.*, 76, 3518–3526.

LaHUE, M. D., J. B. PATE, and J. P. LODGE (1970): Atmospheric nitrous oxide concentrations in the humid tropics. *J. Geophys. Res.*, 75, 2922–2926.

LAMONTAGNE, R. A., J. W. SWINNERTON, and V. J. LINNENBOM (1971): Non-equilibrium of carbon monoxide and methane at the air–sea interface. *J. Geophys. Res.*, 76, 5117–5121.

LAMONTAGNE, R. A., J. W. SWINNERTON, V. J. LINNENBOM, and W. D. SMITH (1973): Methane concentrations in various marine environments. *J. Geophys. Res.*, 78, 5317–5324.

LEIGHTON, P. A. (1961): *Photochemistry of Air Pollution* (Academic Press, New York).

LEIGHTON, R. B. and B. G. MURRAY (1966): Behavior of carbon dioxide and other volatiles on Mars. *Science*, 153, 136–144.

LEOVY, C. and Y. MINTZ (1969): Numerical simulation of the atmospheric circulation and climate of Mars. *J. Atmos. Sci.*, 26, 1167–1190.

LEVINE, J. S. (1976): A new estimate of volatile outgassing on Mars. *Icarus*, 28, 165–169.

LEVINE, J. S. and G. R. RIEGLER (1974): Argon in the Martian atmosphere. *Geophys. Res. Lett.*, 1, 285–287.

LEVY, H. (1971): Normal atmosphere: Large radical and formaldehyde concentrations predicted. *Science*, 173, 141–143.

LEVY, H. (1972): Photochemistry of the lower troposphere. *Planet. Space Sci.*, 20, 919–935.

LEVY, H. (1973a): Photochemistry of minor constituents in the troposphere. *Planet. Space Sci.*, 21, 575–591.

LEVY, H. (1973b): Tropospheric budgets for methane, carbon monoxide and related species. *J. Geophys. Res.*, 78, 5325–5332.

LEWIS, J. S. (1968): An estimate of the surface conditions of Venus. *Icarus*, 8, 434-456.

LEWIS, J. S. (1970): Venus: Atmospheric and lithospheric composition. *Earth Planet. Sci. Lett.*, 10, 73–80.

LEWIS, J. S. (1971): The atmosphere, clouds, and surface of Venus. *Amer. Scientist*, 59, 557–566.

LEWIS, J. S. (1972): Low temperature condensation from the solar nebula. *Icarus*, 16, 241–252.

LEWIS, J. S. (1974): The temperature gradient in the solar nebula. *Science*, 186, 440–443.

LI, Y.-H. (1972): Geochemical mass balance among lithosphere, hydrosphere, and atmosphere. *Amer. J. Sci.*, 272, 119–137.

LI, Y.-H., T. TAKAHASHI, and W. S. BROECKER (1969): Degree of saturation of $CaCO_3$ in the oceans. *J. Geophys. Res.*, 74, 5507–5525.

LINNENBOM, V. J., J. W. SWINNERTON, and R. A. LAMONTAGNE (1973): The ocean as a source for atmospheric carbon monoxide. *J. Geophys. Res.*, 78, 5333–5340.

LISS, P. S. and P. G. SLATER (1974): Flux of gases across the air–sea interface. *Nature*, 247, 181–184.

LIU, S. C. and T. M. DONAHUE (1974a): The aeronomy of hydrogen in the atmosphere of the earth. *J. Atmos. Sci.*, 31, 1118–1136.

LIU, S. C. and T. M. DONAHUE (1974b): Mesospheric hydrogen related to exospheric escape mechanisms. *J. Atmos. Sci.*, 31, 1466–1470.

LIU, S. C. and T. M. DONAHUE (1974c): Realistic model of hydrogen constituents in the lower atmosphere and escape flux from the upper atmosphere. *J. Atmos. Sci.*, 31, 2238–2242.

LIU, S. C. and T. M. DONAHUE (1976): The regulation of hydrogen and oxygen escape from Mars. *Icarus*, 28, 231–246.

LODGE, J. P., P. A. MACHADO, J. P. PATE, D. C. SHEESLEY, and A. F. WARTBURG (1974): Atmospheric trace chemistry in the American humid tropics. *Tellus*, 26, 250–253.

LODGE, J. P. and J. B. PATE (1966): Atmospheric gases and particulates in Panama. *Science*, 153, 408–410.

LOEWUS, M. W. and C. C. DELWICKE (1963): Carbon monoxide production by algae. *Plant Physiol.*, 38, 371–374.

LOVELOCK, J. E. and J. P. LODGE (1972): Oxygen in the contemporary atmosphere. *Atmospheric Environment*, 6, 575–578.

LOVELOCK, J. E. and L. MARGULIS (1974): Homeostatic tendencies of the earth's atmosphere. *Origins of Life.* 5, 93–103.

LUPTON, J. E. (1973): Direct accretion of ^{3}He and ^{3}H from cosmic rays. *J. Geophys. Res.*, 78, 8330–8337.

MCALESTER, A. L. (1968): *The History of Life* (Prentice-Hall, Englewood Cliffs, New Jersey).

MCALESTER, A. L. (1970): Animal extinctions, oxygen consumption, and Atmospheric history. *J. Paleontol.*, 44, 405–409.

MCALESTER, A. L. (1973): Phanerozoic biotic crises. In *The Permian and Triassic Systems and Their Mutual Boundary* (Canadian Society of Petroleum Geologists Memoir No. 2), pp. 11–15.

MCCONNELL, J. C. (1973): Atmospheric ammonia. *J. Geophys. Res.*, 78, 7812–7821.

MCCONNELL, J. C. and M. B. MCELROY (1973): Odd nitrogen in the atmosphere. *J. Atmos. Sci.*, 30, 1465–1480.

MCCONNELL, J. C., M. B. MCELROY, and S. C. WOFSY (1971): Natural sources of atmospheric CO. *Nature*, 233, 187–188.

MACDONALD, G. J. F. (1963): The escape of helium from the earth's atmosphere. *Rev. Geophys.*, 1, 305–349.

MACDONALD, G. J. F. (1964): The escape of helium from the earth's atmosphere. In P. J. Brancazio and A. G. W. Cameron, eds., *The Origin and Evolution of Atmospheres and Oceans* (John Wiley and Sons, New York), pp. 127–182.

MCELROY, M. B. (1967): The upper atmosphere of Mars. *Astrophys. J.*, 150, 1125–1138.

MCELROY, M. B. (1968): The upper atmosphere of Venus in light of the Mariner 5 measurements. *J. Atmos Sci.*, 25, 574–577.

MCELROY, M. B. (1969a): Atmospheric composition of the Jovian planets. *J. Atmos. Sci.*, 26, 798–812.

MCELROY, M. B. (1969b): Structure of the Venus and Mars atmospheres. *J. Geophys. Res.*, 74, 29–41.

MCELROY, M. B. (1972): Mars: An evolving atmosphere. *Science*, 175, 443–445.

MCELROY, M. B. (1974): Comment at a conference on The Atmosphere of Venus (Goddard Institute for Space Studies, New York, October 15–17, 1974).

MCELROY, M. B. and T. M. DONAHUE (1972): Stability of the Martian atmosphere. *Science*, 177, 986–988.

MCELROY, M. B. and D. M. HUNTEN (1969a): The ratio of deuterium to hydrogen in the Venus atmosphere. *J. Geophys. Res.*, 74, 1720–1739.

MCELROY, M. B. and D. M. HUNTEN (1969b): Molecular hydrogen in the atmosphere of Mars. *J. Geophys. Res.*, 74, 5807–5809.

MCELROY, M. B. and J. C. MCCONNELL (1971a): Dissociation of CO_2 in the Martian atmosphere. *J. Atmos. Sci.*, 28, 879–884.

MCELROY, M. B. and J. C. MCCONNELL (1971b): Nitrous oxide: A natural source of stratospheric NO. *J. Atmos. Sci.*, 28, 1095–1098.

MCELROY, M. B., S. C. WOFSY, J. E. PENNER, and J. C. MCCONNELL (1974): Atmospheric ozone: Possible impact of stratospheric aviation. *J. Atmos. Sci.*, 31, 287–303.

McGovern, W. E. (1969): The primitive earth: Thermal models of the upper atmosphere for a methane dominated environment. *J. Atmos. Sci.*, 26, 623–635.

MacGregor, A. M. (1927): The problem of the Precambrian atmosphere. *South African J Sci.*, 24, 155–172.

MacGregor, I. D. and A. R. Basu (1974): Thermal structure of the lithosphere: A petrologic model. *Science*, 185, 1007–1011.

Machta, L. and E. Hughes (1970): Atmospheric oxygen in 1967 to 1970. *Science*, 168, 1582–1584.

Manabe, S. (1970): The dependence of atmospheric temperature on the concentration of carbon dioxide. In S. F. Singer, ed., *Global Effects of Environmental Pollution* (Springer-Verlag, New York), pp. 25–29.

Manabe, S. and R. T. Wetherald (1967): Thermal equilibrium of the atmosphere with a given distribution of relative humidity. *J. Atmos. Sci.*, 24, 241–259.

Margulis, L. (1969): New phylogenies of the lower organisms: Possible relation to organic deposits in Precambrian sediments. *J. Geol.*, 77, 606–617.

Margulis, L. (1970): *Origin of Eukaryotic Cells: Evidence and Research Implications for a Theory of the Origin of Microbial Plant and Animal Cells on the Precambrian Earth* (Yale University Press, New Haven, Conn.).

Margulis, L. (1971a): Microbial evolution on the early earth. In R. Buvet and C. Ponnamperuma, eds., *Chemical Evolution and the Origin of Life* (North-Holland, New York), pp. 480–484.

Margulis, L. (1971b): Symbiosis and evolution, *Scientific American*, 224, no. 8, 49–57.

Margulis, L. (1971c): The origin of plant and animal cells. *Amer. Scientist*, 59, 230–235.

Margulis, L. (1972): Early cellular evolution. In C. Ponnamperuma, ed., *Exobiology* (North-Holland, Amsterdam), pp. 342–368.

Margulis, L. and J. E. Lovelock (1974): Biological modulation of the earth's atmosphere. *Icarus*, 21, 471–489.

Marov, M. Ya. (1972): Venus: A perspective at the beginning of planetary exploration. *Icarus*, 16, 415–461.

Martell, E. A. (1970): Transport patterns and residence times for atmospheric trace constituents vs. altitude. *Advances in Chemistry Series, Number 93* (American Chemical Society, Washington), pp. 138–157.

Martell, E. A. (1973): The distribution of minor constituents in the stratosphere and lower mesosphere. In B. M. McCormac, ed., *Physics and Chemistry of Upper Atmospheres* (D. Reidel, Dordrecht, Holland), pp. 24–33.

Mason, B. J. (1957): *The Physics of Clouds* (Oxford University Press, London).

Mason, B. (1958): *Principles of Geochemistry* (John Wiley and Sons, New York), 2nd ed.

Mason, E. A. and T. R. Marrero (1970): The diffusion of atoms and molecules. In D. R. Bates and I. Esterman, eds., *Advances in Atomic and Molecular Physics* (Academic Press, New York), Vol. 6, pp. 155–232.

MASTENBROOK, H. J. (1968): Water vapor distribution in the stratosphere and high troposphere. *J. Atmos. Sci.*, 25, 299–311.

MASTENBROOK, H. J. (1971): The variability of water vapor in the stratosphere. *J. Atmos. Sci.*, 28, 1495–1501.

MEADOWS, A. J. (1973): The origin and evolution of the atmospheres of the terrestrial planets. *Planet. Space Sci.*, 21, 1467–1474.

MEIER, R. R. and P. MANGE (1973): Spatial and temporal variations of the Lyman-alpha airglow and related atomic hydrogen distributions. *Planet. Space Sci.*, 21, 309–327.

MENARD, H. W. (1961): Some rates of regional erosion. *J. Geol.*, 69, 154–161.

MENARD, H. W. and S. M. SMITH (1966): Hypsommetry of ocean basin provinces. *J. Geophys. Res.*, 71, 4305–4325.

MICHEL, F. C. (1971): Solar wind induced mass loss from magnetic field-free planets. *Planet. Space Sci.*, 19, 1580–1583.

MIGEOTTE, M. and L. NEVEN (1952): Recents progres dans l'observation du spectre infra-rouge du soleil à la station scientifique du Jungfraujoch (Suisse), *Mem. Soc. Roy. Sci. Liège* (4th Ser.), 12, 165–178.

MILLER, S. L. and L. E. ORGEL (1974): *The Origins of Life on Earth* (Prentice-Hall, Englewood Cliffs, New Jersey).

MITCHELL, R. H. and J. H. CROCKET (1971): Diamond genesis—a synthesis of opposing views. *Mineral. Deposita*, 6, 392–403.

MOLINA, M. J. and F. S. ROWLAND (1974): Stratospheric sink for chlorofluoromethanes: chlorine atom-catalysed destruction of ozone. *Nature*, 249, 810–812.

MOORBATH, S., R. K. O'NIONS, and R. J. PANKHURST (1973): Early Archaean age for the Isua Iron Formation, West Greenland. *Nature*, 245, 138–139.

MOORBATH, S., R. K. O'NIONS, R. J. PANKHURST, N. H. GALE, and V. R. MCGREGOR (1972): Further rubidium–strontium age determinations on the very early Precambrian rocks of the Godthaab District, West Greenland. *Nature*, 240, 78–82.

MORGAN, W. J. (1968): Rises, trenches, great faults, and crustal blocks. *J. Geophys. Res.*, 73, 1959–1982.

MOULTON, F. R. (1905): On the evolution of the solar system. *Astrophys. J.*, 22, 165–181.

MUELLER, R. F. (1964): A chemical model for the lower atmosphere of Venus. *Icarus*, 3, 285–298.

MUELLER, R. F. (1969): Planetary probe: Origin of atmosphere of Venus. *Science*, 163, 1322–1324.

MUELLER, R. F. (1970): Dehydrogenation of Venus. *Nature*, 227, 363–364.

MUELLER, R. F. and S. J. KRIDELBAUGH (1973): Kinetics of CO_2 production on Venus. *Icarus*, 19, 531–541.

MUIR, M. (1975): Comment at conference on The Early History of the Earth. (University of Leicester, England).

MURRAY, B. C. and M. C. MALIN (1973): Polar volatiles of Mars. Theory versus observation. *Science*, 182, 437–443.

MURRAY, B. C., W. R. WARD, and S. C. YEUNG (1973): Periodic insolation variations on Mars. *Science*, 180, 638–640.

National Academy of Sciences (1968): *Physics of the Earth in Space* (Report of a study by the Space Science Board).

NEWBURN, R. L. and S. GULKIS (1973): A survey of the outer planets, Jupiter, Saturn, Uranus, Neptune, Pluto, and their satellites. *Space Sci. Rev.*, 14, 179–271.

NEWELL, N. D. (1963): Crises in the history of life. *Scientific American*, 208, no. 2, 76–92.

NEWELL, R. E. (1970): Water vapor pollution in the stratosphere by the supersonic transporter. *Nature*, 226, 70–71.

NICHOLLS, G. D. (1965): The geochemical history of the oceans. In J. P. Riley and G. Skirrow, eds., *Chemical Oceanography* (Academic Press, New York), Vol. 2, pp. 277–294.

NICHOLLS, G. D. (1967): Geochemical studies in the ocean as evidence for the composition of the mantle. In S. K. Runcorn, ed., *Mantles of the Earth and Terrestrial Planets* (Interscience, New York), pp. 285–304.

NIER, A. O. (1950): A redetermination of the relative abundances of the isotopes of neon, krypton, rubidium, xenon, and mercury. *Phys. Rev.*, 79, 450–454.

NIER, A. O., W. B. HANSON, A. SEIFF, M. B. MCELROY, N. W. SPENCER, R. J. DUCKETT, T. C. D. KNIGHT, and W. S. COOK (1976): Composition and structure of the Martian atmosphere: Preliminary results from Viking I. *Science*, 193, 786–788.

NOXON, J. F. (1975): NO_2 in the stratosphere and troposphere by ground-based absorption spectroscopy. *Science*, 189, 547–549.

NORDLIE, B. E. (1968): Calculation of the basaltic gas phase composition (abstr). *Geol. Soc. Amer. Spec. Paper*, 115, 166.

NORDLIE, B. E. (1972): Gases—Volcanic. In R. W. Fairbridge, ed., *The Encyclopedia of Geochemistry and Environmental Sciences* (Van Nostrand, New York), pp. 387–391.

NURSALL, J. R. (1959): Oxygen as a prerequisite to the origin of the Metazoa. *Nature*, 183, 1170–1172.

OEHLER, D. Z., J. W. SCHOPF, and K. A. KVENVOLDEN (1972): Carbon isotopic studies of organic matter in Precambrian rocks. *Science*, 175, 1246–1248.

OLSON, J. M. (1970): The evolution of photosynthesis. *Science*, 168, 438–446.

OPARIN, A. I. (1938): *The Origin of Life* (Dover, New York), 2nd ed.

OPARIN, A. I. (1961): *Life: Its Nature, Origin and Development* (Academic Press, New York).

OPARIN, A. I. (1972): The appearance of life in the universe. In C. Ponnamperuma, ed., *Exobiology, Frontiers of Biology* (North-Holland, Amsterdam), Vol. 23, pp. 1–15.

ORVILLE, P. M. (1974): Crust–atmosphere interactions. Presented at a conference on The Atmosphere of Venus. (Goddard Institute for Space Studies, New York, October 15–17, 1974).

OWEN, T. (1974): Martian climate: An empirical test of possible gross variations. *Science*, 183, 763–764.

OWEN, T. (1976): Volatile inventories on Mars. *Icarus*, 28, 171–177.

OWEN, T. and K. BIEMANN (1976): Composition of the atmosphere at the surface of Mars: Detection of argon-36 and preliminary analysis. *Science*, 193, 801–803.

PALES, J. C. and C. D. KEELING (1965): The concentration of atmospheric carbon dioxide in Hawaii. *J. Geophys. Res.*, 70, 6053–6076.

PALM, A. (1969): The evolution of Venus' atmosphere. *Planet Space Sci.*, 17, 1021–1028.

PARKER, E. N. (1958): Dynamics of the interplanetary gas and magnetic fields. *Astrophys. J.*, 128, 664–676.

PARKER, E. N. (1964): The solar wind. *Scientific American*, 210, 66–76.

PARKER, E. N. (1971): Recent developments in theory of solar wind. *Rev. Geophys. Space Phys.*, 9, 825–835.

PATTERSON, T. N. L. (1966): Atomic and molecular hydrogen in the thermosphere. *Planet. Space Sci.*, 14, 417–423.

PEARSON, F. J. and D. W. FISHER (1971): Chemical composition of atmospheric precipitation in the northeastern United States. *Geological Survey Water Supply Paper* 1535-P (U.S. Government Printing Office, Washington, D.C.).

PEPIN, R. O. and P. SIGNER (1965): Primordial rare gases in meteorites. *Science*, 149, 253–265.

PERRY, E. C., J. MONSTER, and T. REIMER (1971): Sulfur isotopes in Swaziland System barites and the evolution of earth's atmosphere. *Science*, 171, 1015–1016.

PHINNEY, D. (1972): ^{36}Ar, Kr, and Xe in terrestrial materials. *Earth Planet. Sci. Lett.*, 16, 413–420.

PICKWELL, G. V., E. G. BARHAM, and J. W. WILTON (1964): Carbon monoxide production by a bathypelagic siphonophore. *Science*, 140, 860–862.

PIPER, D. Z. and L. A. CODISPOTI (1975): Marine phosphorite deposits and the nitrogen cycle. *Science*, 188, 15–18.

POCHODA, P. and M. SCHWARZSCHILD (1964): Variation of the gravitational constant and the evolution of the sun. *Astrophys. J.*, 139, 587–593.

POLDERVAART, A. (1955): Chemistry of the earth's crust. In A. Poldervaart, ed., *Crust of the Earth (Geol. Soc. Amer. Spec. Paper 62)*, pp. 119–144.

POLLACK, J. B. (1969): A nongrey CO_2–H_2O greenhouse model of Venus. *Icarus*, 10, 314–341.

POLLACK, J. B. (1971): A nongrey calculation of the runaway greenhouse: Implications for Venus' past and present. *Icarus*, 14, 295–306.

POLLACK, J. B., D. PITMAN, B. N. KHARE, and C. SAGAN (1970a): Goethite on Mars: A laboratory study of physically and chemically bound water in ferric oxides. *J. Geophys. Res.*, 75, 7480–7490.

POLLACK, J. B., R. N. WILSON, and G. G. GOLES (1970b): A reexamination of the stability of goethite on Mars. *J. Geophys. Res.*, 75, 7491–7500.

PONNAMPERUMA, C. and N. W. GABEL (1968): Current status of chemical studies on the origin of life. *Space Life Sci.*, 1, 64–96.

POSTGATE, J. R. (1968): The sulphur cycle. In G. Nickless, ed., *Inorganic Sulfur Chemistry* (Elsevier, New York), pp. 259–279.

PRESSMAN, J. and P. WARNECK (1970): The stratosphere as a chemical sink for carbon monoxide. *J. Atmos. Sci.*, 27, 155–163.

PYTKOWICZ, R. M. (1967): Carbonate cycle and the buffer mechanism of recent oceans. *Geochim. Cosmochim. Acta*, 31, 63–73.

RAITT, W. J., R. W. SCHUNK, and P. M. BANKS (1975): A comparison of the temperature and density structure in high and low speed thermal proton flows. *Planet. Space Sci.*, 23, 1103–1117.

RASOOL, S. I. and C. DEBERGH (1970): The runaway greenhouse and the accumulation of CO_2 in the Venus atmosphere. *Nature*, 226, 1037–1039.

RASOOL, S. I. and S. H. SCHNEIDER (1971): Atmospheric carbon dioxide and aerosols: Effects of large increases on global climate. *Science*, 173, 138–141.

RATNER, M. I. and J. C. G. WALKER (1972): Atmospheric ozone and the history of life. *J. Atmos. Sci.*, 29, 803–808.

REDFIELD, A. C. (1958): The biological control of chemical factors in the environment. *Amer. Scientist*, 46, 205–221.

REDFIELD, A. C., B. H. KETCHUM, and F. A. RICHARDS (1963): The influence of organisms on the composition of sea water. In M. N. Hill, ed., *The Sea* (Interscience, New York), Vol. 2, pp. 26–77.

REES, C. E. (1970): The sulphur isotope balance of the ocean: An improved model. *Earth Planet. Sci. Lett.*, 7, 366–370.

REID, G. C., I. S. A. ISAKSEN, T. E. HOLZER, and P. J. CRUTZEN (1976): Influence of ancient solar-proton events on the evolution of life. *Nature*, 259, 177–179.

REITER, E. R. (1971): *Atmospheric Transport Processes* (U.S. Atomic Energy Commission. Division of Technical Information).

RHOADS, D. C. (1973): The influence of deposit-feeding benthos on water turbidity and nutrient recycling. *Amer. J. Sci.*, 273, 1–22.

RHOADS, D. C. and J. W. MORSE (1971): Evolutionary and ecologic significance of oxygen-deficient marine basins. *Lethaia*, 4, 413–428.

RICHARDS, F. A. (1965): Anoxic basins and fjords. In J. P. Riley and G. Skirrow, eds., *Chemical Oceanography* (Academic Press, New York), Vol. 1, pp. 611–645.

RILEY, J. P. and R. CHESTER (1971): *Introduction to Marine Chemistry*. (Academic Press, New York).

RINGWOOD, A. E. (1959): On the chemical evolution and densities of the planets. *Geochim. Cosmochim Acta*, 15, 257–283.

RINGWOOD, A. E. (1966): The chemical composition and origin of the earth. In P. M. Harley, ed., *Advances in Earth Science* (MIT Press, Boston), pp. 287–356.

RIPPERTON, L. A. and F. M. VUKOVICH (1971): Gas phase destruction of tropospheric ozone. *J. Geophys. Res.*, 76, 7328–7333.

ROBINSON, E. and R. C. ROBBINS (1970a): Gaseous nitrogen compound pollutants from urban and natural sources. *J. Air Pollution Control Assoc.*, 20, 303–306.

ROBINSON, E. and R. C. ROBBINS (1970b): Gaseous atmospheric pollutants from urban and natural sources. In S. F. Singer, ed., *Global Effects of Environmental Pollution* (Springer-Verlag, New York), pp. 50–64.

ROEDDER, E. (1965): Liquid CO_2 inclusions in olivine bearing nodules and phenocrysts from basalts. *Amer. Mineral.*, 50, 1746–1782.

ROEDER, R. C. and P. R. DEMARQUE (1966): Solar evolution and Brans–Dicke cosmology. *Astrophys. J.*, 144, 1016–1023.

ROMANKEVICH, YE. A. (1968): Organic carbon and nitrogen deposits in recent and quaternary sediments of the Pacific Ocean. *Oceanology*, 8, 658–672.

RONOV, A. B. (1958): Organic carbon in sedimentary rocks (in relation to the presence of petroleum). *Geochem.*, 510–536.

RONOV, A. B. (1968): Probable changes in the composition of sea water during the course of geological time. *Sedimentol.*, 10, 25–43.

RONOV, A. B. and A. A. YAROSHEVSKIY (1967): Chemical structure of the earth's crust. *Geochem.*, 1041–1066 (trans. from *Geokhimiya*, no. 11, 1285–1309, 1967).

RONOV, A. B. and A. A. YAROSHEVSKIY (1969): Chemical composition of the earth's crust. In P. J. Hart, ed., *The Earth's Crust and Upper Mantle. Amer. Geophys. Union Monograph 13.* (Washington, D.C.), pp. 37–57.

ROSS, J. E. and L. H. ALLER (1976): The chemical composition of the sun. *Science*, 191, 1223–1229.

RUBEY, W. W. (1951): Geologic history of sea water: An attempt to state the problem. *Geol. Soc. Amer. Bull.*, 62, 1111–1147.

RUDERMAN, M. A. (1974): Possible consequences of nearby supernova explosions for atmospheric ozone and terrestrial life. *Science*, 184, 1079–1081.

RUSSEL, H. N. and D. H. MENZEL (1933): The terrestrial abundance of the permanent gases. *Proc. U.S. Nat. Acad. Sci.*, 19, 997–1001.

RUTTEN, M. G. (1971): *The Origin of Life by Natural Causes* (Elsevier, Amsterdam).

RYTHER, J. H. (1969): Photosynthesis and fish production in the sea. *Science*, 166, 72–76.

RYTHER, J. H. and W. M. DUNSTAN (1971): Nitrogen, phosphorus, and eutrophication in the coastal marine environment. *Science*, 171, 1008–1013.

SACKETT, W. M. (1964): The depositional history and isotopic organic carbon composition of marine sediments. *Marine Geol.*, 173–185.

SACKETT, W. M., C. W. POAG, and B. J. EADIE (1974): Kerogen recycling in Ross Sea, Antarctica. *Science*, 185, 1045–1047.

SAGAN, C. (1960): *The Radiation Balance of Venus* (California Institute of Technology, Jet Propulsion Lab., Tech. Rept. No. 32–34).

SAGAN, C. (1962): Structure of the lower atmosphere of Venus. *Icarus*, 1, 151–169.

SAGAN, C. (1967): Origins of atmospheres of earth and planets. In S. K. Runcorn, ed., *International Dictionary of Geophysics* (Pergamon Press, New York), Vol. 1, pp. 97–104.

SAGAN, C. (1973): Ultraviolet selection pressure on the earliest organisms. *J. Theoret. Biol.*, 39, 195–200.

SAGAN, C. and G. MULLEN (1972): Earth and Mars: Evolution of atmospheres and surface temperatures. *Science*, 177, 52–56.

SAGAN, C. and J. B. POLLACK (1974): Differential transmission of sunlight on Mars: Biological implications. *Icarus*, 21, 490–495.

SAGAN, C., O. B. TOON, and P. J. GIERASCH (1973): Climatic change on Mars. *Science*, 181, 1045–1051.

SCEP (1970): *Man's Impact on the Global Environment* (Report of the Study of Critical Environment Problems, MIT Press, Cambridge, Massachusetts).

SCHIDLOWSKI, M., R. EICHMANN, and C. E. JUNGE (1975): Precambrian sedimentary carbonates: Carbon and oxygen isotope geochemistry and implications for the terrestrial oxygen budget. *Precambrian Res.*, 2, 1–69.

SCHOFIELD, K. (1967): An evaluation of kinetic rate data for reactions of neutrals of atmospheric interest. *Planet. Space Sci.*, 15, 643–670.

SCHOLZ, T. G., D. H. EHHALT, L. E. HEIDT, and E. A. MARTELL (1970): Water vapor, molecular hydrogen, methane, and tritium concentrations near the stratopause. *J. Geophys. Res.* 75, 3049–3054.

SCHOPF, J. W. (1972): Precambrian paleobiology. In C. Ponnamperuma, ed., *Exobiology, Frontiers of Biology* (North-Holland, Amsterdam), Vol. 23, pp. 16–61.

SCHOPF, J. W. (1976): Evidence of Archaean life: A brief appraisal. In B. F. Windley, ed., *The Early History of the Earth* (John Wiley and Sons, New York), pp. 589–593.

SCHOPF, J. W., D. Z. OEHLER, R. J. HORODYSKI, and K. A. KVENVOLDEN (1971): Biogenicity and significance of the oldest known stromatolites. *J. Paleontol.*, 45, 477–485.

SCHOPF, T. J. M., A. FARMANFARMAIAN, and J. L. GOOCH (1971): Oxygen consumption rates and their paleontologic significance. *J. Paleontol.*, 45, 247–252.

SCHORN, R. A., C. B. FARMER, and S. J. LITTLE (1969): High-dispersion spectroscopic studies of Mars. III. Preliminary results of 1968–1969 water vapor studies. *Icarus*, 11, 283–288.

SCHUNK, R. W. and J. C. G. WALKER (1970a): Thermal diffusion in the F2-region of the ionosphere. *Planet. Space Sci.*, 18, 535–557.

SCHUNK, R. W. and J. C. G. WALKER (1970b): Minor ion diffusion in the F2-region of the ionosphere. *Planet. Space Sci.*, 18, 1319–1334.

SCHUNK, R. W. and J. C. G. WALKER (1972): Oxygen and hydrogen ion densities above Millstone Hill. *Planet. Space Sci.*, 20, 581–589.

SCHÜTZ, K., C. E. JUNGE, R. BECK, and B. ALBRECHT (1970): Studies of atmospheric N_2O. *J. Geophys. Res.*, 75, 2230–2246.

SHELDON, W. R. and J. W. KERN (1972): Atmospheric helium and geomagnetic field reversals. *J. Geophys. Res.*, 77, 6194–6201.

SHEPPARD, P. A. (1963): Atmospheric tracers and the study of the general circulation of the atmosphere. *Rep. Prog. Phys.*, 26, 213–267.

SIEVER, R. (1968): Sedimentological consequences of steady-state ocean–atmosphere. *Sedimentol.*, 11, 5–29.

SILLÉN, L. G. (1961a): The physical chemistry of sea water. In M. Sears, ed., *Oceanography* (American Association for the Advancement of Science, Washington, D.C.), pp. 549–581.

SILLÉN, L. G. (1961b): The physical chemistry of sea water. In *Treatise on Marine Ecology and Paleoecology* (Geological Society of America, New York), pp. 185–238.

SILLÉN, L. G. (1966): Regulation of O_2, N_2 and CO_2 in the atmosphere; thoughts of a laboratory chemist. *Tellus*, 18, 198–206.

SINCLAIR, A. C. E., J. P. BASART, D. BUHL, W. A. GALE, and M. LIWSCHITZ (1970): Preliminary results of interferometric observations of Venus at 11.1-cm wavelength. *Radio Sci.*, 5, 347–354.

SKINNER, B. J. (1969): *Earth Resources* (Prentice-Hall, Englewood Cliffs, New Jersey).

SMITH, K. C. (1969): Biochemical effects of ultraviolet light on DNA. In F. Urbach, ed., *The Biological Effects of Ultraviolet Radiation* (Pergamon Press, New York), pp. 47–56.

SMITH, F. L. and C. SMITH (1972): Numerical evaluation of Chapman's grazing incidence integral ch(x, χ). *J. Geophys. Res.*, 77, 3592–3597.

SMOLUCHOWSKI, R. (1968): Mars: Retention of ice. *Science*, 159, 1348–1350.

SNYDER, C. W. (1967): Solar wind. In R. W. Fairbridge, ed., *The Encyclopedia of Atmospheric Sciences and Astrogeology* (Reinhold, New York), pp. 903–905.

SPITZER, L. (1949): The terrestrial atmosphere above 300 km. In G. P. Kuiper, ed., *The Atmospheres of the Earth and Planets* (Univ. of Chicago Press, Chicago), pp. 211–247.

STANIER, R. Y., M. DOUDEROFF, and E. A. ADELBERG (1970): *The Microbial World* (Prentice-Hall, Englewood Cliffs, New Jersey), 3rd ed.

STEBBINGS, R. F. and J. A. RUTHERFORD (1968): Low-energy collisions between $O^+(^4S)$ and H(1s). *J. Geophys. Res.*, 73, 1035–1038.

STEDMAN, D. H., W. CHAMEIDES, and J. O. JACKSON (1975): Comparison of experimental and computed values for $j(NO_2)$. *Geophys. Res. Lett.*, 2, 22–25.

STEDMAN, D. H., R. J. CICERONE, W. L. CHAMEIDES, and R. B. HARVEY (1976): Absence of N_2O photolysis in the troposphere. *J. Geophys. Res.*, 81, 2003–2004.

STEIN, J. A. and J. C. G. WALKER (1965): Models of the upper atmosphere for a wide range of boundary conditions. *J. Atmos. Sci.*, 22, 11–17.

STEPHENSON, M. (1949): *Bacterial Metabolism* (Longmans, Green and Co., London. MIT Press 1966), 3rd ed.

STEVENSON, F. J. (1962): Chemical state of the nitrogen in rocks. *Geochim. Cosmochim. Acta*, 26, 797–809.

STOLARSKI, R. S. and R. J. CICERONE (1974): Stratospheric chlorine: A possible sink for ozone. *Can. J. Chem.*, 52, 1610–1615.

STRICKLAND, J. D. H. (1965): Production of organic matter in the primary stages of the marine food chain. In J. P. Riley and G. Skirrow, eds., *Chemical Oceanography* (Academic Press, New York), Vol. 1, pp. 477–610.

STROBEL, D. F. (1971): Odd nitrogen in the mesosphere. *J. Geophys. Res.*, 76, 8384–8393.

STROBEL, D. F., D. M. HUNTEN, and M. B. McELROY (1970): Production and diffusion of nitric oxide. *J. Geophys. Res.*, 75, 4307–4321.

STROBEL, D. F. and G. R. SMITH (1973): On the temperature of the Jovian thermosphere. *J. Atmos. Sci.*, 30, 718–725.

STROBEL, D. F. and E. J. WEBER (1972): Mathematical model of the polar wind. *J. Geophys. Res.*, 77, 6864–6869.

SUTTON, J. and J. V. WATSON (1974): Tectonic evolution of continents in early Proterozoic times. *Nature*, 247, 433–435.

SVERDRUP, H. V., M. W. JOHNSON, and R. H. FLEMING (1942): *The Oceans. Their Physics, Chemistry, and General Biology* (Prentice-Hall, Englewood Cliffs, New Jersey).

SYKES, L. R. (1969): Seismicity of the mid-oceanic ridge system. In P. J. Hart, ed., *The Earth's Crust and Upper Mantle, Geophys. Mon. 13* (American Geophysical Union, Washington, D.C.), pp. 148–153.

SYLVESTER-BRADLEY, P. C. (1972): The geology of juvenile carbon. In C. Ponnamperuma, ed., *Exobiology, Frontiers of Biology* (North-Holland, Amsterdam), Vol. 23, pp. 62–94.

TAYLOR, H. A. and W. J. WALSH (1972): The light-ion trough, the main trough, and the plasmasphere. *J. Geophys. Res.*, 77, 6716–6723.

THIMANN, K. V. (1963): *The Life of Bacteria* (Macmillan, New York), 2nd ed.,

THOMPSON, B. A., P. HARTECK, and R. R. REEVES (1963): Ultraviolet absorption coefficients of CO_2, CO, O_2, H_2O, N_2O, NH_3, NO, SO_2, and CH_4 between 1850 and 4000 A. *J. Geophys. Res.*, 68, 6431–6436.

TINSLEY, B. A. (1974): Hydrogen in the upper atmosphere. *Fundamentals of Cosmic Physics*, 1, 201–300.

TOOMS, J. S., C. P. SUMMERHAYES, and D. S. CRONAN (1969): Geochemistry of marine phosphate and manganese deposits. *Oceanogr. Mar. Biol.*, 7, 49–100.

TORR, M. R., J. C. G. WALKER, and D. G. TORR (1974): Escape of fast oxygen from the atmosphere during geomagnetic storms. *J. Geophys. Res.*, 79, 5267–5271.

TOWE, K. M. (1970): Oxygen collagen priority and the early Metazoan record. *Proc. U.S. Natl. Acad. Sci.*, 65, 781–788.

TRENDALL, A. F. and J. G. BLOCKLEY (1970): The iron formations of the Precambrian Hamersley Group, Western Australia. *Geol. Survey Western Australia Bull.* 119.

TUREKIAN, K. K. (1964): Degassing of argon and helium from the earth. In P. J. Brancazio and A. G. W. Cameron, eds., *The Origin and Evolution of Atmospheres and Oceans* (John Wiley and Sons, New York), pp. 74–85.

TUREKIAN, K. K. (1968): *Oceans* (Prentice-Hall, Englewood Cliffs, New Jersey).

TUREKIAN, K. K. (1971): Rivers, tributaries, and estuaries. In D. W. Hood, ed., *Impingement of Man on the Oceans* (John Wiley and Sons, New York), pp. 9–73.

TUREKIAN, K. K. (1972): *Chemistry of the Earth* (Holt, Rinehart and Winston, New York).

TUREKIAN, K. K. and S. P. CLARK (1969): Inhomogeneous accumulation of the earth from the primitive solar nebula. *Earth Planet. Sci. Lett.*, 6, 346–348.

TUREKIAN, K. K. and S. P. CLARK (1975): The nonhomogeneous accumulation model for terrestrial planet formation and the consequences for the atmosphere of Venus. *J. Atmos. Sci.*, 32, 1257–1261.

TURNER, J. S. and E. G. BRITTAIN (1962): Oxygen as a factor in photosynthesis. *Biol. Rev.*, 37, 130–170.

URBACH, F. (1969): *The Biologic Effects of Ultraviolet Radiation* (Pergamon, New York).

U.S. Standard Atmosphere Supplement (1966): Prepared under sponsorship of ESSA, NASA, and USAF (U.S. Government Printing Office, Washington, D.C.).

UZZELL, T. and C. SPOLSKY (1974): Mitochondria and plastids as endosymbionts: A revival of special creation. *Amer. Scientist*, 62, 334–343.

VACCARO, R. F. (1965): Inorganic nitrogen in sea water. In J. P. Riley and G. Skirrow, eds., *Chemical Oceanography* (Academic Press, New York), Vol. 1, pp. 365–408.

VALENTINE, J. W. (1973): Phanerozoic taxonomic diversity: A test of alternate models. *Science*, 180, 1078–1079.

VALENTINE, J. W. and E. M. MOORES (1972): Global tectonics and the fossil record. *J. Geol.*, 80, 167–184.

VAN HOUTEN, F. B. (1973): Origin of red beds. A review, 1961–1972. In *Annual Review of Earth and Planetary Sciences* (Annual Review, Inc., Palo Alto, California), Vol. 1, pp. 39–61.

VAN VALEN, L. (1971): The history and stability of atmospheric oxygen. *Science*, 171, 439–443.

VEEH, H. H., W. C. BURNETT, and A. SOUTAR (1973): Contemporary phosphorites on the continental margin of Peru. *Science*, 181, 844–845.

VERNIANI, F. (1966): The total mass of the Earth's atmosphere. *J. Geophys. Res.*, 71, 385–391.

VIDAL-MADJAR, A., J. E. BLAMONT, and B. PHISSAMAY (1973): Solar Lyman-alpha changes and related hydrogen density distribution at the earth's exobase (1969–1970). *J. Geophys. Res.*, 78, 1115–1144.

VON ZAHN, U. (1970): Neutral air density and composition at 150 kilometers. *J. Geophys. Res.*, 75, 5517–5527.

WALKER, J. C. G. (1965): Analytic representation of upper atmosphere densities based on Jacchia's static diffusion models. *J. Atmos. Sci.*, 22, 462–463.

WALKER, J. C. G. (1967a): Upper atmosphere. In R. W. Fairbridge, ed., *The*

Encyclopedia of Atmospheric Sciences and Astrogeology (Reinhold, New York), pp. 1064–1073.

WALKER, J. C. G. (1967b): Atmospheric nomenclature. In R. W. Fairbridge, ed, *The Encyclopedia of Atmospheric Sciences and Astrogeology* (Reinhold, New York), pp. 80–81.

WALKER, J. C. G. (1974): Stability of atmospheric oxygen. *Amer. J. Sci.*, 274, 193–214.

WALKER, J. C. G. (1975a): Atmospheric physics. In B. M. McCormac, ed., *Atmospheres of Earth and Planets* (D. Reidel, Dordrecht, Holland).

WALKER, J. C. G. (1975b): Evolution of the atmosphere of Venus. *J. Atmos. Sci.*, 32, 1248–1256.

WALKER, J. C. G. (1976a): Implications for atmospheric evolution of the inhomogeneous accretion model of the origin of the earth. In B. F. Windley, ed., *The Early History of the Earth* (John Wiley and Sons, New York), pp. 537–546.

WALKER, J. C. G. (1976b): Formation of the inner planets. *Monthly Notes Astron. Soc. Southern Africa*, 35, 2–8.

WALKER, J. C. G., K. K. TUREKIAN, and D. M. HUNTEN (1970): An estimate of the present-day deep-mantle degassing rate from data on the atmosphere of Venus. *J. Geophys. Res.*, 75, 3558–3561.

WALLACE, L. (1969): Analysis of the Lyman-alpha observations of Venus made from Mariner 5. *J. Geophys. Res.*, 74, 115–131.

WARD, W. R. (1973): Large-scale variations in the obliquity of Mars. *Science*, 181, 260–262.

WARD, W. R. (1974): Climatic variations on Mars. 1. Astronomical theory of insolation. *J. Geophys. Res.*, 79, 3375–3386.

WARD, W. R., B. C. MURRAY, and M. C. MALIN (1974): Climatic variations on Mars. 2. Evolution of carbon dioxide atmosphere and polar caps. *J. Geophys. Res.*, 79, 3387–3395.

WEINSTOCK, B. and H. NIKI (1972): Carbon monoxide balance in nature. *Science*, 176, 290–292.

WESTLAKE, D. F. (1963): Comparisons of plant productivity. *Biol. Rev.*, 38, 385–425.

WHALEN, B. A., J. R. MILLER, and I. B. McDIARMID (1971): Evidence for a solar wind origin of auroral ions from low-energy ion measurements. *J. Geophys. Res.*, 76, 2406–2418.

WHITE, D. E. and G. A. WARING (1963): Volcanic emanations. *U.S. Geol. Survey Prof. Paper 440-K*.

WHITFIELD, M. (1974): Accumulation of fossil CO_2 in the atmosphere and in the sea. *Nature*, 247, 523–525.

WIJLER, J. and C. C. DELWICHE (1954): Investigations on the denitrifying process in soil. *Plant and Soil*, 5, 155–169.

WILLIAMS, P. M. (1971): The distribution and cycling of organic matter in the ocean. In S. J. Faust and J. V. Hunter, eds., *Organic Compounds in Aquatic Environments* (Marcel Dekker, New York), pp. 145–163.

WILLIAMS, R. T. and A. E. BAINBRIDGE (1973): Dissolved CO, CH_4, and H_2 in the southern ocean. *J. Geophys. Res.*, 78, 2691–2694.

WILLIAMSON, S. J. (1973): *Fundamentals of Air Pollution* (Addison-Wesley, Reading, Mass.).

WILSON, D. F., J. W. SWINNERTON, and R. A. LAMONTAGNE (1970): Production of carbon monoxide and gaseous hydrocarbons in seawater: Relation to dissolved organic carbon. *Science*, 168, 1577–1579.

WINDLEY, B. F. (1976): *The Early History of the Earth* (John Wiley and Sons, New York).

WITTENBERG, J. B. (1960): The source of carbon monoxide in the float of the Portuguese Man of War. *J. Exp. Biol.*, 37, 698–705.

WOFSY, S. C., J. C. McCONNELL, and M. B. McELROY (1972): Atmospheric CH_4, CO, and CO_2. *J. Geophys. Res.*, 77, 4477–4493.

WOFSY, S. C. and M. B. McELROY (1973): On vertical mixing in the upper stratosphere and lower mesosphere. *J. Geophys. Res.*, 78, 2619–2624.

WOFSY, S. C. and M. B. McELROY (1974): HO_X, NO_X, and ClO_X: Their role in atmospheric photochemistry. *Can. J. Chem.*, 52, 1582–1591.

WOOD, J. A. (1967): Meteorites. In R. W. Fairbridge, ed., *The Encyclopedia of Atmospheric Sciences and Astrogeology* (Reinhold, New York), pp. 561–564.

YEH, T. (1970): A three-fluid model of solar winds. *Planet. Space Sci.*, 18, 199–215.

INDEX

307